Tributes
Volume 30

Liber Amicorum Alberti
A Tribute to Albert Visser

Volume 19
From Quantification to Conversation. Festschrift for Robin Cooper on the occasion of his 65th birthday
Staffan Larsson and Lars Borin, eds.

Volume 20
The Goals of Cognition. Essays in Honour of Cristiano Castelfranchi
Fabio Paglieri, Luca Tummolini, Rino Falcone and Maria Miceli, eds.

Volume 21
From Knowledge Representation to Argumentation in AI, Law and Policy Making. A Festschrift in Honour of Trevor Bench-Capon on the Occasion of his 60th Birthday
Katie Atkinson, Henry Prakken and Adam Wyner, eds.

Volume 22
Foundational Adventures. Essays in Honour of Harvey M. Friedman
Neil Tennant, ed.

Volume 23
Infinity, Computability, and Metamathematics. Festschrift celebrating the 60th birthdays of Peter Koepke and Philip Welch
Stefan Geschke, Benedikt Löwe and Philipp Schlicht, eds.

Volume 24
Modestly Radical or Radically Modest. Festschrift for Jean Paul Van Bendegem on the Occasion of his 60th Birthday
Patrick Allo and Bart Van Kerkhove, eds.

Volume 25
The Facts Matter. Essays on Logic and Cognition in Honour of Rineke Verbrugge
Sujata Ghosh and Jakub Szymanik, eds.

Volume 26
Learning and Inferring. Festschrift for Alejandro C. Frery on the Occasion of his 55th Birrthday
Bruno Lopes and Talita Perciano, eds.

Volume 27
Why is this a Proof? Festschrift for Luiz Carlos Pereira
Edward Hermann Haeusler, Wagner de Campos Sanz and Bruno Lopes, eds.

Volume 28
Conceptual Clarifications. Tributes to Patrick Suppes (1922-2014)
Jean-Yves Béziau, Décio Krause and Jonas R. Becker Arenhart, eds.

Volume 29
Computational Models of Rationality. Essays Dedicated to Gabriele Kern-Isberner on the Occasion of her 60th Birthday
Christoph Beierle, Gerhard Brewka and Matthias Thimm, eds.

Volume 30
Liber Amicorum Alberti. A Tribute to Albert Visser
Jan van Eijck, Rosalie Iemhoff and Joost J. Joosten, eds.

Tributes Series Editor
Dov Gabbay   dov.gabbay@kcl.ac.uk

Liber Amicorum Alberti
A Tribute to Albert Visser

edited by

Jan van Eijck

Rosalie Iemhoff

and

Joost J. Joosten

© Individual authors and College Publications 2016. All rights reserved.

ISBN 978-1-84890-204-6

College Publications
Scientific Director: Dov Gabbay
Managing Director: Jane Spurr

http://www.collegepublications.co.uk

Original artwork for cover by Jasmijn Visser: Phaistos (2005) graphic pen on paper 150cm x 180cm.
Cover design by Laraine Welch

Printed by Lightning Source, Milton Keynes, UK

All rights reserved. No part of this publication may be reproduced, stored in a retrieval system or transmitted in any form, or by any means, electronic, mechanical, photocopying, recording or otherwise without prior permission, in writing, from the publisher.

Foreword

The first time that I witnessed a presentation by Albert must have been on November 8th, 2007. It was on a Thursday, and I remember trying to find my way to Plantage Muidergracht, living under the impression that I would be able to discern between those heading towards "A Day of Mathematical Logic" and those going elsewhere. Regrettably, I recall nothing of the lecture, but I do remember an attendee make a smooth transition, in a matter of seconds, from what looked like deep slumber to applause for the lecturer. A seasoned scientist.

On Friday, July 16th, 2010, I was facing a three-headed committee. It was lunchtime, and one of the heads was Albert's. The committee decided to appoint me to the position of PhD candidate under (ultimate) supervision of Albert. In the years that followed, Albert and I had many conversations. Mostly over lunch.

In my first month, the group moved from the "Bestuursgebouw" to Janskerkhof. At the end of the afternoon of our final day at the Uithof, Albert was listening to Pink Floyd, and explained that he wished to induce a feeling of nostalgia. I was not attached to the old location, so it was hard for me to have nostalgic feelings then. It is much easier today.

Albert had people come to him all the time for consultation, so he needed an office of his own. Only, the building design made this hard to realize, so in compromise Albert handled his job from a sizable glass cuboid. Some referred to this as the aquarium, but I preferred to call it the "Visserkom".

Together, we worked on the closed fragments of provability logics for constructive theories such as Heyting's arithmetic enriched with Markov's principle. Many years prior, in a paper dedicated to Smoryński, Albert had proven some property or other of this system. The argument leaned heavily on the finite model property on the arithmetic side. For our purposes, we had to steer away from this dependency, in order to generalize to enrichments with Church's thesis. I am sad to say that we never succeeded. But I did learn how to properly use "*quod non*" in proofs. Not only does this make me a better mathematician, but also a better man.

Monday, July 11th, 2012, was the first day of the Logic Colloquium, held that year in Barcelona. My first international conference. Albert introduced me to assorted people. I had yet to learn how to determine which talks I ought to attend. As a proxy for good judgment, I used Albert's. This worked wonderfully.

The day is Thursday, July 12th, 2013, Manchester. Another Logic Colloquium. Albert and me joined a pack of logicians in pursuit of lunch. The front-runners hunted for a particular vegetarian restaurant, and the remainder of the pack duly followed. "Where is this restaurant?" I asked Albert. Albert, visibly agitated: "I have no clue, but apparently very far away." Even for someone with Albert's experience, it is stressful to have to give a presentation right after lunch.

In my papers, I often thank Albert for his insightful comments. Albert has a keen eye, and his comments help to improve whatever this eye happens to fall on. Only, throughout these years, I have never been able to reliably predict just what this might be. But variance wanes as volume grows, so I made it a point to consult Albert as often as possible. I wonder whether Albert knew why.

Eight thirty in the morning on Saturday, 19th July, 2014, the Logic Colloquium in Vienna. Albert and me have never been more similar. Together we listened to Dana Scott's lecture on point-free geometry. Everybody wanted to be there; even the hallway was packed. It was magnificent, and impossible to beat. I skipped Albert's lecture later that afternoon. And he skipped mine. Our talks coincided, by the way.

Albert loves jokes, and so do I. Our conversations in his Visserkom could drift substantially from the intended topic. As often as not, a decent conversation seamlessly degenerated into an attempt of Albert to find something on internet. Nothing is less interesting than watching someone search the internet, hence I would slyly leave Albert's office at this point. On Tuesday, September 15th, 2015, another pleasant conversation transformed itself into an attempt of Albert to find a picture on the internet, so I prepared for immediate departure. Albert: "When I wish to indicate the end of a conversation with you, all I have to do is search for something, and you're gone." Good joke.

Albert, congratulations! I remember once when you finished three papers over a Christmas holiday, so we can have high expectations of your retirement. Let me close off with a reference to one of the many mathematical discoveries named after you: Visser rules!

Jeroen Goudsmit

Preface

How can we characterize Albert Visser as a logician, as a philosopher, as a university professor, as a colleague, as a friend? Many attributes come to mind: original, deep, wise, versatile, sincere, modest, friendly, humorous. Albert is mild-mannered, but he is also known for his acumen in spotting the weak points in an argument or the flaws in a proof, and he can be tenacious as a critic of sloppy thinking. In public defences of PhD theses, the defendants fear Albert's questions the most. Albert can be extremely funny and sometimes delightfully playful and irresponsible, as when you meet up to discuss a paper, but instead are lured by Albert into spending a pleasant afternoon comparing different versions of Händel arias that can be found on the internet.

Albert's interests range widely, from theories of arithmetic to Heidegger-style philosophy ("Das Nichts nichtet", and so on). We believe that the great variety of qualities that is apparent from Albert's research and teaching also shines through in this book tribute. There are contributions by logicians, by computer scientists and by philosophers, in a wide spectrum of styles, ranging from austere to frivolous, with all shades in between.

The miracle that allowed us to put together this high-quality volume in next to no time is explained by the size and quality of the network that connects Albert to the *haut monde* of the most eminent logicians and philosophers throughout the world. When we asked around for contributions we had to make sure we did not cast the net too widely, for fear that a single volume would not be enough to collect the harvest of contributions. We decided to restrict our invitations to potential authors in close touch with Albert, either as colleagues or as former students.

If we overlooked someone with our invitations to contribute, we are to blame, and we apologize. But apart from our feelings of guilt about our rather haphazard procedure of sending out invitations, editing this book was a joyful experience. In the process of putting this volume together, while reading and editing, we also learned quite a bit. And there is more to be learned: the treasure trove is far from exhausted. Studying this book one can learn new facts and theorems, and explore new avenues in logic; at a different level, the book also tells the story of Albert's diversity as a scientist and scholar. Many of the contributions are clearly inspired by Albert's work; they explore variations of Visserian themes, or they prove generalizations of Visser theorems.

The *Logic Group Preprint Series* of Utrecht University lists sixty-eight preprints by Albert Visser. Add to this an extensive collection of course notes, all containing a wealth of material and showing a thoroughly original mind at work (one of us recalls Albert's survey of the key discoveries and inventions of mankind; the list included 'invention of the dog'), and it becomes clear that Albert has been very productive indeed. The papers in the preprint list have truly Albertian titles: *Jumping in Arithmetic*; *Peano Basso and Peano Corto*; *Cardinal arithmetic in the style of Baron von Münchhausen*; *Prolegomena to the categorical study of interpretations*; *No escape from Vardanyan's Theorem*; *Faith & Falsity, A study of faithful interpretations and false Σ^0_1-sentences*; *The Donkey and the Monoid: Dynamic Semantics with Control Elements*; *Dynamic Negation, The One and Only*; *Lazy & quarrelsome brackets*; and *Peano's smart children: A provability logical study of systems with built-in consistency*.

Confronted with such a display of wit, what to do about the name of the present book? Should we pay homage to Albert by trying to imitate him in the art of title invention? If so, how? We discussed various possibilities: *Visser bites logic*; *Liber Visserorum*; *Summa Visseraica*; *Liber Amicorum, Lieber Albert* and *The Visser Volume*. In the end we decided to give up our attempts to outperform

Albert in a game of which he is the true master, and we settled for "Book from his friends for Albert", *Liber Amicorum Alberti*, a title that duly describes this book as what it is.

Our heartfelt thanks to all authors who contributed, not only for the effort they put into their papers, but also for their flexibility and good cheer in handling our numerous and sometimes ill timed typesetting and proofreading requests. Jörg Endrullis and Volodya Shavrukov helped us with chasing and removing (or circumventing) some tenacious LaTeX bugs. Jane Spurr of College Publications, swift, cheerful and competent as ever, was invaluable in guiding us through the production process. Thanks to Utrecht University for the financial help that made the appearance of this book possible. We also thank Jasmijn Visser for sharing her artwork for the cover.

Last but not least, congratulations to Albert with a splendid career where he produced a continuous stream of deep research in logic and foundations of mathematics, and was able to combine these scientific activities with being one of the key contributors to the Cognitive Artificial Intelligence Programme (nowadays called the Artificial Intelligence Programme) of Utrecht University. One of the reasons that this program, in which three faculties have to work together, is successful, is because of Albert's *polder* approach: in a pleasant and informal way he made sure that the different communities all had their equal share in the program. For some time, Albert was even called *mister CKI* by students and staff.

We value Albert as a colleague, as a teacher and as a friend, and we hope he will find as much joy in reading this book as the authors and editors derived from putting it together.

Jan van Eijck, Rosalie Iemhoff, and Joost J. Joosten
Editors

LIST OF AUTHORS

HENK BARENDREGT
Institute for Computing and Information Sciences, Radboud University, Nijmegen
Url: http://www.cs.ru.nl/~henk

JOHAN VAN BENTHEM
Institute for Logic, Language and Computation (ILLC), Amsterdam
Url: http://staff.fnwi.uva.nl/j.vanbenthem

LEV BEKLEMISHEV
Steklov Mathematical Institute, Moscow, Moscow M.V. Lomonosov State University, National Research University Higher School of Economics, Moscow, Russia
Url: http://www.mi.ras.ru/~bekl

JAN BERGSTRA
Informatics Institute, Faculty of Science, University of Amsterdam
Url: https://staff.fnwi.uva.nl/j.a.bergstra

INGE BETHKE
Informatics Institute, Faculty of Science, University of Amsterdam
Url: https://staff.fnwi.uva.nl/i.bethke

MARC BEZEM
Department of Informatics, University of Bergen, Norway
Url: http://www.ii.uib.no/~bezem

GURAM BEZHANISHVILI
Department of Mathematical Sciences, New Mexico State University, USA
Url: http://sierra.nmsu.edu/gbezhani

NICK BEZHANISHVILI
Institute for Logic, Language and Computation (ILLC), Amsterdam
Url: https://staff.fnwi.uva.nl/n.bezhanishvili

MARTA BÍLKOVÁ
Institute of Computer Science, CAS, Prague, Czech Republic
Url: http://web.ff.cuni.cz/~bilkmaff

HANS VAN DITMARSCH
LORIA, Vandoeuvre-lès-Nancy, France
Url: https://sites.google.com/site/hansvanditmarsch

JAN VAN EIJCK
Centrum voor Wiskunde en Informatica (CWI), Amsterdam
Institute for Logic, Language and Computation (ILLC), Amsterdam
Url: https://homepages.cwi.nl/~jve

ALI ENAYAT
Department of Philosophy, Linguistics, and Theory of Science
University of Gothenburg, Sweden
Url: http://flov.gu.se/english/contact/staff/ali-enayat

JÖRG ENDRULLIS
Department of Theoretical Computer Science, Vrije Universiteit, Amsterdam
Url: http://joerg.endrullis.de

CLEMENS GRABMAYER
Department of Theoretical Computer Science, Vrije Universiteit, Amsterdam
Url: http://www.few.vu.nl/~cgr600

TUOMAS HAKONIEMI
Department of Logic, History and Philosophy of Science, University of Barcelona, Spain
Url: https://thakon.wordpress.com

VOLKER HALBACH
New College, Oxford, UK
Url: http://users.ox.ac.uk/~sfop0114

ROSALIE IEMHOFF
Department of Philosophy, Utrecht University
Url: https://www.phil.uu.nl/~iemhoff

JULIA ILIN
Institute for Logic, Language and Computation (ILLC), Amsterdam
Url: https://staff.fnwi.uva.nl/j.ilin

DICK DE JONGH
Institute for Logic, Language and Computation (ILLC), Amsterdam
Url: https://staff.science.uva.nl/d.h.j.dejongh

JOOST JOOSTEN
Department of Logic, History and Philosophy of Science, University of Barcelona, Spain
Url: http://www.phil.uu.nl/~jjoosten

JAN WILLEM KLOP
Department of Theoretical Computer Science, Vrije Universiteit, Amsterdam
Centrum voor Wiskunde en Informatica (CWI), Amsterdam
Url: http://www.cs.vu.nl/~jwk

JAN KRAJÍČEK
Faculty of Mathematics and Physics, Charles University, Prague, Czech Republic
Url: http://www.karlin.mff.cuni.cz/~krajicek

LESZEK KOŁODZIEJCZYK
University of Warsaw, Warszawa, Poland
Url: http://www.mimuw.edu.pl/~lak

JOOP LEO
Department of Philosophy, Utrecht University
Url: http://joopleo.com

ROBERT NIEUWENHUIS
Barcelogic.com and Technical University of Catalonia (UPC), Barcelona, Spain
Url: https://www.cs.upc.edu/~roberto

PERE PARDO
LORIA, Vandoeuvre-lès-Nancy, France
Url: https://sites.google.com/site/perepardoventura

ANDREW POLONSKY
Institut Galilée, Université Paris 13, France
Url: https://www.researchgate.net/profile/Andrew_Polonsky

RAHIM RAMEZANIAN
Sharif University of Technology, Tehran, Iran
Url: https://sites.google.com/site/rahimramezanian

FRANÇOIS SCHWARZENTRUBER
Ecole Normale Superieure (ENS), Rennes, France
Url: http://people.irisa.fr/Francois.Schwarzentruber

DANIYAR SHAMKANOV
Steklov Mathematical Institute of Russian Academy of Sciences, and National Research University Higher School of Economics, Moscow, Russia
Url: https://www.hse.ru/en/org/persons/63969405

VOLODYA SHAVRUKOV
Email: v.yu.shavrukov@gmail.com

ANNE TROELSTRA
Institute for Logic, Language and Computation (ILLC), Amsterdam
Email: a.s.troelstra@uva.nl

FAN YANG
Applied Logic Group, Delft University of Technology
Url: https://sites.google.com/site/fanyanghp

DOMENICO ZAMBELLA
Dipartimento di Matematica, Università di Torino, Italy
Email: domenico.zambella@unito.it

CONTENTS

Foreword   v

Preface   vii

List of authors   ix

Memories
Henk Barendregt   1

Tales From an Old Manuscript
Johan van Benthem   5

Some Abstract Versions of Gödel's Second Incompleteness Theorem Based on Non-Classical Logics
Lev Beklemishev and Daniyar Shamkanov   15

A Negative Result on Algebraic Specifications of the Meadow of Rational Numbers
Jan A. Bergstra and Inge Bethke   31

Completeness of Cutting Planes Revisited
Marc Bezem and Robert Nieuwenhuis   37

An Algebraic Approach to Filtrations for Superintuitionistic Logics
Guram Bezhanishvili and Nick Bezhanishvili   47

Uniform Interpolation in Provability Logics
Marta Bílková   57

Gossip in Dynamic Networks
Hans van Ditmarsch, Jan van Eijck, Pere Pardo, Rahim Ramezanian, François Schwarzentruber   91

Variations on a Visserian Theme
Ali Enayat   99

Reduction Cycles in Lambda Calculus and Combinatory Logic
Jörg Endrullis, Jan Willem Klop, Andrew Polonsky   111

Linear Depth Increase in Leftmost-Outermost Rewrite Sequences
Clemens Grabmayer   125

Labelled Tableaux for Interpretability Logics
Tuomas Hakoniemi and Joost J. Joosten   141

The Root of Evil – A Self-Referential Play in One Act
Volker Halbach   155

Reasoning in Circles
Rosalie Iemhoff   165

NNIL Axioms Have the Finite Model Property
Julia Ilin, Dick de Jongh, Fan Yang   177

A Reflection on Collection
Leszek A. Kołodziejczyk   187

Expansions of Pseudofinite Structures and Circuit and Proof Complexity
Jan Krajíček   195

Zero or Hero?
Joop Leo   205

Dense Chains of Σ_n Sentences with Strong Conservativity Properties
V. Yu. Shavrukov   209

Checking Your History
A.S. Troelstra   223

On the Diameter of Lascar Strong Types
Domenico Zambella   231

Memories

HENK BARENDREGT

Some free associations about and for Albert Visser.

1. About 40 years ago a young student from the University of Twente came to me at the Mathematical Institute in Utrecht, saying he wanted to study logic (Dirk van Dalen was then in Oxford). That was you, Albert. You asked what subjects were given in that field and whether you could get an exemption for what you already had studied in Twente. Your interest in logic had been stimulated by Louk Fleischhacker, one of your teachers in Twente, which I had shortly met at the Montessori Lyceum Amsterdam in my last year at that school. Not long after that, you were officially enrolled to study logic at Utrecht University.

2. In a break during a seminar you said that it probably would be difficult to obtain a research position towards a PhD. Jan Bergstra then said something like: "If you have such kind of thoughts, then indeed it will be difficult to obtain a PhD." Therefore you should not think that way, was the implication. Five years later you had your PhD. There were several strong results in your thesis. The ADN theorem I am trying to formulate by heart.

DEFINITION (Ershov). An equivalence relation $\sim$ on the natural numbers is called *pre-complete* if every partial computable function ψ can be made total modulo $\sim$ (your terminology), i.e. for every ψ there is a total computable function f such that for every n, if $\psi(n)$ is defined, then $\psi(n) \sim f(n)$.

PROPOSITION (Ershov). Let $\sim$ be a pre-complete equivalence relation on the natural numbers. Then for every total computable function g there exists a natural number n such that $g(n) \sim n$.

THEOREM (Anti Diagonal Normalization Theorem (ADN), Visser). Let $\sim$ be a pre-complete equivalence relation. Then every partial computable function ψ can be made total modulo $\sim$ in such a way that one can still say something about the n such that $\psi(n)$ is not defined. In fact, for every partial computable function δ satisfying a particular condition, every partial computable ψ can be made total by a computable f modulo $\sim$ such that moreover if $\psi(n)$ is not defined, then $\delta(f(n))$ is undefined. Of course this can't be the case for any partial computable function: take for example a total computable δ. In your result you showed that a sufficient condition is that for every n such that $\delta(n)$ is defined one has $\delta(n) \not\sim n$.

The proof is very short, but practically unintelligible, even after a second reading (as is the case, but less so, with the proof of the recursion theorem of which your ADN is a considerable strengthening). Your theorem has many applications. Bernardi and Sorbi used in [2] a construction in which the ADN theorem has been applied $\omega + 1$ times to get a nice result about Peano Arithmetic. Through an application of ADN by Statman it follows that one can represent a partial computable functions ψ

by a lambda term Ψ, where one can prescribe properties of the term $\Psi^\ulcorner n\urcorner$ in case $\psi(n)$ is undefined [1]. Ten years earlier, in my own PhD thesis, it took some effort to cover the case that $\Psi^\ulcorner n\urcorner$ had to be unsolvable, in case $\psi(n)$ was undefined [0].

3. Not long after your ADN theorem was published, a colleague of Ershov in Siberia came to Utrecht and was looking for you. You were off that day and that was a disappointment for him (the next day he had to go back to Novosibirsk in the former Soviet Union with its permafrost).

4. On the lambda terms you have introduced a topology: $O \subseteq \Lambda$ is *open* if O is closed under β-conversion and moreover O (after coding) a Π_1^0 set (equivalently $\Lambda - O$ is computably enumerable). You, but also Statman and myself [3,4], have given powerful applications.

5. You use, as a working logician, in Dutch an attractive linguistic expression. 'Consider a number n' is usually rendered as 'Laat n een getal zijn'. You state and write 'Bezie een getal n', which is shorter (only by one word, but it has a one-one correspondence with the English version and is more pretty: it 'runs more smoothly in the mind').

6. Another good memory was a difference of opinion between you and Karst Koymans. You insisted that natural deduction logic is entirely without axioms, only having deduction rules. Karst insisted that this is impossible, one always needs a beginning in a proof. I agreed with Karst: you have to assume that if A is in the set Γ, then $\Gamma \vdash A$; this then can be regarded as an axiom. Your dispute took place during a long walk in a forest. I was more focussed on the large quantities of lushious blackberries for which both of you had no eyes at that moment.

7. Once upon a time I was paying a visit to your home. My winter hat, which my good father had brought me as a present from China, lay besides me on your sofa. Because the hat was made of fur (already then barely politically acceptable) you thought it was your black cat, until I got up ready to leave and to your horror the 'cat' was put on my head *way too fast*!

8. In those days I learned from you to make a nice and pretty dessert: warm homemade cranberry compote with cold yogurt without sugar and whipped cream at room temperature. It is now one of my favorites and Lidia and I still prepare it every now and then.

9. Then there are some more private memories that are better not told here. We will speak about

these in due time.

Finally, I'm sure you will enjoy your retirement.
Wishing you and Karin all the best!

References

[0] H. Barendregt [1971] Some extensional term models for λ-calculi and combinatory logics. PhD Thesis, Utrecht University.
URL: <repository.ubn.ru.nl/handle/2066/27329>.

[1] H. Barendregt [1992] Representing 'undefined' in lambda calculus. *J. Funct Programming*, 2(3), 367-374.
URL: <repository.ubn.ru.nl/handle/2066/17243>.

[2] C. Bernardi and A. Sorbi [1983] Classifying Positive Equivalence Relations. *J. Symbolic Logic*, 48(3), 529-538.

[3] R. Statman and H. Barendregt [1999] Applications of Plotkin terms: partitions and morphisms for closed terms. *J. Funct. Programming*, 9(5), 565-575.
URL: <repository.ubn.ru.nl/handle/2066/17267>.

[4] R. Statman and H. Barendregt [2005] Böhm's Theorem, Church's Delta, Numeral Systems, and Ershov Morphisms. In: *Processes, terms and cycles: steps on the road to infinity: essays dedicated to Jan Willem Klop on the occasion of his 60th birthday*, Springer, 40-54.
URL: <repository.ubn.ru.nl/handle/2066/33258>.

Tales From an Old Manuscript

JOHAN VAN BENTHEM

Abstract

'Possibility semantics' for classical logics is on the rise these days, and it throws interesting new light on the borderline between classical and intuitionistic logic. This note is about a source from 1981 that is attracting some attention, a handwritten Tech Report from the Groningen Institute of Mathematics, never officially published.[1] I will present its content, and discuss what issues this raises that may be of current relevance.

1 Albert Visser in a nutshell

First things first. The main points of this short note are not logical theorems but three empirical observations, based on long experience. Knowing Albert Visser and interacting with him is a pleasure, corresponding with Albert Visser is engaging, and working together with Albert Visser is an educational experience that is one of the best self-improvement methods known to mankind. This concludes the main content of this note, the rest is aftermath.

I have enjoyed all three mentioned dimensions, but my theme for this piece is some thoughts touching on a general issue that has colored my contacts with Albert. My heart beats to the rhythms of *classical logic*, but his (I suspect) to those of *intuitionistic logic*. What follows is a possible new perspective on the interface between these two lifestyles – though I hasten to add that Albert is a master at both, and knows much more about their interfaces than I do.

2 Possible worlds semantics for classical logic

Possibilities Around 1980 possible worlds semantics for modal logics was under attack from various sides. There was a growing philosophical interest in an ontology of smaller, less elaborate 'possibilities' or 'situations', but there were also related developments in the temporal semantics of natural language where intervals, rather than fully specified points, were emerging as the indices of evaluation (cf. Barwise and Perry 1983, van Benthem 1983). While standard possible worlds-based modal logic has continued to flourish (cf. Blackburn, van Benthem & Wolter, eds. 2006), models with possibilities ordered by inclusion, too, return all the time. For instance, working in an epistemological setting where the options to be considered are often less fine-grained than worlds, Holliday 2014, Holliday 2015 develop a framework of possibilities that allows for a functional view of beliefs, returning to the classic paper Humberstone 1981.[2]

---

[1] An online facsimile can be found at http://dare.uva.nl/cgi/arno/show.cgi?fid=621016.

[2] For instance, coming from mereology, Bochmann 2015, too, develops new possibilities models and rich matching languages. Rumfitt 2015 uses a possibilities-based paradigm to mount a sustained philosophical defense of classical logic.

Interestingly, going against a prejudice that ontological change must also lead to logic change, say toward some partial, intuitionistic or hyper-intensional framework, the models elaborated in this line naturally support classical logic.[3] And this brings us to the topic of this paper.

An old handwritten tech report As Holliday notes, the possibilities models that he ends up with are close to those proposed in an unpublished report van Benthem 1981, with a short summary included in the published paper van Benthem 1986. However, my motivations in 1981 were different from the modern possibilities program, and this gives me the theme for this contribution. Why did I propose these structures (called 'possible worlds models for classical logic' – not a very good name), what results were found, and what relevant logical program might still emerge?

The first thing to do was reading the original report from the Mathematical Institute in Groningen. This was not always easy, apart from the fading ink. The author is often terse and apodictic: something which may be all-right to inflict on others, but which seems really objectionable when doing it to *oneself*, even if 35 years later. What follows is my current reconstruction of things written back then, with a selection of some key results, proofs that I have checked, plus some comments on the enterprise as they occur to me now. I hope that this still has some current value: for possibility semantics, and perhaps beyond.

Intuitionistic and classical logic As a student in Amsterdam, I was exposed first-hand to a lot of intuitionistic logic, represented by some of its finest minds: Anne Troelstra, Dick de Jongh, and later on, Albert Visser. Still, probably out of a revolutionary 1960s desire to be different, my heart went out to classical logic. My dissertation on modal logic was squarely in that tradition, and my favorite tool was classical model theory, not constructive proof theory or the like. But of course, for a working logician, things are never that clear-cut. I used proof theory in my later work on categorial grammar and substructural logics, and as for intuitionistic logic, I have always felt a fascination for what it has to offer semantically. In particular, intuitionistic models for coming to know things have always appealed to me, especially their links with information structure and information dynamics (van Benthem 1989, 2009).

More than that, the 1981 report even shows an unease with classical logic. It starts with the complaint that classical completeness proofs have the inelegant feature that one needs to pick 'some' maximally consistent extension of a given set of formulas to create a model, and likewise, that basic model-theoretic constructions such as ultraproducts need an ultrafilter going beyond what is often just a filter containing the relevant initial structure. The report then proposes what it calls an 'intuitionistic' alternative, though it notes that intuitionistic completeness proofs still have a problem, as they use prime filters that split disjunctions in an arbitrary manner, often going beyond the original input filter.

A modal semantics So, we take models **M** for the language of first-order logic that are tuples $(W, \leq, D, I)$ where W is a set of worlds (or stages, or information pieces, or anything suitable that appeals to you), $\leq$ is a partial order (inclusion between stages), D is a map assigning to each world w a set $D(w)$ of objects (the domain of that world), and I is an interpretation function sending predicates P and worlds w to sets of tuples of the appropriate arity with objects chosen from $D(w)$. In doing so, we impose some further constraints on our structures. The first of these is standard modal:

---

[3] Actually, possibility models are also modal models in the usual style. Many differences that seem vast in the literature are matters of philosophical interpretation, not mathematical substance.

Cumulation for domains: if $w \leq v$, then $D(w) \subseteq D(v)$.

We also impose, as in models for intuitionistic logic,

Persistence for atomic facts: if Pd holds at w and $w \leq v$, then Pd holds at v.

However, in line with our classical orientation, we also impose

Cofinality for atomic facts: if for all $v \geq w$, there exists a $u \geq v$ with Pd true at u, then Pd is already true at w.

The report motivates the latter condition as follows, while pointing out how it embodies the negation law $\neg\neg\phi \to \phi$. If the truth of a formula ϕ is 'inevitable' cofinally in the model, then we might as well call ϕ true now. This sounds like the Sure Thing Principle of decision theory: if every available current action of yours will result in believing that ϕ, believe ϕ right now.[4]

REMARK. *Later literature on possibility- or stage-based models, in natural language semantics, metaphysics, or epistemology, has added a different intuition motivating cofinality. This time, it appears as a desirable form of 'refinability': if a stage w fails to make ϕ true, then there must still be some more refined state $v \geq w$ that refutes ϕ, in the sense of making $\neg\phi$ true. An early source for this line of motivation is the 'data semantics' of Veltman 1984.*

The following truth definition now interprets the standard first-order language, with a little benign sloppiness in notation. In what follows, in our exposition, we restrict attention to just three logical operations, viz. conjunction, negation, and universal quantification:[5]

$\mathbf{M}, w, \mathbf{d} \models Px$ iff   the tuple of objects $\mathbf{d}$ is in $I(P, w)$

$\mathbf{M}, w, \mathbf{d} \models \phi \wedge \psi$ iff $\mathbf{M}, w, \mathbf{d} \models \phi$ and $\mathbf{M}, w, \mathbf{d} \models \psi$

$\mathbf{M}, w, \mathbf{d} \models \neg\phi$ iff   for no $v \geq w$, $\mathbf{M}, w, \mathbf{d} \models \phi$

$\mathbf{M}, w, \mathbf{d} \models \forall x\phi$ iff   for all $v \geq w$, for all $d \in D(v)$, $\mathbf{M}, v, \mathbf{d}d \models \phi$

More formally, instead of tuples of objects $\mathbf{d}$, one can use assignment functions s in a format $\mathbf{M}, w, s \models \phi$ where s sends the free variables in ϕ to objects in the domain $D(w)$. Domain Cumulation makes sure this remains well-defined when we go up to worlds higher in the inclusion order.

Further operations are considered defined, with $\phi \vee \psi$ viewed as $\neg(\neg\phi \wedge \neg\psi)$, and $\exists x\phi$ as $\neg\forall x\neg\phi$.

With the language interpreted in this way, it can be shown by a straightforward induction that the earlier two semantic properties of persistence and cofinality lift to all formulas.

---

[4]Cofinality is weaker than total 'inevitability' of a formula defined as truth on some barrier across the inclusion graph, the second-order truth condition employed in Beth's original semantics for intuitionistic logic. The difference raises some interesting technical complexities on infinite trees, but we omit them here.

[5]Also, mainly for simplicity, we will only consider only variables and constants in our language, a countably infinite set of each (for technical reasons to become clear shortly), but no complex term-forming function symbols.

Completeness by one canonical model With models like this, the completeness proof for a classical first-order axiom system becomes straightforward, referring to a unique canonical model. Consider the set of all *consistent deductively closed* sets Σ of formulas in our language, where we assume that each set Σ has an associated language $L(\Sigma)$ which involves only finitely many individual constants.[6] When we refer to inclusion in what follows, we mean inclusion of sets of formulas and associated languages. Now we observe that the following standard decomposition principles hold for all formulas and consistent sets in the relevant associated languages:

FACT 2.1.

$$\begin{aligned}
\phi \wedge \psi \in \Sigma &\quad \textit{iff} \quad \phi \in \Sigma \textit{ and } \psi \in \Sigma \\
\neg \phi \in \Sigma &\quad \textit{iff} \quad \textit{for all } \Delta \supseteq \Sigma, \phi \notin \Delta \\
\forall x \phi \in \Sigma &\quad \textit{iff} \quad \textit{for all } \Delta \supseteq \Sigma \textit{ and all constants } c, [c/x]\phi \in \Delta
\end{aligned}$$

Proof. The proof of these equivalences involves only minimal properties of any proof system. In particular, in the argument for the negation clause, it is not assumed that the underlying proof system is classical. The analogy of this decomposition result with the three earlier truth conditions will be clear. ⊣

We can exploit these decomposition properties of consistent sets by defining a canonical model **M** as consisting of all consistent language-indexed deductively closed sets ordered by inclusion, and deriving its interpretation of atoms as usual, directly from the atomic formulas that are explicitly present in the sets. The key feature of this model is given by the usual equivalence of membership and truth, for all sets Σ in the model and all sentences $[c/x]\phi$ with all free variables in the formula ϕ replaced by constants in the language $L(\Sigma)$:

LEMMA 2.2 (Truth). $\mathbf{M}, \Sigma \models [c/x]\phi$ iff $[c/x]\phi \in \Sigma$.

Proof. This is shown by a straightforward induction on formulas. ⊣

The further property of Persistence is immediate from the use of set inclusion, but Cofinality does ask for something classical, namely, the derivability of the double negation law $\neg\neg\phi \to \phi$.

In this light, completeness for classical logic does not need any arbitrary choices of maximally consistent sets. But what is more, the above proof does not assume any form of the Axiom of Choice, and hence it suggests a more constructive version of standard meta-theory.[7]

Back to classical models: generic branches Now, as pointed out in our source text, this simple and minimal completeness proof comes at a price. For, it does not tell us that a non-derivable formula has a counterexample that is 'standard', being just a single-world model of the usual sort. However, a route toward such models is provided by what the report calls the 'usual method' of generic branches, borrowed from set-theoretic forcing.

Consider any countable possibility model as defined above, and look at the branches in it – i.e., the maximal linearly ordered subsets. If a branch ends at some finite stage, we have an obvious classical model for all formulas true there. But such models can also be found for infinite *generic branches* B that satisfy the following two properties:

---

[6] The finiteness restriction ensures that we can extend our sets with new witnesses for false universal formulas. Alternative solutions are possible.

[7] It is of interest to compare this analysis with that of the intuitionistically valid completeness proofs for intuitionistic logic due to Veltman and De Swart in the 1970s: see the exposition in Troelstra & van Dalen 1988.

(a) for each formula ϕ, either ϕ or $\neg\phi$ is true at some stage on B,

(b) for each formula $\neg\forall x\phi$ true somewhere on the branch B, some instance $\neg[c/x]\phi$ is also true somewhere on that branch for some individual constant c.

Each branch B in any model naturally induces a classical model $\mathbf{M}(B)$ whose domain consists of all objects occurring in worlds on B, with the interpretation of predicate atoms copied from what is true on the branch. Persistence makes sure that this stipulation is well-defined. Moreover, for the special case of generic branches that force choices and provide witnesses for false universal formulas, we can prove the following fact by induction on formulas:

FACT 2.3. $\mathbf{M}(B) \models \phi$ *iff there is some world w on B such that $\mathbf{M}, w \models \phi$.*

Proof. We use induction on ϕ. For conjunctions, the key point is that having two formulas true at different points on a branch makes both of them true at the greater of the two points, by persistence. For negation, again persistence is crucial. In one direction, if any point on a branch makes ϕ true, then no other point on the branch can make $\neg\phi$ true. The opposite direction uses clause (a) of genericity, plus the inductive hypothesis. Finally, the right to left direction of the universal quantifier case follows immediately by persistence and domain cumulation, while deriving the opposite direction makes use of clause (b) of genericity plus cofinality. ⊣

What makes these facts useful is the following observation.

FACT 2.4. *Any countable model contains generic branches, and such branches even start from every point.*

Proof. Generic branches can be found by a model-internal analogue of the Henkin completeness construction. We enumerate formulas and pick a branch, making sure that all instances of formulas occur with all objects occurring in points on the branch.[8] Using this enumeration as our scheduling, to continue the branch so far, we choose a further point that either makes the scheduled formula or its negation true, and if the latter case is of the form $\neg\forall x\phi$, we use the fact that $\forall x\phi$ fails at the point, and continue to a possibly even further point where some instance $\neg[c/x]\phi$ is true, which must exist by the truth definition plus cofinality. ⊣

3 Taking the framework further

Possibility models for a classical setting may look like 'beautiful noise' induced by purely philosophical qualms about standard models of first-order logic. However, the remainder, and in fact the bulk of my old report was devoted to taking stock of what a switch to such models would do to working model theory. In particular, new versions of old notions emerge, while we can also use notions and techniques from modal logic. Here is a key example.

From ultraproducts to filter products Consider the fundamental notion of an *ultraproduct*, used in the well-known proof of the compactness theorem for constructing a model for a finitely satisfiable set Σ of formulas out of models for its finite subsets. This proof uses an arbitrary maximal extension of some initial filter carrying all the key information – here, the 'regular filter' of all sets $\{\Delta \subseteq \Sigma \mid \Sigma_0 \subseteq \Delta\}$ with Σ_0 running over all finite subsets of Σ.

---

[8] I use the countability restriction as a precaution, which need not be necessary for a best result. Also, the precise enumeration pattern requires some care, the details of which are omitted here.

The following notion makes the compactness construction unique, while preserving the basic behavior of ultraproducts with respect to first-order formulas according to Los' Theorem.

Consider any family $\{\mathbf{M}_i \mid i \in I\}$ of possibility models in our sense, and fix some filter F on I. We define the *filter product* of this family *with respect to F* as the following new possibility model $\mathbf{M}$. Let the variable G run over all non-empty filters on I that extend F, and for the worlds, collect all world-like objects in the products $\Pi^G \mathbf{M}_i$ of the family $\{\mathbf{M}_i \mid i \in I\}$ in the usual model-theoretic sense: each world is an equivalence class of functions from indices i to worlds in $\mathbf{M}_i$, divided out by G. As for objects belonging to such worlds w, these are equivalence classes of functions from indices to objects in $D(w)$ divided out by the relevant filter.[9]

The interpretation of atomic predicates on these functional objects is given locally in the usual way. For the sake of illustrating this product setting and the role of filters, we display one clause:

P holds of a tuple of equivalence classes $\mathbf{d}^G$ in $\Pi^G \mathbf{M}_i$ iff
$\{i \in I \mid$ in $\mathbf{M}_i, P$ holds of the tuple of coordinates $(\mathbf{d})_i\} \in G$.

Analogously, for the stage inclusion order, we put $(w)^G \leq (v)^{G'}$ if and only if $\{i \in I \mid$ in $\mathbf{M}_i, w_i \leq v_i\} \in G'$.[10]

REMARK. *There is a small but interesting issue here with identity across worlds. Worlds $(w)^G$ at stage G need not occur as such at later stages, since an extended filter G' may identify more points in the product of the models $\mathbf{M}_i$ than G did. Even so, there is a family of well-defined functions $f_{GG'}$ taking worlds $(w)^G$ at earlier stages to worlds $(w)^{G'}$ at later stages (and the same is true for objects at those worlds) – where the well-definedness uses the crucial intersection property of filters. Thus, we have to generalize the earlier models to the case where domain cumulation can run via arbitrary functions, not necessarily the identity. This is a minor but significant modification of our earlier setting, with independent motivation in current semantics of modal predicate logic, for instance, in removing cases of incompleteness.[11] In order to keep notation simple, in what follows, we do not display the maps $f_{GG'}$ explicitly.*

Now we can prove the following generalized Los Theorem for filter products $\mathbf{M}$ of a family of models $\{\mathbf{M}_i \mid i \in I\}$ with respect to F. For any world $(w)^G$, formula ϕ, and tuple of objects $(\mathbf{d})^G$:

FACT 3.1. $\mathbf{M}, (w)^G, (\mathbf{d})^G \models \phi$ *iff* $\{i \in I \mid \mathbf{M}_i, w_i, (\mathbf{d})_i \models \phi\} \in G$.

Proof. The proof for this equivalence is a simple induction like that for the Los Theorem itself, but now using our definition of the logical constants referring to filter extensions. (a) The atomic clause works by definition. (b) The case of a conjunction uses upward closure plus the intersection property of filters in a straightforward manner. (c1) For a negation $\neg \phi$, first assume that we do not have the set $\{i \in I \mid \mathbf{M}_i, w_i, (\mathbf{d})_i \models \neg \phi\}$ in the filter G. Then, by standard reasoning, there must be an extended filter G' that contains $\{i \in I \mid \mathbf{M}_i, w_i, (\mathbf{d})_i \models \neg \phi\}$. Unpacking this by the truth definition plus the upward closure of filters, we have that the set of indices $\{i \in I \mid$ for some $v_i \geq_i w_i, \mathbf{M}_i, v_i, (\mathbf{d})_i \models \phi\}$ is in G'. But then we can collect all these worlds v_i into one functional object v in the domain for the product of the models $\mathbf{M}_i$ – where it will not matter what we do on

---
[10] By way of motivation, note that this definition gives the intuitively correct result for inclusion if the stage does not change, i.e., $G' = G$.

[11] Van Benthem 2010 discusses the literature on generalized functional models for modal predicate logic, where objects at one world only need to have 'counterparts' in other worlds.

indices outside the preceding set, given the fact that we have divided out by the filter G'.[12] Now, by the inductive hypothesis, $(v)^{G'}$ makes ϕ true for the relevant objects $\mathbf{d}$, while by our definition of filter products, $(v)^{G'}$ is a stage-inclusion successor of $(w)^G$. Unpacking the truth definition for negation, it follows that not $\mathbf{M}, (w)^G, (\mathbf{d})^G \models \neg\phi$. (c2) As for the opposite direction, suppose that not $\mathbf{M}, (w)^G, (\mathbf{d})^G \models \neg\phi$. So there is a $G' \supseteq G$ and $(v)^{G'} \geq (w)^G$ such that $\mathbf{M}, (v)^{G'}, (\mathbf{d})^{G'} \models \phi$. By the inductive hypothesis, then, $\{i \in I \mid \mathbf{M}_i, v_i, (\mathbf{d})_i \models \phi\}$ is in G' (#). It follows that the set $\{i \in I \mid \mathbf{M}_i, w_i, (\mathbf{d})_i \models \neg\phi\}$ cannot be in G. For, by the truth definition for negation, it is contained in $\{i \in I \mid \text{not } \mathbf{M}_i, w_i, (\mathbf{d})_i \models \phi\}$, which would then also be in the filter G and therefore in G'. But this contradicts (#). (d) The case of the universal quantifier can be proved by similar reasoning. We merely note one direction where a small additional move is needed. If the set $\{i \in I \mid \text{not } \mathbf{M}_i, w_i, (\mathbf{d})_i \models \forall x\phi\}$ is in some filter $G' \supseteq G$, then we can create a further product stage v as we did before for the negation case, but now also with objects d attached to its coordinate worlds such that $\phi(d)$ fails. And we can do still better: by applying cofinality to the relevant coordinates, and then collecting the worlds obtained into a set in G', we can even make sure that $\neg\phi(d)$ holds for the matching product world.

Now we are almost done. We still must check that filter products are truly possibility models. Here Persistence is a simple consequence of our product definition and the truth definition, and it amounts technically to saying that the earlier domain-inclusion maps $f_{GG'}$ are relational homomorphisms. Checking Cofinality for the product model requires a little bit more care (as is in fact done in the original report), using cofinality inside the coordinate models $\mathbf{M}_i$ to pass to upward worlds where the negation of the relevant atomic formula holds, and then collecting these worlds into a member of a suitable filter extending F. ⊣

As a corollary, we get the compactness theorem for possibility semantics by a canonical construction without arbitrary extensions to ultrafilters. Taking a filter product of models for finite subsets with respect to the regular filter will give a model for the whole set.

Further possibility-based model theory The original 1981 report has more material than what we cover here. It defines analogues for all the major modal model constructions: generated submodels, disjoint unions, and in particular also, a notion of 'zigzag invariance' that combines bisimulation for stages with potential isomorphism for objects. In addition, we have the above filter products, plus an inverse operation of taking a 'filter base'. The upshot of the analysis is a complex argument leading up to the following statement:

CLAIM 3.2. *A class of possibility models is definable by a set of first-order sentences if and only if it is closed under the formation of generated submodels, disjoint unions, zigzag-invariant images, filter products and filter bases.*

Given that I have not checked this proof in detail yet, the status of this claim is at present a conjecture. However, in this connection, it is worth noting that Holliday 2015 uses a notion like the above filter product to develop a duality between classical and possibilities models.

Excursion: submodel preservation Instead of pursuing the preceding theme, I will briefly raise a further one, not in the original report: namely, what happens to an old interest of Albert and mine (and our friends): preservation theorems for modal and intuitionistic logic, and in particular,

---

[12] *A methodological caveat.* Here we use the Axiom of Choice, and this seems to be a sin against our earlier aim of reducing our dependence on such principles. It would be of interest to see if we can circumvent the present use.

characterizing preservation under *submodels*. Which special syntactic forms arise on possibility models, and what is the relation to the classical Los-Tarski Theorem?

We can think in two natural directions here. One way is to start from the usual syntactic target class of 'universal' sentences of the form $\forall$-prefix + matrix with only $\{\neg, \wedge, \text{atoms}\}$, and ask for its characteristic preservation behavior in the setting of possibility models. This turns out to involve going to 'submodels' of possibility models in the following sense:

> We leave the stage structure the same and only throw away objects from domains of worlds, in such a way that the basic constraints of cumulation, persistence, and cofinality continue to hold.

I believe that a preservation theorem of this kind can be proved by a variant of the standard compactness-based argument for first-order logic, but now using filter models, rather than standard models.

But there is also an opposite direction of thought. We can investigate what natural notions of submodel arise in the setting of possibility models, and then ask what syntactic preservation classes match these. Then at least three natural options arise for operating on models:

> One option is dropping *objects* from world domains as above,
> a second is dropping *stages*, and a third: stages plus objects.

This time, one has to be even more cautious, since dropping stages can easily endanger the truth of cofinality: concrete counter-examples are easy to find.[13] But assuming that we take care of this desideratum in some way (only taking 'admissible' submodels), it will be clear that we now need strong syntactic restrictions on the matrix part of formulas with only $\neg, \wedge,$ and atoms to ensure persistence, since the negation is really a modality over the stage order.

I have no preservation theorem of this sort yet, but now that possibilities semantics is on the table again, I submit that these are interesting questions. Moreover, to get to such a result, I would urge the reader to visit an obvious inspiration point. Preservation theorems for passing to submodels in modal or intuitionistic propositional logic, with appropriate proof techniques for that setting, can be found in van Benthem, de Jongh, Renardel & Visser 1995.

Further directions Finally, here are two lines of inquiry not studied in any detail in the original document discussed here, but which make good sense in the current literature on possibility models.

Language extensions Any serious generalization of a semantics naturally raises the question whether the old vocabulary of logical operators in the original restricted model class is still adequate for capturing essential distinctions in the new broader setting. Indeed, intuitionistic logic itself involves a language extension, as it has two natural disjunctions. One is the classical 'slow disjunction' of our semantics, saying that a choice between the disjuncts is made cofinally – where not always the same disjunct needs to be chosen. Another option is the constructive 'fast disjunction', where we have to choose the disjunct right now at the present stage. In this case, clearly, we must drop the cofinality assumption governing our classical semantics. But the origins of possibility models also suggest other options, with languages that drop even persistence.

In terms of, say, the interval models of the early 1980s, where points may be viewed limits of histories of ever descending subsets, the conjunction of persistence and cofinality amounts to a

---

[13] One technical way to see the problem is that cofinality is not a statement that itself can be formulated in syntactic universal form.

strong assumption of 'reducibility' of truth at intervals to truth in the points that they contain.[14] Such simply reducible notions clearly fail to exploit the full sui generis structure of intervals, such as the 'collective properties' that hold on them, or the significant *merges* (or lowest upper bounds in the inclusion order) that can be made of them. If we do go this way, we get additional notions, such as a 'sum modality' $\phi + \psi$ true at those stages that are lowest upper bounds of a stage where ϕ is true and one where ψ is true. This new notion supports an interesting logic of its own (van Benthem 1989). Thus, possibility semantics may need richer languages to unleash its full potential, whether the resulting logic is classical or not.

Translations Another natural issue arising in the present setting is syntactic *translation* of languages and proof calculi. It is well-known that merely looking at surface syntax does not tell us the true identity of logical systems: we need to look at their systematic connections.[15] Now the possibility semantics for classical logic presented here is reminiscent of the double-negation based Gödel-Gentzen translation of classical logic into intuitionistic logic.

What this suggests is a dual perspective on what we have been doing in this article: as either changing a semantics for a given logic, or as giving a translation into some other logic. Moreover, translating our classical logic into intuitionistic logic is not the end of the road. The Gödel translation of intuitionistic logic into classical modal *S4* can also be applied here (using ideas going back to McKinsey and Tarski) to rephrase our semantics and logic inside a *bimodal* classical logic, with one modality for the inclusion structure and modalities for the first-order quantifiers. A thorough study of bimodal translational perspectives is made for modal logic itself in van Benthem, Bezhanishvili & Holliday 2015. Thus, we may either use the new semantics, or move up to a standard semantics for an enriched language for possibility models.

4 Conclusion

This admittedly sketchy revival of an old manuscript may have shown that possibility semantics for classical logic is of conceptual interest in its own right – though by now, this point has also been made quite persuasively in other literature (Humberstone 1981, Holliday 2015). As an additional benefit, a model theory for classical logic going for bare essentials, such as unique filter-based constructions for completeness and compactness with no appeal to the Axiom of Choice, seems a worthwhile enterprise (cf. Andréka, Bezhanishvili, Németi & van Benthem 2014). Perhaps this will merely restate in a more austere style what is already known by other means, but there may also be more to it. Our discussion of preservation theorems at least suggests how the playing field of classical model theory may get richer in the process. I have more to say on this, but my text already seems a lot for a short Festschrift piece.

Let me end on a personal note, the same way I started. The date on the handwritten report discussed here is December 1981. Thinking about this gave me a shock, and again made me realize what an alien mind wrote this old text. December 1981 is the very month my second son was born, so what was this author thinking? Did he have nothing better to do in that period than write up logic? I am not sure I know all the ins and outs of Albert's mind and its priorities, but one thing I do know. He would never cruelly abandon little babies like that.

---

[14]In natural language, this form of reducibility would restrict attention to the special linguistic class of 'distributive' predicates.

[15]By the way, as the cognoscenti will know, translations between logical systems are of course a recurrent theme throughout Albert's oeuvre.

BIBLIOGRAPHY

H. Andréka, N. Bezhanishvili, I. Németi & J. van Benthem, 2014, 'Changing a Semantics: Opportunism or Courage?', in M. Manzano, I. Sain and E. Alonso, eds., *The Life and Work of Leon Henkin*, Birkhaüser Verlag, 307–337.

J. Barwise and J. Perry, 1983, *Situations and Attitudes*, The MIT Press, Cambridge Mass.

J. van Benthem, 1981, 'Possible Worlds Semantics for Classical Logic', Report ZW 8018, Mathematical Institute, University of Groningen.

J. van Benthem, 1983, *The Logic of Time*, Reidel, Dordrecht.

J. van Benthem, 1986, 'Partiality and Non-Monotonicity in Classical Logic', *Logique et Analyse* 29, 225–247.

J. van Benthem, 1989, 'Semantic Parallels in Natural Language and Computation', in H-D. Ebbinghaus et al., eds., *Logic Colloquium 1987*, North-Holland, Amsterdam, 331–375.

J. van Benthem, 2009, 'The Information in Intuitionistic Logic', *Synthese* 167:2, 251–270.

J. van Benthem, 2010, *Modal Logic for Open Minds*, CSLI Publications, Stanford University.

J. van Benthem, N. Bezhanishvili & W. Holliday, 2015, 'A Bimodal Perspective on Possibility Semantics', ILLC, University of Amsterdam and Department of Philosophy, UC Berkeley.

J. van Benthem, D. de Jongh, G. Renardel & A. Visser, 1995, 'NNIL, A Study in Intuitionistic Propositional Logic', in A. Ponse, M. de Rijke & Y. Venema, eds., *Modal Logic and Process Algebra*, CSLI Lecture Notes, Stanford University, 289–326.

P. Blackburn, J. van Benthem & F. Wolter, eds., 2006, *Handbook of Modal Logic*, Elsevier Science, Amsterdam.

A. Bochmann, 2015, 'Classical Scott Consequence Relations and their Mereological Semantics', Holon Institute of Technology, Israel.

W. Holliday, 2014, 'Partiality and Adjointness in Modal Logic', in R. Goré, B. Kooi & A. Kurucz, eds., *Advances in Modal Logic*, Volume 10, University of Groningen, 313–332.

W. Holliday, 2015, 'Possibility Frames and Forcing for Modal Logic', Department of Philosophy & Group in Logic and the Methodology of Science, University of California, Berkeley.

L. Humberstone, 1981, 'From Worlds to Possibilities', *Journal of Philosophical Logic* 10(3), 313–339.

I. Rumfitt, 2015, *The Boundary Stones of Thought. An Essay in the Philosophy of Logic*, Oxford University Press, Oxford.

A. Troelstra & D. van Dalen, 1988, *Constructivism in Mathematics: An Introduction*, North-Holland Publishers, Amsterdam.

F. Veltman, 1984, 'Data Semantics', in J. Groenendijk, T. Janssen, and M. Stokhof, eds., *Truth, Interpretation and Information,* Foris, Dordrecht, 43–63.

A. Visser, 2014, 'Johan van Benthem and Löb's Logic', Lecture at J65 Retirement Symposium, University of Amsterdam.

Y. Wang, 2015, *Notes on Mathematical Modal Logic*, Lectures by Johan van Benthem at the Berkeley-Stanford Logic Circle in San Francisco, Department of Philosophy, Stanford University.

Some Abstract Versions of Gödel's Second Incompleteness Theorem Based on Non-Classical Logics

LEV BEKLEMISHEV AND DANIYAR SHAMKANOV

To Albert Visser, a remarkable logician and a dear friend, whose papers and conversations are a source of constant inspiration

Abstract

We study abstract versions of Gödel's second incompleteness theorem and formulate generalizations of Löb's derivability conditions that work for logics weaker than the classical one. We isolate the role of the contraction rule in Gödel's theorem and give toy examples of systems based on modal logic without contraction invalidating Gödel's argument.

1 Introduction

One of the topics that have been fascinating logicians over the years is Gödel's second incompleteness theorem (G2). Both mathematically and philosophically G2 is well known to be more problematic than his first incompleteness theorem (G1). G1 and Rosser's Theorem are well understood in the context of recursion theory. Abstract logic-free formulations have been given by Kleene [12] ('symmetric form'), Smullyan [20] ('representation systems') and others. Sometimes G2 is considered as a minor addition to G1, whose role is to exhibit a specific form of the sentence independent from a given theory, namely its consistency assertion. However, starting with the work of Kreisel, Orey, Feferman, and others, who provided various nontrivial uses of G2, it has been gradually understood that the two results are of a rather different nature and scope. G2 has more to do with the (modal-logical) properties of the provability predicate and the phenomenon of self-reference in sufficiently expressive systems. A satisfactory general mathematical context for G2, however, still seems to be lacking.

The main difficulties in G2 are due to the fact that we cannot easily delineate a class of formulas that 'mean' consistency. Thus, the most intuitively appealing formulation of G2 — *sufficiently strong consistent theories cannot prove their own consistency* — remains non-mathematical. For a concrete formal system, such as Peano arithmetic PA, one can usually write out a specific 'natural' formula Con_{PA} and declare it to be the expression of consistency. This approach is rather common in mathematics but has several deficiencies: Firstly, it ties the statement to a very particular formula, coding mechanism etc., and provides no clue why this choice is better than other ones. Secondly, instead of a general theorem working uniformly for a wide class of theories, we only obtain a specific statement for an individual theory such as PA. We do not know what is the natural consistency assertion for an arbitrary extension of PA. Thus, we have a problem with translating our informal intuition into strict mathematical terms.

The way to better understand G2 is through investigating its range and generalizations. A lucky circumstance is that G2 also holds for larger syntactically defined classes of consistency formulas, some of which are apparently intensionally correct (adequately express consistency), but some are not. Thus, it is still possible to formulate mathematical results in certain important aspects *more* (rather than less) general than the broad intuitive formulation of G2 above.

A universally accepted approach to general formulations of G2 appeared in the fundamental paper by Feferman [3] who showed, among other things, that G2 holds for all consistency assertions defined by Σ_1-numerations. Feferman deals with first-order theories T in the language containing that of PA and specified by recursively enumerable (r.e.) sets of axioms. Feferman assumes fixed some natural Gödel numbering of the syntax of T as well as some specific axiomatization of first order logic. A Σ_1-formula $\alpha(x)$ defining the set of Gödel numbers of axioms of T in the standard model of PA is called a Σ_1-*numeration* of T.[1] It determines the provability formula $\text{Prov}_\alpha(x)$ and the corresponding consistency assertion Con_α. Feferman's statement of G2 is that for all consistent theories T given by Σ_1-numerations α and containing a sufficiently strong fragment of PA, the formula Con_α is unprovable in T.

This theorem is considerably more general than any specific instance of G2 for an individual theory T. However, it also presupposes quite a lot: first order logic and its axiomatization, Gödel numbering, the way formula Prov_α is built from α.

Exploring bounds to G2 leads to relaxing various assumptions involved in Feferman's statement:

- One can weaken the axioms of arithmetic (for a representative selection see Bezboruah–Shepherdson [2], Pudlák [14], Wilkie–Paris [24], Adamowicz–Zdanowski [1], Willard [25, 26]).

- One can consider theories modulo interpretability. This approach started with the work of Feferman [3]. In the recent years it lead to particularly attractive coding-free formulations of generalizations of G2 due to Harvey Friedman and Albert Visser (see [21–23]).

- One can weaken the requirements on the proof predicate aka derivability conditions (see Feferman [3], Löb [13], Jeroslow [10, 11]).

- One can weaken the logic.

It is the latter two aspects, less studied in the literature, that we are going to comment on in this note. Firstly, let us briefly recall the history of derivability conditions.

Gödel [5] gave a sketch of a proof of G2 and a promise to provide full details in a subsequent publication. This promise has not been fulfilled, and a detailed proof of this theorem — for a system Z related to first-order arithmetic PA — only appeared in a monograph by Hilbert and Bernays [9]. In order to structure a rather lengthy proof, Hilbert and Bernays formulated certain conditions on the proof predicate in Z sufficient for the proof of G2. Later Martin Löb [13] gave an elegant form to these conditions by stating them fully in terms of the provability predicate $\text{Pr}(x)$ and obtained an important strengthening of G2 known as Löb's Theorem. Essentially the same properties of the provability predicate were earlier noted by Gödel in his note [6], where he proposed to treat the provability predicate as a connective $\Box$ in modal logic, though the idea that these conditions

---

[1] Feferman deals with the notion of r.e. formula rather than with the equivalent notion of Σ_1-formula more common today.

constitute necessary requirements on a provability predicate most likely only appeared later. For the sake of brevity we call the Gödel–Hilbert–Bernays–Löb conditions simply *Löb's conditions* below.

A traditional proof of G2 (for arithmetical theories) consists of a derivation of G2 from the fixed point lemma using Löb's conditions (see e.g. [17]). An accurate justification of these conditions is technically not so easy, and textbooks rarely provide enough details here, however see Smoryński [19] and Rautenberg [15] for readable expositions.

Löb's conditions are applicable to formal theories at least containing the connective of implication and closed under the *modus ponens* rule. Here we give more general abstract formulations of G2 which presuppose very little about logic. They are rather close in the spirit and the level of generality to the recursion-theoretic formulations of G1 due to Smullyan. When a good implication is added to the language one essentially obtains the familiar Löb's conditions. However, we show that Gödel's argument presupposes admissibility of the contraction rule restricted to □-formulas in the logic under consideration. Moreover, the uniqueness of Gödelian fixed point is based on the similarly restricted form of weakening.

In the last part of the paper we present a system invalidating a formalized version of G2. We consider a version of propositional modal logic K4 based on the contraction-free fragment of classical logic extended by fixed point operators (defined for any formulas modalized in the fixed point variables). By means of a cut-elimination theorem for this system we establish the failure of G2 and some other properties such as the infinity of the Gödel- and Henkin-type fixed points.

2 Abstract Provability Structures

DEFINITION 2.1. Let us call *an abstract consequence relation* a structure $S = (L_S, \leq_S, \top, \bot)$, where L_S is a set of *sentences* of S, $\leq_S$ is a transitive reflexive relation on L_S, $\top$ and $\bot$ are distinguished elements of L_S ('axiom' and 'contradiction'). A sentence $x \in L_S$ is called *provable in S*, if $\top \leq_S x$, and *refutable in S*, if $x \leq_S \bot$. Sentences x, y are called *equivalent in S*, if $x \leq_S y$ and $y \leq_S x$. The equivalence of x and y will be denoted $x =_S y$.

The structure S represents syntactical (rather than semantical) data about the theory in question. In a typical case, for example, for arithmetical theories S, the relation $x \leq_S y$ denotes the provability of y from hypothesis x, whereas $\top$ and $\bot$ are some standard provable and refutable formulas, respectively, e.g., $0 = 0$ and $0 \neq 0$.

In concrete situations we can enrich this structure by additional data, for example, by the conjunction and the implication connectives. Notice that we do not assume either $\bot \leq_S x$ or $x \leq_S \top$, nor do we assume the existence of any logical connectives (such as negation) in S.

S is called *inconsistent* if $\top \leq_S \bot$, otherwise it is called *consistent*. By transitivity, if S is consistent then no sentence is both provable and refutable. S is called *complete* if every $x \in L_S$ is either provable or refutable. S is called *r.e.*, if L_S is recursive and $\leq_S$ is r.e. (as a binary relation). T is called an *extension* of S if $L_T = L_S$ and $\leq_S$ is contained in $\leq_T$.

Let P_S and R_S denote the sets of provable and of refutable sentences of S, respectively. If S is consistent and r.e., then P_S and R_S is a pair of disjoint r.e. sets. We say that S *separates pairs of disjoint r.e. sets* if for each such pair (A, B) there is a total computable function f such that

$$\forall n \in A\ f(n) \in P_S \text{ and } \forall n \in B\ f(n) \in R_S.$$

The following statement is a natural version of G1 and Rosser's theorem for abstract consequence relations (á la Kleene and Smullyan); we omit the standard proof.

PROPOSITION 2.2. *1. If S is r.e., consistent and complete, then both P_S and R_S are decidable.*

2. If S is r.e. and separates disjoint pairs of r.e. sets, then every consistent extension of S is incomplete and undecidable.

Next, we introduce two operators $\Box, \boxtimes : L_S \to L_S$ representing provability and refutability predicates in S.

DEFINITION 2.3. *Provability* and *refutability operators* for an abstract consequence relation S are functions $\Box, \boxtimes : L_S \to L_S$ satisfying the following conditions, for all $x, y \in L_S$:

C1. $x \leq_S y \Rightarrow \Box x \leq_S \Box y$ and $\boxtimes y \leq \boxtimes x$.

C2. $\top \leq_S \boxtimes \bot$;

C3. $x \leq_S \Box y$ and $x \leq_S \boxtimes y \Rightarrow x \leq_S \boxtimes \top$;

C4. $\boxtimes x \leq_S \Box \boxtimes x$.

The structure $(L_S, \leq_S, \top, \bot, \Box, \boxtimes)$ is called an *abstract provability structure* (APS).

Intuitively, $\Box x$ is the sentence expressing the provability of a sentence x, whereas $\boxtimes x$ expresses its refutability in S. Condition C1 means that provability of y follows from provability of x whenever y is derivable from x; similarly, refutability of y implies refutability of x. Conditions C2 and C3 are axioms for contradiction: according to C2, refutability of $\bot$ is provable in S; according to C3, $\top$ is refutable if some sentence y is both provable and refutable. Finally, Condition 4 means that the refutability of x can be formally checked in S. It is an analogue of Löb's condition L2 (see below).

Note that we consider the refutability operator on a par with the provability operator, since we do not assume that the logic of S necessarily has a well-defined operation of negation, that is, we cannot always define $\boxtimes x$ as $\Box \neg x$.

REMARK. It is rather natural to additionally require that $\Box \bot =_S \boxtimes \top$: refutability of $\top$ and provability of $\bot$ are expressed by the same statement $\top \leq_S \bot$. Yet, it is not, strictly speaking, needed in this very abstract context, and we take $\boxtimes \top$ as our default expression of inconsistency.

DEFINITION 2.4. We say that an abstract provability structure S has a *Gödelian fixed point* if there is a sentence $p \in L_S$ such that $p =_S \boxtimes p$. Similarly, S has a *Henkinian fixed point* if there is a sentence $p \in L_S$ such that $p =_S \Box p$.

Notice that Gödel considered a dual sentence q expressing its own unprovability in S. R. Jeroslow [11] noticed that the sentence stating its own refutability allows to prove G2 under somewhat more general conditions than those of Löb. In our formalism the sentence q is not expressible, therefore we are using Jeroslow's idea.

A very abstract version of G2 can now be stated as follows.

THEOREM 2.5. *Suppose an APS S has a Gödelian fixed point.*

1. If S is consistent, then $\boxtimes \top$ is not refutable in S;

2. $\boxtimes \boxtimes \top \leq_S \boxtimes \top$, that is, Statement (i) is formalizable in S.

Proof. Let $p =_S \boxtimes p$. First we prove Statement (ii) omitting the subscript $_S$ everywhere:

1. $\boxtimes p \leq \Box \boxtimes p \leq \Box p$ by C4 and C1;

2. $p = \boxtimes p \leq \boxtimes \top$ by C3 (since $\boxtimes p \leq \boxtimes p$);

3. $\boxtimes \boxtimes \top \leq \boxtimes p = p \leq \boxtimes \top$ by C1.

Proof of Statement (i): Assume $\boxtimes \top \leq \bot$. By the previous argument $p \leq \boxtimes \top$, hence $p \leq \bot$. By C1, $\boxtimes \bot \leq \boxtimes p = p \leq \bot$. Therefore, by C2, $\top \leq \boxtimes \bot \leq \bot$. ⊣

The following statement shows that under some additional condition the Gödelian–Jeroslowian fixed point is unique modulo equivalence in S and coincides with the inconsistency assertion for S. Therefore, the existence of such a fixed point is not only sufficient but also necessary for the validity of (a formalized version of) G2. The additional condition is

C5. $x \leq_S \top$, for all $x \in L_S$.

THEOREM 2.6. *Assume C5 holds for S. Then $p =_S \boxtimes \top$ for all Gödelian fixed points p and (if such a sentence exists)*
$$\boxtimes \boxtimes \top =_S \boxtimes \top.$$

Proof. We know that $p \leq \boxtimes \top$. Since $p = \boxtimes p \leq \top$ we obtain $\boxtimes \top \leq \boxtimes p = p$. Hence $p = \boxtimes \top$ and therefore $\boxtimes \boxtimes \top = \boxtimes \top$. ⊣

3 Consequence Relations with Implication

Classical Löb's conditions emerge for APSs with an implication. A decent implication can be defined for consequence relations representing derivability of a sentence from a (multi)set of assumptions. In other words, we now go to a more general but less symmetric format $\Gamma \vdash \phi$, where Γ is a finite multiset and ϕ an element of a given set L_S. In order to avoid confusion we use the more standard notation $\vdash$ instead of $\leq$ and will follow the standard conventions of sequential proof format. In particular, Γ, ϕ denotes the result of adjoining $\phi \in L_S$ to a multiset of sentences Γ, and Γ, Δ denotes the multiset union of Γ and Δ.[2]

DEFINITION 3.1. *A consequence relation with an implication is a structure $S = (L_S, \vdash, \rightarrow, \top, \bot)$ where $\vdash$ is a binary relation between finite multisets of elements of L_S and elements of L_S; $\rightarrow$ is a binary operation on L_S; $\top$ and $\bot$ are distinguished elements of L_S such that the following conditions hold:*

I1. $\phi \vdash \phi$;

I2. if $\Gamma, \psi \vdash \phi$ and $\Delta \vdash \psi$ then $\Gamma, \Delta \vdash \phi$;

I3. $\Gamma, \phi \vdash \psi \iff \Gamma \vdash \phi \rightarrow \psi$;

I4. $\Gamma, \top \vdash \phi \iff \Gamma \vdash \phi$.

---

[2] Our strive for generality does not go as far as to consider lists of formulas rather than multisets.

Notice that Conditions I1 and I2 generalize reflexivity and transitivity of $\leq_S$. Setting $\phi \leq_S \psi$ as $\phi \vdash \psi$ yields an abstract consequence relation in the sense of Definition 2.1. Condition I3 speaks for itself. Condition I4 conveniently stipulates that provability from the empty multiset of assumptions is the same as provability from $\top$. It also implies $\top \to \bot =_S \bot$.

Similarly to the implication one can consider consequence relations with other additional connectives of which we are mostly interested in conjunction.

DEFINITION 3.2. *Conjunction* is a binary operator $\otimes : L_S^2 \to L_S$ satisfying
$$\Gamma, \phi, \psi \vdash \theta \iff \Gamma, \phi \otimes \psi \vdash \theta.$$

Clearly, such an operation is associative and commutative, though not necessarily idempotent. If conjunction is available, then $\phi_1, \ldots, \phi_n \vdash \psi$ holds in S if and only if $\phi_1 \otimes \cdots \otimes \phi_n \vdash \psi$, disregarding the ordering of brackets. Hence, in the presence of conjunction in S the relation $\leq_S$ uniquely determines the corresponding multiset consequence relation $\Gamma \vdash \phi$.

For consequence relations with an implication we can define negation $\neg \phi$ by $\phi \to \bot$.[3] The following simple lemma shows that the implication respects the deductive equivalence relation in S and the negation satisfies the contraposition principle.

LEMMA 3.3. 1. *If* $\Gamma \vdash \phi \to \psi$ *and* $\Delta \vdash \phi$, *then* $\Gamma, \Delta \vdash \psi$;

2. $\phi_1 =_S \phi_2$ *and* $\psi_1 =_S \psi_2$ *implies* $(\phi_1 \to \psi_1) =_S (\phi_2 \to \psi_2)$;

3. $\Gamma, \phi \vdash \psi$ *implies* $\Gamma, \neg \psi \vdash \neg \phi$.

Next we turn to the derivability conditions. Assume S is a consequence relation with an implication.

DEFINITION 3.4. $\Box : L_S \to L_S$ satisfies Löb's derivability conditions for S if

L1. $\Box(\phi \to \psi) \vdash \Box \phi \to \Box \psi$;

L2. $\Box \phi \vdash \Box \Box \phi$;

L3. $\vdash \phi$ implies $\vdash \Box \phi$.

LEMMA 3.5. *For any consequence relation with an implication the following statements are equivalent:*

1. $\Box$ *satisfies Löb's conditions for* S;

2. $\Box$ *satisfies L2 and* S *is closed under the rule*
$$\frac{\Gamma \vdash \phi}{\Box \Gamma \vdash \Box \phi};$$

3. S *is closed under the rule*
$$\frac{\Gamma, \Box \Delta \vdash \phi}{\Box \Gamma, \Box \Delta \vdash \Box \phi}.$$

---

[3] In general, it is possible to consider other, for example, stronger, forms of negation.

REMARK. Notice that the last rule is formulated slightly differently from the more standard rule for modal logic K4:
$$\frac{\Gamma, \Box\Gamma \vdash \phi}{\Box\Gamma \vdash \Box\phi}.$$
The latter has a form of built-in contraction that we are not assuming here.

It is natural to define refutability $\boxtimes\phi$ as provability of negation $\Box\neg\phi$. Notice that since $\bot =_S \top \to \bot$ we have $\boxtimes\top =_S \Box\bot$, whenever L1 holds for $\Box$. However, as the example in Section 4 shows, this translation does not always yield an APS in the sense of Definition 2.3. To sort things out we need to consider two additional conditions on the consequence relation.

DEFINITION 3.6. A consequence relation with an implication

- *satisfies contraction* if $\Gamma, \phi, \phi \vdash \psi$ implies $\Gamma, \phi \vdash \psi$;

- *satisfies weakening* if $\Gamma \vdash \psi$ implies $\Gamma, \phi \vdash \psi$, for any ϕ.

The first condition intuitively means that any hypothesis can be used several times in a derivation. Recall that for Girard's linear logic this condition is not met, however it is postulated, for example, for relevant logics. It turns out that a certain amount of contraction is essential for the proof of G2.

The second condition corresponds to the requirement $x \leq_S \top$ that was needed to guarantee that $\boxtimes\top$ is a Gödelian fixed point and that such a fixed point is unique.

For consequence relations with an implication we have the following proposition.

PROPOSITION 3.7. *Suppose S satisfies contraction, $\Box : L_S \to L_S$ satisfies Löb's conditions for S and $\boxtimes\phi := \Box(\phi \to \bot)$. Then $(L_S, \leq_S, \Box, \boxtimes, \top, \bot)$ is an APS.*

Proof. By Lemma 3.3 $\phi \vdash \psi$ implies $\neg\psi \vdash \neg\phi$. This yields Conditions C1 and C4. Condition C2 obviously follows from Condition 1 for a good consequence relation. Let us prove C3. By Lemma 3.3(i) we have: $\phi, \neg\phi, \top \vdash \bot$. Hence, $\phi, \neg\phi \vdash \top \to \bot$, therefore $\Box\phi, \Box\neg\phi \vdash \Box\neg\top$ by Condition 1. The rules of transitivity and contraction imply that, if $\Gamma \vdash \Box\phi$ and $\Gamma \vdash \Box\neg\phi$, then $\Gamma \vdash \Box\neg\top$. ⊣

Thus, from Proposition 3.7 we obtain the following expected corollary, parallel to Theorem 2.5, for consequence relations satisfying contraction.

THEOREM 3.8. *Suppose S satisfies contraction and $\Box$ satisfies Löb's conditions for S. Then Theorem 2.5 holds for S.*

For an analogue of Theorem 2.6 on the uniqueness of a Gödelian fixed point we also need a weakening property.

THEOREM 3.9. *Suppose S satisfies contraction and weakening and $\Box$ satisfies Löb's conditions for S. Then all Gödelian fixed points in S (if exist) are equivalent to $\boxtimes\top =_S \Box\bot$.*

REMARK. As it turns out, contraction and weakening for S, though natural, are somewhat excessive requirements for the validity of Theorems 3.8 and 3.9. A consequence relation with an implication

- *satisfies $\Box$-contraction* if $\Gamma, \Box\phi, \Box\phi \vdash \psi$ implies $\Gamma, \Box\phi \vdash \psi$;

- *satisfies □-weakening* if $\Gamma \vdash \phi$ implies $\Gamma, \Box\psi \vdash \phi$, for any ψ.

Conditions C3 and C5 of APS can also be weakened to

C3'. $\boxtimes x \leq_S \Box y$, $\boxtimes x \leq_S \boxtimes y \Rightarrow \boxtimes x \leq_S \boxtimes\top$;

C5'. $\boxtimes x \leq_S \top$.

With these modifications, the proofs of Theorems 2.5 and 2.6 stay the same, which in turn yields more general versions of Theorems 3.8 and 3.9 for consequence relations satisfying only □-contraction and □-weakening.

The property of □-contraction actually holds for some meaningful arithmetical systems lacking general contraction rule, for example, for a version of Peano arithmetic based on affine predicate logic considered by the second author of this paper (as yet, unpublished).

4 A Non-Gödelian Theory with Fixed Points

In view of Theorems 3.8 and 3.9 it is natural to ask whether the assumptions of □-contraction and □-weakening are substantial for these results. More specifically, two questions immediately present themselves:

1. Does there exist a consequence relation with an implication satisfying Löb's conditions for □ in which a Gödelian fixed point exists, but G2 fails? (The failure of G2 can be understood in two different senses — as a failure of its formalized version, and as a failure of its non-formalized version. Our examples will show the failure of both versions.)

2. Do Gödelian fixed points in such a system S have to be unique, even if S satisfies weakening?

In this section we provide an example showing that the answer to the first question is positive and to the second one negative. Moreover, we formulate a system in which there are many more fixed points than are officially required for a proof of G2. Our system S is a version of modal logic K4 based on the multiplicative $\{\to, \otimes, \bot\}$ fragment of a classical logic without contraction. It also has a built-in fixed point operator where the expression $\mathsf{fp}\, x.A(x)$ denotes some fixed point of $A(x)$ for formulas A modalized in the variable x. Thus, one will be able to derive

$$\mathsf{fp}\, x.A(x) =_S A(\mathsf{fp}\, x.A(x)),$$

for each formula $A(x)$ modalized in x. Let us now turn to the exact definitions.

Consider the set of formulas Fm_0 given by the grammar:

$$A ::= p \mid x \mid \bot \mid (A \to A) \mid \Box A,$$

where p stands for *atomic propositions* and x stands for *variables* (the alphabets of atomic propositions and variables are disjoint). We define the *set of formulas of* S by extending the set Fm_0 by a new constructor: if A is a formula and all free occurrences of x in A are within the scope of modal operators, then $\mathsf{fp}\, x.A$ is a formula, and $\mathsf{fp}\, x$ binds all free occurrences of x. A formula B is *closed* if it does not contain any free occurrences of variables. For a closed formula B, we denote by $A[B//x]$

the result of replacing all free occurrences of x in A by B. We also put $\neg A := A \to \bot$, $\top := \neg \bot$ and $A \otimes B := \neg(A \to \neg B)$.

A *sequent* is an expression of the form $\Gamma \Rightarrow \Delta$, where Γ and Δ are finite multisets of closed formulas. The sequent calculus S is defined in the standard way by the following initial sequents and inference rules:

$$\Gamma, A \Rightarrow A, \Delta \qquad \Gamma, \bot \Rightarrow \Delta$$

$$(\text{fix}_L) \frac{\Gamma, A[\text{fp}\, x.\, A //x] \Rightarrow \Delta}{\Gamma, \text{fp}\, x.\, A \Rightarrow \Delta} \qquad (\text{fix}_R) \frac{\Gamma \Rightarrow A[\text{fp}\, x.\, A //x], \Delta}{\Gamma \Rightarrow \text{fp}\, x.\, A, \Delta}$$

$$(\to_L) \frac{\Gamma, B \Rightarrow \Delta \quad \Sigma \Rightarrow A, \Pi}{\Gamma, \Sigma, A \to B \Rightarrow \Pi, \Delta} \qquad (\to_R) \frac{\Gamma, A \Rightarrow B, \Delta}{\Gamma \Rightarrow A \to B, \Delta}$$

$$(\Box) \frac{\Sigma, \Box \Pi \Rightarrow A}{\Gamma, \Box \Sigma, \Box \Pi \Rightarrow \Box A, \Delta}\ .$$

Explicitly displayed formulas in the conclusions of the rules are called *principal formulas* of the corresponding inferences. In the rules (fix$_L$), (fix$_R$), ($\to_L$) and ($\to_R$), the elements of Γ, Δ, Σ and Π are called *side formulas*. In initial sequents and in applications of the rule ($\Box$), the elements of Γ and Δ are *weakening formulas*. We call the elements of $\Box\Sigma$ and $\Box\Pi$ in the corresponding applications of ($\Box$) *active formulas*. In addition, explicitly displayed formulas in initial sequents are called *axiomatic formulas*.

A *proof in* S is a finite tree whose nodes are marked by sequents and leaves are marked by initial sequents that is constructed according to the rules of the sequent calculus. A sequent $\Gamma \Rightarrow \Delta$ is *provable in* S if there is a proof with the root marked by $\Gamma \Rightarrow \Delta$.

We associate with S a consequence relation with an implication and conjunction in the usual way by letting $\Gamma \vdash_S \phi$ iff $\Gamma \Rightarrow \phi$ is provable in S. The main thing we need to prove about S is the closure of S under the cut rule, which would show that $\Gamma \vdash_S \phi$ is indeed a well-defined consequence relation (see Theorem 5.2 below).

Since S is cut-free, the following propositions are easy to establish. Firstly, we obtain the failure of formalized G2.

PROPOSITION 4.1. *The sequent* $\Box(\Box\bot \to \bot) \Rightarrow \Box\bot$ *is not provable in* S.

Recall that an inference rule is called *admissible* (for a given proof system) if, for every instance of the rule, the conclusion is provable whenever all premises are provable.

PROPOSITION 4.2. *The Löb rule and the Henkin rule*

$$(\text{Löb}) \frac{\Box A \Rightarrow A}{\Rightarrow A} \qquad (\text{Hen}) \frac{\Box A \Rightarrow A \quad A \Rightarrow \Box A}{\Rightarrow A}$$

are not admissible in S.

Proof. Consider the Henkin fixed point $\text{fp}\, x.\, \Box x$. The sequent $\Rightarrow \text{fp}\, x.\, \Box x$ is not provable in S. Hence, the Henkin rule is not admissible and neither is the stronger Löb rule. ⊣

PROPOSITION 4.3. *There are infinitely many Henkinian and Gödelian fixed points in* S.

Proof. The routine of bound variables in S is such that the formulas $\mathsf{fp}\, x_i.\Box x_i$ for graphically distinct variables x_i are all inequivalent. (There is no rule of bound variables renaming and, in fact, it is easy to convince oneself that there are no cut-free proofs in S of the sequents $\mathsf{fp}\, x_i.\Box x_i \Rightarrow \mathsf{fp}\, x_j.\Box x_j$, for $i \neq j$.) The same holds for the Gödelian fixed points of S: the formulas $\mathsf{fp}\, x_i.\Box\neg x_i$ are pairwise inequivalent in S. ⊣

5 Cut-admissibility for S

For a proof of the cut-admissibility theorem for S we need the following standard lemma. Let the *size* $\|\pi\|$ *of a proof* π be the number of nodes in π.

LEMMA 5.1. *The weakening rule*

$$(\text{weak})\ \frac{\Gamma \Rightarrow \Delta}{\Sigma, \Gamma \Rightarrow \Delta, \Pi}$$

is admissible for S, *and its conclusion has a proof of at most the same size as the premise.*

THEOREM 5.2. *The cut rule*

$$(\text{cut})\ \frac{\Gamma \Rightarrow \Delta, A \quad A, \Sigma \Rightarrow \Pi}{\Gamma, \Sigma \Rightarrow \Pi, \Delta},$$

is admissible for S. *Moreover, if* π_1 *and* π_2 *are proofs of the premises of* (cut), *then the conclusion of* (cut) *has a proof with the size being less than* $\|\pi_1\| + \|\pi_2\|$.

Proof. Assume we have an inference

$$(\text{cut})\ \frac{\begin{array}{c}\pi_1 \\ \vdots \\ \Gamma \Rightarrow \Delta, A\end{array} \quad \begin{array}{c}\pi_2 \\ \vdots \\ A, \Sigma \Rightarrow \Pi\end{array}}{\Gamma, \Sigma \Rightarrow \Pi, \Delta},$$

where π_1 and π_2 are proofs in S. We proof by induction on $\|\pi_1\| + \|\pi_2\|$ that for any formula A there exists a proof $\mathcal{E}_A(\pi_1, \pi_2)$ of $\Gamma, \Sigma \Rightarrow \Pi, \Delta$ with the size being less than $\|\pi_1\| + \|\pi_2\|$.

Consider the final inference in π_1. If the formula A is in a position of a weakening formula in it, then we erase A in π_1 and extend the sequent $\Gamma \Rightarrow \Delta$ to $\Gamma, \Sigma \Rightarrow \Pi, \Delta$ by adding new weakening formulas. This transformation of π_1 defines $\mathcal{E}_A(\pi_1, \pi_2)$. Moreover, we have $\|\mathcal{E}_A(\pi_1, \pi_2)\| = \|\pi_1\| < \|\pi_1\| + \|\pi_2\|$.

Suppose the formula A is an axiomatic formula in the final inference of π_1. Then the proof π_1 consists of an initial sequent and the multiset Γ has the form Γ_0, A. We obtain $\mathcal{E}_A(\pi_1, \pi_2)$ by applying the admissible rule (weak):

$$(\text{weak})\ \frac{\begin{array}{c}\pi_2 \\ \vdots \\ A, \Sigma \Rightarrow \Pi\end{array}}{\Gamma_0, A, \Sigma \Rightarrow \Pi, \Delta}.$$

We have $\|\mathcal{E}_A(\pi_1, \pi_2)\| \leqslant \|\pi_2\| < \|\pi_1\| + \|\pi_2\|$.

Now suppose the formula A is a side formula. Then the final inference in π_1 can be (fix_L), (fix_R), ($\to_L$) or ($\to_R$).

In the case of ($\to_R$), the proof π_1 has the form

$$\begin{array}{c} \pi_1' \\ \vdots \\ (\to_R) \dfrac{\Gamma, B \Rightarrow C, \Delta_0, A}{\Gamma \Rightarrow B \to C, \Delta_0, A} \end{array},$$

where $\Delta = B \to C, \Delta_0$. We define $\mathcal{E}_A(\pi_1, \pi_2)$ as

$$\mathcal{E}_A(\pi_1', \pi_2)$$
$$\vdots$$
$$(\to_R) \dfrac{\Gamma, B, \Sigma \Rightarrow \Pi, C, \Delta_0}{\Gamma, \Sigma \Rightarrow \Pi, B \to C, \Delta_0}.$$

The proof $\mathcal{E}_A(\pi_1', \pi_2)$ is defined by the induction hypothesis for π_1' and π_2. We also have $\|\mathcal{E}_A(\pi_1, \pi_2)\| = \|\mathcal{E}_A(\pi_1', \pi_2)\| + 1 < \|\pi_1'\| + \|\pi_2\| + 1 = \|\pi_1\| + \|\pi_2\|$.

In the case of (fix_R), the proof π_1 has the form

$$\begin{array}{c} \pi_1' \\ \vdots \\ (\text{fix}_R) \dfrac{\Gamma \Rightarrow B[\text{fp}\, x.\, B//x], \Delta_0, A}{\Gamma \Rightarrow \text{fp}\, x.\, B, \Delta_0, A} \end{array},$$

where $\Delta = \text{fp}\, x.\, B, \Delta_0$. We define $\mathcal{E}_A(\pi_1, \pi_2)$ as

$$\mathcal{E}_A(\pi_1', \pi_2)$$
$$\vdots$$
$$(\text{fix}_R) \dfrac{\Gamma, \Sigma \Rightarrow \Pi, B[\text{fp}\, x.\, B//p], \Delta_0}{\Gamma, \Sigma \Rightarrow \Pi, \text{fp}\, x.\, B, \Delta_0}.$$

The proof $\mathcal{E}_A(\pi_1', \pi_2)$ is defined by the induction hypothesis, and $\|\mathcal{E}_A(\pi_1, \pi_2)\| = \|\mathcal{E}_A(\pi_1', \pi_2)\| + 1 < \|\pi_1'\| + \|\pi_2\| + 1 = \|\pi_1\| + \|\pi_2\|$.

The remaining cases of ($\to_L$) and (fix_L) can be analyzed analogously, so we omit them.

Now consider the final inference in π_2. If the formula A is a weakening, an axiomatic or a side formula in it, then we can define $\mathcal{E}_A(\pi_1, \pi_2)$ in a similar way to the previous cases.

Suppose that the formula A is a principal or an active formula in the final inferences of π_1 and π_2. Then A has the form $\text{fp}\, x.\, A_0$, $A_0 \to A_1$ or $\Box A_0$.

If $A = \Box A_0$, then π_2 has one of the two forms

$$\begin{array}{c} \pi_2' \\ \vdots \\ (\Box) \dfrac{A_0, \Sigma_1, \Box\Sigma_2 \Rightarrow D}{\Sigma_0, \Box A_0, \Box\Sigma_1, \Box\Sigma_2 \Rightarrow \Box D, \Pi_0} \end{array} \qquad \begin{array}{c} \pi_2' \\ \vdots \\ (\Box) \dfrac{\Sigma_1, \Box A_0, \Box\Sigma_2 \Rightarrow D}{\Sigma_0, \Box\Sigma_1, \Box A_0, \Box\Sigma_2 \Rightarrow \Box D, \Pi_0} \end{array},$$

where $\Sigma = \Sigma_0, \Box\Sigma_1, \Box\Sigma_2$ and $\Pi = \Box D, \Pi_0$. In addition, the proof π_1 has the form

$$(\Box) \frac{\begin{array}{c} \pi_1' \\ \vdots \\ \Gamma_1, \Box\Gamma_2 \Rightarrow A_0 \end{array}}{\Gamma_0, \Box\Gamma_1, \Box\Gamma_2 \Rightarrow \Box A_0, \Delta},$$

where $\Gamma = \Gamma_0, \Box\Gamma_1, \Box\Gamma_2$. If π_2 has the first form, then we define $\mathcal{E}_A(\pi_1, \pi_2)$ as

$$(\Box) \frac{\begin{array}{c} \mathcal{E}_{A_0}(\pi_1', \pi_2') \\ \vdots \\ \Gamma_1, \Box\Gamma_2, \Sigma_1, \Box\Sigma_2 \Rightarrow D \end{array}}{\Gamma_0, \Box\Gamma_1, \Box\Gamma_2, \Sigma_0, \Box\Sigma_1, \Box\Sigma_2 \Rightarrow \Box D, \Pi_0, \Delta}.$$

We have $\|\mathcal{E}_A(\pi_1, \pi_2)\| = \|\mathcal{E}_{A_0}(\pi_1', \pi_2')\| + 1 < \|\pi_1'\| + \|\pi_2'\| + 1 < \|\pi_1\| + \|\pi_2\|$. If π_2 has the second form, then we define $\mathcal{E}_A(\pi_1, \pi_2)$ as

$$(\Box) \frac{\begin{array}{c} \mathcal{E}_A(f(\pi_1), \pi_2') \\ \vdots \\ \Box\Gamma_1, \Box\Gamma_2, \Sigma_1, \Box\Sigma_2 \Rightarrow D \end{array}}{\Gamma_0, \Box\Gamma_1, \Box\Gamma_2, \Sigma_0, \Box\Sigma_1, \Box\Sigma_2 \Rightarrow \Box D, \Pi_0, \Delta},$$

where $f(\pi_1)$ is the proof obtained by erasing multisets Γ_0 and Δ from the conclusion of π_1. We have $\|\mathcal{E}_A(\pi_1, \pi_2)\| = \|\mathcal{E}_A(f(\pi_1), \pi_2')\| + 1 < \|f(\pi_1)\| + \|\pi_2'\| + 1 = \|\pi_1\| + \|\pi_2\|$.

In the case of $A = \mathsf{fp}\, x.\, A_0$, the proofs π_1 and π_2 have the form

$$(\mathsf{fix}_\mathsf{R}) \frac{\begin{array}{c} \pi_1' \\ \vdots \\ \Gamma \Rightarrow \Delta, A_0[\mathsf{fp}\, x.\, A_0//x] \end{array}}{\Gamma \Rightarrow \Delta, \mathsf{fp}\, x.\, A_0} \qquad (\mathsf{fix}_\mathsf{L}) \frac{\begin{array}{c} \pi_2' \\ \vdots \\ A_0[\mathsf{fp}\, x.\, A_0//x], \Sigma \Rightarrow \Pi \end{array}}{\mathsf{fp}\, x.\, A_0, \Sigma \Rightarrow \Pi}.$$

We put $\mathcal{E}_A(\pi_1, \pi_2) = \mathcal{E}_{A_0[\mathsf{fp}\, x.\, A_0//x]}(\pi_1', \pi_2')$ and see that $\|\mathcal{E}_A(\pi_1, \pi_2)\| = \|\mathcal{E}_{A_0[\mathsf{fp}\, x.\, A_0//x]}(\pi_1', \pi_2')\| < \|\pi_1'\| + \|\pi_2'\| < \|\pi_1\| + \|\pi_2\|$.

If $A = A_0 \to A_1$, then the proofs π_1 and π_2 have the form

$$(\to_\mathsf{R}) \frac{\begin{array}{c} \pi_1' \\ \vdots \\ A_0, \Gamma \Rightarrow \Delta, A_1 \end{array}}{\Gamma \Rightarrow \Delta, A_0 \to A_1} \qquad (\to_\mathsf{L}) \frac{\begin{array}{cc} \pi_2' & \pi_2'' \\ \vdots & \vdots \\ A_1, \Sigma_1 \Rightarrow \Pi_1 & \Sigma_0 \Rightarrow \Pi_0, A_0 \end{array}}{\Sigma_0, A_0 \to A_1, \Sigma_1 \Rightarrow \Pi_0, \Pi_1},$$

where $\Sigma = \Sigma_0, \Sigma_1$ and $\Pi = \Pi_0, \Pi_1$. By the induction hypothesis, $\mathcal{E}_{A_0}(\pi_2'', \pi_1')$ is defined and $\|\mathcal{E}_{A_0}(\pi_2'', \pi_1')\| < \|\pi_2''\| + \|\pi_1'\|$. Since $\|\mathcal{E}_{A_0}(\pi_2'', \pi_1')\| + \|\pi_2'\| < \|\pi_2''\| + \|\pi_1'\| + \|\pi_2'\| < \|\pi_1\| + \|\pi_2\|$, then $\mathcal{E}_{A_1}(\mathcal{E}_{A_0}(\pi_2'', \pi_1'), \pi_2')$ is defined by the induction hypothesis. We put $\mathcal{E}_A(\pi_1, \pi_2) = \mathcal{E}_{A_1}(\mathcal{E}_{A_0}(\pi_2'', \pi_1'), \pi_2')$. In addition, we have $\|\mathcal{E}_A(\pi_1, \pi_2)\| = \|\mathcal{E}_{A_1}(\mathcal{E}_{A_0}(\pi_2'', \pi_1'), \pi_2')\| < \|\mathcal{E}_{A_0}(\pi_2'', \pi_1')\| + \|\pi_2'\| < \|\pi_1\| + \|\pi_2\|$. ⊣

6 Conclusions and Future Work

The preliminary results presented in this paper indicate the following conclusions:

- Derivability conditions can be stated in a way not assuming much about logic. However,

- Gödel's argument presupposes a certain amount of contraction for the logic under consideration.

The role of contraction rule here is similar to its role in Liar-type paradoxes including Russell's paradox in set theory. Thus, Vyacheslav Grishin (see [7, 8]) pioneered the study of set theory with full comprehension based on a logic without contraction. He demonstrated that the pure comprehension scheme is consistent in this logic. He also showed, however, that the extensionality principle allows for this system to actually *prove* contraction even if there is no postulated contraction in the logic.

One can also consider systems of arithmetic based on contraction-free logic, see e.g. Restall [16, Chapter 11]. For one such system, considered by the second author of this paper, the rule of $\Box$-contraction is admissible, which according to our results still yields G2. Thus, we are still missing convincing examples of mathematical theories based on weak logics for which G2 would fail.

- For consequence relations with an implication and with $\Box$ satisfying Löb's conditions, the existence of appropriately many fixed points does not imply their uniqueness. Nor does it imply formalized versions of G2 and Löb's theorem $\Box(\Box\phi \to \phi) \vdash \Box\phi$.

This shows that the move from *diagonalized algebras* in the sense of R. Magari, i.e., Boolean algebras with $\Box$ satisfying Löb's conditions and having fixed points, to *diagonalizable algebras* (modal algebras satisfying Löb's identity) is, in general, not possible for logics without contraction and weakening. See Smoryński [18, 19] for a nice exposition of the original setup.

- One can also show that the admissibility of Löb's rule does not, in general, imply a formalized version of G2.

A system S^∞ witnessing this property can be obtained by extending the notion of proof in the system S to possibly non-well-founded proof trees. Infinite proofs may arise because of the presence of the fixed point rules. For S^∞, unlike S, one can show that Löb's rule is admissible and, moreover, all fixed points are unique. Yet, formalized G2 is still underivable. The analysis of S^∞ is based on another cut-admissibility theorem, which we postpone to a later publication.

We remark that the system S does not provide a counterexample to the non-formalized version of G2, since $\Rightarrow \neg\Box\bot$ is not provable. Such counterexamples can be obtained by extending S by additional rules. Let T be obtained by adding to S the rule

$$\frac{\Gamma, \phi \Rightarrow \Delta}{\Gamma, \Box\phi \Rightarrow \Delta}.$$

In the presence of this rule the modal rule ($\Box$) can be simplified to

$$\frac{\Box\Gamma \Rightarrow \phi}{\Pi, \Box\Gamma \Rightarrow \Box\phi, \Sigma}.$$

Obviously, in T the reflection principle $\Box\phi \Rightarrow \phi$ and the consistency assertion $\Rightarrow \neg\Box\bot$ are provable. However, the cut-admissibility theorem holds for T with essentially the same proof as for S (one should only check one additional case which is easy). Hence, T is consistent and provides a counterexample to the non-formalized version of G2. Nor does it prove the formalized version of G2.

The modal axioms of T correspond to those of the modal logic S4. There is an even simpler example of the same kind corresponding to the maximal modal logic Triv (in which the modality degenerates) given by the axiom $\phi \leftrightarrow \Box\phi$. Consider the fragment Triv of S in the language without modality axiomatized by all the rules of S except for ($\Box$). In particular, fixed-point operators $\mathsf{fp}\, x.A(x)$ can now be applied to any formula $A(x)$ without modalities. As usual, the cut-admissibility theorem holds for this logic by (a fragment of) the same proof. For the corresponding consequence relation one can now *define* the operation $\Box$ by letting $\Box\phi := \phi$. Then it is easy to check that Löb's conditions are satisfied. Yet, the logic is consistent, despite the fact that it contains a formula $c := \mathsf{fp}\, x.\neg x$ for which $c =_{\mathsf{Triv}} \neg c$. This is similar to the well-known consistency of the Liar sentence in contraction-free framework (see, e.g., [16]).

Acknowledgements

The authors would like to thank Johan van Benthem for useful comments and questions. This work is supported by the Russian Foundation for Basic Research, grant 15-01-09218a, and by the Presidential council for support of leading scientific schools.

BIBLIOGRAPHY

[1] Z. Adamowicz and K. Zdanowski. Lower bounds for the provability of herbrand consistency in weak arithmetics. *Fundamenta Mathematicae*, 212(3):191–216, 2011.
[2] A. Bezboruah and J. C. Shepherdson. Gödel's second incompleteness theorem for Q. *The Journal of Symbolic Logic*, 41(2):503–512, 1976.
[3] S. Feferman. Arithmetization of metamathematics in a general setting. *Fundamenta Mathematicae*, 49:35–92, 1960.
[4] S. Feferman, J.R. Dawson, S.C. Kleene, G.H. Moore, R.M. Solovay, and J. van Heijenoort, editors. *Kurt Gödel Collected Works, Volume 1: Publications 1929–1936*. Oxford Univeristy Press, 1996.
[5] K. Gödel. Über formal unentscheidbare Sätze der Principia Mathematica und verwandter Systeme I. *Monatshefte für Mathematik und Physik*, 38:173–198, 1931.
[6] K. Gödel. Eine Interpretation des intuitionistischen Aussagenkalkuls. *Ergebnisse Math. Kolloq.*, 4:39–40, 1933. English translation in [4], pages 301–303.
[7] V.N. Grishin. On some non-standard logic and its application to set theory. In *Investigations on formalized languages and non-classical logics*, pages 135–171. Nauka, Moscow, 1974. In Russian.
[8] V.N. Grishin. Predicate and set-theoretic calculi based on logic without contractions. *Mathematics of the USSR-Izvestiya*, 18(1):41–59, 1982.
[9] D. Hilbert and P. Bernays. *Grundlagen der Mathematik, Vols. I and II, 2d ed.* Springer-Verlag, Berlin, 1968.
[10] R.G. Jeroslow. Consistency statements in formal theories. *Fundamenta Mathematicae*, 72:2–39, 1970.
[11] R.G. Jeroslow. Redundancies in the Hilbert–Bernays derivability conditions. *The Journal of Symbolic Logic*, 38(3):359–367, 1973.
[12] S.C. Kleene. A symmetric form of Gödel's theorem. *Indagationes Mathematicae*, 12:244–246, 1950.
[13] M.H. Löb. Solution of a problem of Leon Henkin. *The Journal of Symbolic Logic*, 20:115–118, 1955.
[14] P. Pudlák. Cuts, consistency statements and interpretations. *The Journal of Symbolic Logic*, 50:423–441, 1985.
[15] W. Rautenberg. *A Concise Introduction to Mathematical Logic*. Springer, second edition, 2006.
[16] G. Restall. *On Logics Without Contraction*. PhD thesis, The University of Queensland, 1994. http://consequently.org/papers/onlogics.pdf.
[17] C. Smoryński. The incompleteness theorems. In J. Barwise, editor, *Handbook of Mathematical Logic*, pages 821–865. North Holland, Amsterdam, 1977.
[18] C. Smoryński. Fixed point algebras. *Bull. Amer. Math. Soc.*, 6(3):317–356, 1982.

[19] C. Smoryński. *Self-Reference and Modal Logic*. Springer-Verlag, Berlin, 1985.
[20] R.M. Smullyan. *Diagonalization and Self-Reference*. Oxford Logic Guides 27. Oxford University Press, 1994.
[21] A. Visser. Unprovability of small inconsistency. *Archive for Math. Logic*, 32:275–298, 1993.
[22] A. Visser. Can we make the Second Incompleteness Theorem coordinate free? *Journal of Logic and Computation*, 21(4):543–560, 2011.
[23] A. Visser. The Second Incompleteness Theorem and bounded interpretations. *Studia Logica*, 100(1–2):399–418, 2012.
[24] A. Wilkie and J. Paris. On the scheme of induction for bounded arithmetic formulas. *Annals of Pure and Applied Logic*, 35:261–302, 1987.
[25] D. Willard. Self-verifying systems, the incompleteness theorem and the tangibility reflection principle. *The Journal of Symbolic Logic*, 66:536–596, 2001.
[26] D. Willard. A generalization of the Second Incompleteness Theorem and some exceptions to it. *Annals of Pure and Applied Logic*, 141:472–496, 2006.

A Negative Result on Algebraic Specifications of the Meadow of Rational Numbers

JAN A. BERGSTRA AND INGE BETHKE

Abstract

$\mathbb{Q}_0$—the involutive meadow of the rational numbers—is the zero-totalized expansion field of the rational numbers where the multiplicative inverse operation is made total by imposing $0^{-1} = 0$. In this note, we prove that $\mathbb{Q}_0$ cannot be specified by the usual axioms for meadows augmented by a finite set of axioms of the form $(1 + \cdots + 1 + x^2) \cdot (1 + \cdots + 1 + x^2)^{-1} = 1$.

Dedication: Jan Bergstra acknowledges many years of cooperation with Albert as close colleagues in the former Faculty of Philosophy of Utrecht University as well as recent joint activities in the context of KNAW concerning the aims and scope of mathematics teaching. We share an interest for the logic of elementary number systems.

1 Introduction

This note contributes to the algebraic datatype specification of $\mathbb{Q}_0$—the rational numbers equipped with a total inverse operation. Advantages and disadvantages of dividing by zero in various ways have been amply discussed in the mathematics and logic literature (see e.g. [7, 9]) and we do not wish to add to those matters here. The same holds for issues regarding the origins of various approaches to division by zero. But we believe that the observation made in [4] that $\mathbb{Q}_0$ has a finite initial algebra specification was at that time original. Here we elaborate on this theme.
In [4] it is shown that

$$\mathbb{Q}_0 \cong \mathcal{I}(\Sigma_{\mathsf{Md}}, \mathsf{Md} + L_4)$$

where $\mathcal{I}(\Sigma_{\mathsf{Md}}, \mathsf{Md} + L_4)$ is the initial algebra of the theory Md of meadows given in Table 1 enriched with the axiom L_4 given in Table 2 for $n = 4$. The characterization above has been sharpened in [3] where it is shown that

$$\mathbb{Q}_0 \cong \mathcal{I}(\Sigma_{\mathsf{Md}}, \mathsf{Md} + L_2).(\ddagger)$$

In [1] it is proved that every finite specification of $\mathbb{Q}_0$ can be given in the form $\mathsf{Md} + e$—the meadow axioms enriched with a single equation. Moreover, observe that both L_4 and L_2 do not hold in the presence of the imaginary unit i. In general, every finite algebraic specification of $\mathbb{Q}_0$ contains an equation not valid in $\mathbb{C}_0$—the zero-totalized expansion field of the complex numbers. Again, this is proved in [1].

In the sequel we denote by $(\mathbb{Z}/p\mathbb{Z})_0$ the zero-totalized expansion of the finite prime field of order p. Moreover, we define for $n \in \mathbb{N}$ the numerals $\underline{n}$ by $\underline{0} := 0$ and $\underline{n+1} := \underline{n} + 1$. A necessary and sufficient condition for an initial algebra specification of $\mathbb{Q}_0$ is given in the following theorem.

$$(x + y) + z = x + (y + z)$$
$$x + y = y + x$$
$$x + 0 = x$$
$$x + (-x) = 0$$
$$(x \cdot y) \cdot z = x \cdot (y \cdot z)$$
$$x \cdot y = y \cdot x$$
$$1 \cdot x = x$$
$$x \cdot (y + z) = x \cdot y + x \cdot z$$
$$(x^{-1})^{-1} = x$$
$$x \cdot (x \cdot x^{-1}) = x$$

Table 1. The set Md of axioms for meadows

$$(1 + x_1^2 + \cdots + x_n^2) \cdot (1 + x_1^2 + \cdots + x_n^2)^{-1} = 1 \quad (L_n)$$

Table 2. The axiom schema L_n

THEOREM 1.1. *Let E be a set of meadow equations. Then*

$$\mathbb{Q}_0 \cong \mathcal{I}(\Sigma_{\mathsf{Md}}, \mathsf{Md} + E) \text{ if and only if for all prime numbers } p, (\mathbb{Z}/p\mathbb{Z})_0 \not\models E.$$

Proof. Assume $\mathbb{Q}_0 \cong \mathcal{I}(\Sigma_{\mathsf{Md}}, \mathsf{Md}+E)$. Then $\mathsf{Md}+E \vdash p \cdot p^{-1} = 1$ for all prime numbers p. Hence $(\mathbb{Z}/p\mathbb{Z})_0 \not\models E$ for all primes p. For the converse, assume $(\mathbb{Z}/p\mathbb{Z})_0 \not\models E$ for all primes p. Recall that every minimal meadow is a subdirect product of minimal zero-totalized expansion fields (see e.g. [5]). The minimal zero-totalized expansion fields are the zero-totalized prime fields $(\mathbb{Z}/p\mathbb{Z})_0$ and $\mathbb{Q}_0$. In particular, the initial algebra of $\mathsf{Md} + E$ is such a subdirect product. Since every $\mathbb{Z}/p\mathbb{Z}_0$ is not a model of E, it follows that $\mathcal{I}(\Sigma_{\mathsf{Md}}, \mathsf{Md} + E)$ must be isomorphic to $\mathbb{Q}_0$. ⊣

An application of this theorem yielding a positive result is given below. First we recall a few facts from the theory of numbers (see e.g. [6], Ch. 3). For every odd prime p one of the two congruences $p \equiv 1 \mod 4$ or $p \equiv 3 \mod 4$ hold. Given a prime p and a natural number $0 < n < p$, n is called a *quadratic residue* of p if there exists a natural number x such that $x^2 \equiv n \mod p$. If the congruence is insoluble, n is said to be a *quadratic non-residue*. Every prime p has quadratic residues since $1^2 \equiv 1 \mod p$, but for $p > 3$ there are more: e.g. if $p = 19$, the quadratic residues are 1, 4, 5, 6, 7, 9, 11, 16 and 17. In general, every odd prime p has $\frac{p-1}{2}$ quadratic residues. If $p \equiv 1 \mod 4$ then the lists of quadratic residues and quadratic non-residues are both symmetrical in the sense that if n is a (non-)quadratic residue then $p - n$ is one. On the other hand, if $p \equiv 3 \mod 4$, then n is a quadratic residue if and only if $p - n$ is a quadratic non-residue (see e.g. the case of $p = 19$). The quadratic residues and non-residues have a simple *multiplicative property*: the product of two residues or of two non-residues is a residue.

EXAMPLE 1.2. *There also exist finite initial algebra specifications of $\mathbb{Q}_0$ of the form* Md$+e$ *where e is a single variable equation. Consider the equation* $f(x) \cdot f(x)^{-1} = 1$ *where* $f(x) = (x^2-2)(x^2-3)(x^2-6)$. *Inspection shows that $f(x)$ has no rational root. Thus* $\mathbb{Q}_0 \models f(x) \cdot f(x)^{-1} = 1$. *On the other hand, $f(x)$ has a root modulo every prime number p:*

- *If $p = 2$ then $f(x)$ has a root modulo p for $x = 0$.*

- *If $p > 2$ we apply the multiplicative property of quadratic residues and non-residues. If $(x^2 - 2)$ or $(x^2 - 3)$ have a root modulo p, then $f(x)$ has a root modulo p. If neither $(x^2 - 2)$ nor $(x^2 - 3)$ has a root modulo p then both 2 and 3 are non-residues of p. Hence 6 is a residue of p, i.e. $(x^2 - 6)$ has a root modulo p and thus $f(x)$ has a root modulo p.*

It follows that $(\mathbb{Z}/p\mathbb{Z})_0 \not\models f(x) \cdot f(x)^{-1} = 1$ for every prime number p. By the above theorem we may conclude that Md $+ f(x) \cdot f(x)^{-1} = 1$ is an initial algebra specification of $\mathbb{Q}_0$.

In the next section we apply Theorem 1.1 in order to give a negative result.

2 A Negative Result

A general question concerning the specification of the rationals is whether there exists a logical weakest initial algebra specification. In this section we show that the weakening from L_4 to L_2 cannot be prolonged in a straightforward way.

$$(1 + \underline{n} + x^2) \cdot (1 + \underline{n} + x^2)^{-1} = 1 \quad (H_n)$$
$$(1 + \underline{n}) \cdot (1 + \underline{n})^{-1} = 1 \quad (C_n)$$

Table 3. The axiom schemas H_n and C_n

Substituting 0 for x in H_n in Table 3, we obtain the axiom C_n. In [4] it is proved that $\{C_n \mid n \in \mathbb{N}\}$ together with Md specify the rational numbers. The question then arises whether $\mathbb{Q}_0$ can be specified by Md $+ \Gamma$ for some finite subset $\Gamma \subset \{H_n \mid n \in \mathbb{N}\}$. We give a negative answer below.

Consider the function $f : \mathbb{N} \to \mathbb{N}$ defined by

$$f(n) = \begin{cases} 0 & \text{if } n \leq 1 \text{ or } n \text{ is composite,} \\ n - max\{i \mid i \text{ is a quadratic residue of } n\} & \text{if } n \text{ is prime.} \end{cases}$$

E.g. $f(19) = 2$. In [8], a table for the values of $f(n)$ can be found for the first 10^5 natural numbers (see entry A088192). The occurring values increase very slowly: the largest value found in that table is 43.

In [10], Wright proved the following theorem (see Theorem 2.3).

THEOREM 2.1. *Every nonempty finite subset of $\mathbb{N}^+$ is a set of quadratic residues for infinitely many primes.*

As a corollary we obtain that the function f is unbounded.

COROLLARY 2.2. *f is unbounded.*

Proof. For $n \in \mathbb{N}$ with $n > 2$ consider the set $A = \{1, 2, 3, \ldots, n\}$. By the previous theorem we can pick a prime p such that every element of A is a quadratic residue of p. In particular, 2 is a quadratic residue of p. It follows from Gauss's lemma that $p \equiv 7 \mod 8$ and hence $p \equiv 3 \mod 4$. Thus, since $1, 2, 3, \ldots, n$ are all quadratic residues, $p - 1, p - 2, p - 3, \ldots, p - n$ are all quadratic non-residues. So $max\{i \mid i$ is a quadratic residue of $p\} < p - n$. Therefore $f(p) > n$. ⊣

We now prove that for every finite $\Gamma \subset \{H_n \mid n \in \mathbb{N}\}$ there exists a prime p with $(\mathbb{Z}/p\mathbb{Z})_0 \models \Gamma$.

PROPOSITION 2.3. *Let $\Gamma \subset \{H_n \mid n \in \mathbb{N}\}$ be finite. Then there exists a prime p such that $(\mathbb{Z}/p\mathbb{Z})_0 \models \Gamma$.*

Proof. Say $\Gamma = \{H_0, \ldots, H_n\}$. Pick a prime p such that $f(p) > n + 1$. Suppose $(\mathbb{Z}/p\mathbb{Z})_0 \not\models H_m$ for some $0 \leq m \leq n$. We derive a contradiction as follows. By the assumption, there exists $0 \leq x < p$ with $1 + m + x^2 = 0$, i.e. $x^2 \equiv p - (m+1) \mod p$. Hence $p - (m+1)$ is a quadratic residue of p. Therefore $p - (m+1) \leq max\{i \mid i$ is a quadratic residue of $p\}$ and hence

$$\begin{aligned} m + 1 &= p - (p - (m+1)) \geq p - max\{i \mid i \text{ is a quadratic residue of } p\} \\ &= f(p) > n + 1 \geq m + 1. \end{aligned}$$

⊣

We can now show that $\mathbb{Q}_0$ cannot be specified by a finite set of H_n-axioms.

THEOREM 2.4. *Let $\Gamma \subset \{H_n \mid n \in \mathbb{N}\}$ be finite. Then $\mathbb{Q}_0 \not\cong \mathcal{I}(\Sigma_{\mathsf{Md}}, \mathsf{Md} + \Gamma)$.*

Proof. Immediately by the preceding proposition and Theorem 1.1. ⊣

COROLLARY 2.5. $\mathbb{Q}_0 \not\cong \mathcal{I}(\Sigma_{\mathsf{Md}}, \mathsf{Md} + L_1)$

REMARK 2.6. *Observe that $\mathcal{I}(\Sigma_{\mathsf{Md}}, \mathsf{Md} + L_1)$ is a non-cancellation meadow of characteristic 0 which does not contain $\mathbb{Q}_0$ as a subalgebra since $\mathbb{Q}_0 \not\cong \mathcal{I}(\Sigma_{\mathsf{Md}}, \mathsf{Md} + L_1)$. The existence of such a structure has already been proved in [2] (see Theorem 2.1). However, the proof given there relies on the compactness theorem.*

3 An Open Question

Open questions arise when we extend the rational numbers. E.g. consider the meadow of Gaussian rationals—denoted $\mathbb{Q}_0(i)$—obtained by adjoining the imaginary number i to the meadow of rationals. It is not difficult to see that the polynomial given in the example in Section 1 also yields an initial algebra specification of the Gaussian rationals. Since $f(x) = (x^2 - 2)(x^2 - 3)(x^2 - 6)$ has only real roots, there exist no complex rational ones. It follows that $\mathbb{Q}_0(i) \models f(x) \cdot f(x)^{-1} = 1$. Moreover, working in $\mathsf{Md} + \{i^2 + 1 = 0\}$ it can be shown that every closed term over $\Sigma_{\mathsf{Md}} \cup \{i\}$ is provable equal to a term of the form $l \cdot m^{-1} + p \cdot q^{-1} \cdot i$ where l, m, p, q are numerals. Hence $\mathsf{Md} + \{f(x) \cdot f(x)^{-1} = 1, i^2 + 1 = 0\}$ yields an initial algebra specification of $\mathbb{Q}_0(i)$.

Up to isomorphism, there exists only one simple extension of the rational numbers generated by a transcendental number t. We are unable to prove or disprove the existence of a finite initial algebra specification of $\mathbb{Q}_0(t)$ with t a new constant interpreted as a trancendental element.

Acknowledgement: We are indebted to Rob Tijdeman for valuable discussions including a proposal for proving Corollary 2.2 and the suggestion to make use of [10].

BIBLIOGRAPHY

[1] Bergstra, J.A. and Bethke, I. (2015). *Subvarieties of the variety of meadows.* arXiv: 1510.0402 [math.RA].
[2] Bergstra, J.A., Bethke, I. and Ponse, A. (2015). Equations for formally real meadows. *Journal of Applied Logic*, 13(2):1–23.
[3] Bergstra, J.A. and Middelburg, C.A. (2011). Inversive meadows and divisive meadows. *Journal of Applied Logic*, 9(3):203–220.
[4] Bergstra, J.A. and Tucker, J.V. (2007). The rational numbers as an abstract data type. *Journal of the ACM*, 54(2), Article 7.
[5] Bethke, I. and Rodenburg, P. (2010). The initial meadows. *Journal of Symbolic Logic*, 75(3): 888–895.
[6] Davenport, H. (1952). *The Higher Arithmetic*, Cambridge University Press.
[7] Komori, Y. (1975). Free algebras over all fields and pseudo-fields. Report 10, pp. 9 - 15, Faculty of Science, Shizuoka University, Japan.
[8] *The on-line encyclopedia of integer sequences.* http://oeis.org
[9] Ono, H. (1983). Equational theories and universal theories of fields. *J. Math. Soc. Japan*, 35:289–306.
[10] Wright, S. (2007). Patterns of quadratic residues and nonresidues for infinitely many primes. *Journal of Number Theory*, 123:120–132.

Completeness of Cutting Planes Revisited

MARC BEZEM AND ROBERT NIEUWENHUIS

Abstract

Most of the theoretical background behind integer linear programming (ILP) and 0-1 ILP methods has historically been based on general (rational) Linear Programming.

Motivated by recent practical interest in SAT-related techniques for ILP, here we re-visit cutting planes in a more self-contained way.

We give short and –we think– insight-providing completeness proofs for general bounded ILPs as well as 0-1 ones, analyzing stronger and weaker rule systems, and giving (counter) examples.

Among the insights gained are new completeness results for less prolific and hence more easily automatizable systems including ordered ones.

1 Introduction

The theory of Integer Linear Programming, in part based on the theory of Linear Programming (LP) in the rationals, is well-developed and all results are rigorously proved. It takes, however, considerable effort to understand the sometimes deep proofs. Examples include the proof of completeness of the cutting planes method in [13, Theorem 2.14] and [19, Corollary 23.2b]. This is of course not meant as criticism, but we believe that shorter and more self-contained proofs, if they can be found, may be useful to the community.

Indeed, there has been a considerable recent practical interest in techniques for ILP that are not based on LP-relaxations. On the one hand, there is a large body of work on Pseudo-Boolean (aka 0-1 ILP) solvers based on conflict-driven constraint-learning in a similar way to CDCL clause-learning solvers for SAT (see [18] and its numerous references). On 0-1 ILP problems, such solvers frequently outperform the best commercial LP-relaxation-based tools such as Cplex or Gurobi. On the other hand, also new CDCL-like techniques for full ILP over $\mathbb{Z}$ have recently appeared: Cutsat [12] and IntSat [14]. The latter one also appears to be competitive with these commercial solvers at least on certain full ILP problem classes.

Our motivation for this work is to gain new insights in the cutting-plane inference systems underlying these new CDCL solvers. Since these solvers optimize by solving a sequence of feasibility problems, we here focus on *refutation* completeness. We give short and –we think– insight-providing completeness proofs for general bounded ILPs as well as for 0-1 ones, analyzing stronger and weaker rule systems, and giving (counter) examples. Among the insights gained are new (to the best of our knowledge) completeness results for less prolific and hence more easily automatizable systems including ordered ones.

We address general bounded ILPs in Section 3, and 0-1 ones in Section 4. Several (counter) examples are given in Section 5. Sections 6 and 7 discuss related and future work and conclude.

2 Basic Definitions and Notation

Let $X = \{x_n \ldots x_1\}$ be a set of variables over $\mathbb{Z}$. Throughout this paper we will use the following definitions. An *(integer linear) monomial* is an expression of the form $a_i x_i$ where $a_i \in \mathbb{Z}$ ($1 \leq i \leq n$) is called the *coefficient* of x_i. A *(homogeneous integer linear) polynomial*, denoted by $p, q, \ldots$ is a sum $a_n x_n + \ldots + a_1 x_1$ of n monomials. We sometimes use standard shorthands, omitting, e.g., unit coefficients, or monomials with coefficient zero. If $a_i = 0$ for all i with $1 \leq i \leq n$ then the polynomial is called *empty* and denoted by 0. If p is $a_n x_n + \ldots + a_1 x_1$ and q is $b_n x_n + \ldots + b_1 x_1$, then $p+q$ denotes the polynomial $c_1 x_1 + \ldots + c_n x_n$ where $c_i = a_i + b_i$ ($1 \leq i \leq n$), and if $c \in \mathbb{Z}$ then cp denotes the polynomial $(ca_n)x_n + \ldots + (ca_1)x_1$. Given $d \in \mathbb{N}^+$, we write $d|p$ to denote that d divides all a_i for $1 \leq i \leq n$. If $d|p$ then p/d denotes $(a_n/d)x_n + \ldots + (a_1/d)x_1$.

An *(integer linear) constraint*, denoted by $C, D, \ldots$ is an inequality $p \geq c$ with $c \in \mathbb{Z}$. A *contradiction* is a constraint of the form $0 \geq c$ with $c > 0$. We denote sets of constraints by the letter S. An *integer assignment* is a function $M \colon \{x_n, \ldots, x_1\} \to \mathbb{Z}$. If p is a polynomial $a_n x_n + \ldots + a_1 x_1$ then $M(p)$ is the integer $a_n M(x_n) + \ldots + a_1 M(x_1)$. M *satisfies* a constraint $p \geq c$ if $M(p) \geq_\mathbb{Z} c$, denoted $M \models C$. We write $M \models S$ if M satisfies all constraints in S. Then M is a *solution* or a *model* of S and S is called *feasible*.

A *lower bound for* x is a constraint of the form $x \geq l$ and an *upper* bound for x is a constraint of the form $-x \geq -u$. Then the integers l and u are also called *lower (resp. upper) bounds for* x. We say that a set S of constraints is *bounded* if for each variable there is both a lower and an upper bound in S.

We will use the ordering on variables induced by their subindices, as follows. We always write polynomials with variables in descending order. When we write a constraint $ax_i + p \geq c$ this means that $a \neq 0$ and moreover that all monomials bx_j in p with $b \neq 0$ have $j < i$, and we call x_i *maximal* in this constraint. Note that any constraint with non-empty polynomial has a unique maximal variable.

If S is a set of constraints, we denote by S_i the subset of S of constraints having no variable larger than x_i, i.e., in S_i all variables of $\{x_{i+1} \ldots x_n\}$ have zero coefficients. Formally, given an S, for each $i = 0, \ldots, n$ we define $S_i = \{ (a_n x_n + \ldots + a_1 x_1 \geq c) \in S \mid a_j = 0 \text{ for all } j > i \}$. We clearly have that $S_n = S$, that $S_i \subset S_{i+1}$ ($0 \leq i < n$) and that S_0 includes all contradictions in S (if any).

The following inference rules for constraints are well-known ($\lceil . \rceil$ denotes the standard rounding-up ceiling function and gcd the greatest common divisor).

$$\textbf{Combine}: \quad \frac{p \geq c \quad q \geq d}{ap + bq \geq ac + bd} \quad \text{where} \quad a, b \in \mathbb{N}$$

$$\textbf{Divide}: \quad \frac{a_n x_n + \ldots + a_1 x_1 \geq c}{(a_n/d)x_n + \ldots + (a_1/d)x_1 \geq \lceil c/d \rceil} \quad \text{where} \quad d = \gcd(a_n, \ldots, a_1)$$

These rules are also well-known to be correct, that is, any set of constraints S has the same solutions in $\mathbb{Z}$ as $S \cup \{C\}$ if C is a conclusion obtained by **Combine**-ing or **Divide**-ing with premise(s) in S.

3 Completeness for ILP

The following result is not new, see, e.g., [13, Cor. 2.14]. In fact, it is known to be true even if S is not bounded [19, Corollary 23.2b]. But the proof we provide is new, concise and self-contained. Moreover, the proof allows us to analyse which instances of Combine and Divide are actually used. This will allow us to obtain stronger results.

THEOREM 3.1. *If a bounded set of constraints is infeasible then its closure under Combine and Divide yields a contradiction.*

Proof. Consider any bounded set of constraints, let S be its closure under Combine and Divide, and assume there is no contradiction in S. We prove that S is feasible, by actually building a solution M of S. Recall that S_i is the subset of all constraints in S where the variables in $x_n, \ldots, x_{i+1}$ have coefficient 0. We build a solution M_i for each S_i by induction on i. The base case $i = 0$ is trivial: the empty M_0 is a solution of S_0 since S has no contradictions. For the induction step $i > 0$, assume M_{i-1} is a solution for S_{i-1} (I.H.). We extend M_{i-1} to M_i by defining $M_i(x_i) = \max\{\, c - M_{i-1}(p) \mid x_i + p \geq c \text{ in } S_i \,\}$. This is well-defined: we take the maximum of a non-empty set (since there is a lower bound $x_i \geq l$ in S) that is upper bounded by u, the upper bound for x_i in S. The latter is true since S is closed under Combine and each $x_i + p \geq c$ in S_i can be combined with $-x_i \geq -u$ to give the constraint $p \geq c - u$ in S_{i-1} and, by I.H., $c - M_{i-1}(p) \leq u$.

We now prove $M_i \models S_i$. Constraints C in $S_i \setminus S_{i-1}$ can have the following form:

A) C is of the form $x_i + p \geq c$. Then by construction of M_i we have $M_i \models C$.

B) C is of the form $-ax_i + p \geq c$ with $a > 0$. By construction, $M_i(x_i) = d - M_i(q)$ for some $x_i + q \geq d$ in S_i. By a Combine of a times $x_i + q \geq d$ with C we get $aq + p \geq c + ad$, which is in S_{i-1}, hence by I.H. $aM_i(q) - M_i(p) \geq c + ad$, so $M_i \models -ax_i + p \geq c$.

C) C is of the form $ax_i + p \geq c$ with $a > 1$.

  C1) If $a|p$ then, by Divide, $x_i + p/a \geq \lceil c/a \rceil$ is in S_i and by construction $M_i(x_i) \geq \lceil c/a \rceil - M_i(p/a)$ and hence $M_i(ax_i + p) \geq c$, so $M_i \models C$.

  C2) Otherwise, let x_j be the maximal variable in p with coefficient b not divisible by a. Let $x_j + q \geq d$ in S_j be the constraint causing $M_i(x_j)$ to be $d - M_i(q)$.

   C2a) If $b < 0$, do a Combine of $-b$ times $x_j + q \geq d$ and C (eliminating x_j).

   C2b) If $b > 0$, let r be the smallest natural number such that $b + r$ is divisible by a and do a Combine of r times $x_j + q \geq d$ and C.

  In both cases we get a constraint D in S_i, where x_j's coefficient is divisible by a, and such that $M_i \models D$ iff $M_i \models C$. By repeating this at most $n - 1$ times (where each time the maximal variable with coefficient not divisible by a becomes smaller), finally we get a constraint as in case C1), proving $M_i \models C$. ⊣

One inconvenient aspect of the previous result is that Combine is difficult to control (and hence of little practical value): in fact *any* pair of constraints $p \geq c$ and $q \geq d$ can be combined with *any* pair of *factors* $a, b \in \mathbb{N}$ to obtain $ap + bq \geq ac + bd$. Also, while Divide is fully deterministic, it is not clear when it is useful to apply this rule.

However, a closer look at our proof reveals that in fact only very particular cases of **Combine** and **Divide** are used. The **Divide** rule is applied only in case C1), where the premise is of the form $ax_i + p \geq c$ (i.e., by our notation, x_i is its maximal variable) where $a > 1$, and where *all* coefficients in p are divisible by a, and where indeed the divisor d of **Divide** is a; this is achieved by the **Ordered Exact Divide** rule given below. The **Combine** rule has three kinds of uses in the proof:

- In case B) and before A), **Combine** is used to *eliminate* x_i, the maximal variable in *both* premises, and moreover in one of the premises x_i has coefficient 1. Hence in the other premise x_i's coefficient $-a$ is negative and the factors are fully determined: they must be a and 1, respectively. This is done by the **Ordered VE Combine** rule given below (where VE stands for Variable-Eliminating).

- In cases C2a) and C2b) premise C must be of the form $ax_i + p + bx_j + p' \geq c$ where $a>1$, $a|p$ and $a \nmid b$ (note that by our notation, x_i is its maximal variable and x_j is its largest variable whose coefficient b is not divisible by a). The other premise C' is of the form $x_j + q \geq d$.

  - In case C2a), we have $b < 0$ and **Combine** is again used only to *eliminate* x_j, by adding $-b$ times C' to C, i.e., the factors must be $-b$ and 1.
    This is the **Ordered VE Preparing Combine** rule given below (called like this as it prepares for the **Ordered Exact Divide** step of case C1).

  - In case C2b), we have $b > 0$ and **Combine** must be used to add r times C' to C (i.e., factors r and 1), where r is the smallest natural number such that a divides $b + r$. This is the **Ordered Preparing Combine** rule given below.

Altogether, this proves that the following strongly restricted versions of **Combine** and **Divide** suffice.

Ordered VE Combine :
$$\frac{-ax_i + p \geq c \qquad x_i + q \geq d}{p + aq \geq c + ad} \qquad \text{where } a > 0$$

Ordered VE Preparing Combine :
$$\frac{ax_i + p + bx_j + p' \geq c \qquad x_j + q \geq d}{ax_i + p + p' - bq \geq c - bd} \qquad \text{where } a>1,\ b<0,\ a|p \text{ and } a \nmid b$$

Ordered Preparing Combine :
$$\frac{ax_i + p + bx_j + p' \geq c \qquad x_j + q \geq d}{ax_i + p + (b+r)x_j + p' + rq \geq c + rd} \qquad \text{where } b>0,\ a|p,\ a|(b+r),\ 0<r<a$$

Ordered Exact Divide :
$$\frac{ax_i + p \geq c}{x_i + p/a \geq \lceil c/a \rceil} \qquad \text{where } a>1 \text{ and } a|p$$

THEOREM 3.2. *If a bounded set of constraints is infeasible then its closure under* **Ordered VE Combine, Ordered VE Preparing Combine, Ordered Preparing Combine,** *and* **Ordered Exact Divide** *yields a contradiction.*

Some remarks related to the proof:

1. Bounds are crucial to make the definition of the model correct, ensuring that we take the maximum of a non-empty, bounded set of integers. Without bounds, finiteness of the initial constraint set is needed: the closure of $\{\, x \geq c \mid c \in \mathbb{N} \,\}$ under Combine and Divide has no contradiction but is not feasible. When the set of constraints is finite, bounds are not needed (see, e.g., [19, Corollary 23.2b]; [19, Corollary 17.1c] ensures that bounds of size polynomial in the size of the input data always exist).

2. The ordering of the variables is arbitrary.

3. The model in the above proof is built by taking for each x_i its smallest possible value, given the model built so far for $x_{i-1} \ldots x_1$. There is a dual model based on largest possible values. One can even choose per variable which of the two. The duality should not come as a surprise: if $l \leq x \leq u$, then $l \leq l + u - x \leq u$.

4. The constraints $x_i + p \geq c$ used in the model construction are called *tight* in [12].

5. Finiteness of the set of variables is not essential. The proof above can easily be generalized to a well-ordered set of variables.

6. Theorem 3.2 is a stronger completeness result than Theorem 3.1 since the rules are restricted. These weaker rules are easier to automate. On the other hand, weaker rules could make some proofs longer, for example, proofs of pigeon hole principles. This is a delicate balance: a stronger proof system is only useful for automated theorem proving if shorter proofs can indeed be found. So far this is not the case.

4 0-1 Solutions

This section is on Pseudo-Boolean constraint solving, also known as 0-1 integer linear programming. We consider *0-1 bounded* sets of constraints, i.e., containing *Boolean* bounds $x \geq 0$ and $-x \geq -1$ for every variable. We define the following two new rules that are special cases of Combine in which one of the premises is such a Boolean bound, and that replace the two Preparing rules of the previous section:

Upper 1 Combine :
$$\frac{ax_i + p + bx_j + p' \geq c \qquad -x_j \geq -1}{ax_i + p + (b-r)x_j + p' \geq c - r} \quad \text{where } a|p,\ a|(b-r),\ 0<r<a$$

Lower 0 Combine :
$$\frac{ax_i + p + bx_j + p' \geq c \qquad x_j \geq 0}{ax_i + p + (b+r)x_j + p' \geq c} \quad \text{where } a|p,\ a|(b+r),\ 0<r<a$$

Note that these rules do not include inferences with the stronger bounds $x \geq 1$ and $-x \geq 0$ stating that x is 1 or 0 (initial ones nor derived ones). These are used only by Ordered VE Combine.

THEOREM 4.1. *If a 0-1 bounded set of constraints is infeasible then its closure under* **Ordered VE Combine, Upper 1 Combine, Lower 0 Combine** *and* **Ordered Exact Divide** *yields a contradiction.*

Proof. Consider any 0-1 bounded set of constraints, let S be its closure under the four rules and assume there is no contradiction in S. We prove that S is feasible, by actually building a solution M_i for each S_i by induction on i. The base case $i = 0$ is trivial: the empty M_0 is a solution of S_0 since S has no contradictions. For the induction step $i > 0$, assume M_{i-1} is a solution for S_{i-1} (I.H.). We extend M_{i-i} to M_i by defining

- $M_i(x_i) = 1$ if there is a constraint of the form $x_i + p \geq c$ in S_i with $1 + M_{i-1}(p) = c$;
- $M_i(x_i) = 0$ otherwise.

Note that in both cases $M_i(x_i) = c - M_{i-1}(p)$ for some constraint of the form $x_i + p \geq c$ in S_i (in the second case the lower bound $x_i \geq 0$ is of this form). We now prove that $M_i \models C$ for all constraints C in $S_i \setminus S_{i-1}$. There are three cases, depending on C:

A) C is of the form $x_i + p \geq c$. Then by **Ordered VE Combine** with the upper bound $-x_i \geq -1$, we get $p \geq c - 1$ in S_{i-1} and hence, since $M_{i-1} \models S_{i-1}$, we have $1 + M_{i-1}(p) \geq c$. If $1 + M_{i-1}(p) = c$ then by construction we have $M_i(x_i) = 1$ and $M_i \models C$. If $1 + M_{i-1}(p) > c$ then $M_{i-1}(p) \geq c$ and even if $M_i(x_i) = 0$ we have $M_i \models C$.

B) C is of the form $-ax_i + p \geq c$ with $a > 0$. By construction $M_i(x_i) = d - M_i(q)$ for some $x_i + q \geq d$ in S_i, so we need to prove $-ad + aM_i(q) + M_i(p) \geq c$. By a **Ordered VE Combine** of a times $x_i + q \geq d$ and C we get that $aq + p \geq ad + c$ is in S_{i-1}, hence by I.H. $aM_i(q) + M_i(p) \geq c + ad$.

C) C is of the form $ax_i + p \geq c$ with $a > 1$.

   C1) If $a | p$ then by **Ordered Exact Divide** $x_i + p/a \geq \lceil c/a \rceil$ is in S_i. By case A) $M_i(x_i) + M_i(p/a) \geq \lceil c/a \rceil$ and hence $aM_i(x_i) + aM_i(p/a) \geq c$ and hence $M_i \models C$.

   C2) Otherwise, let x_j be the maximal variable in p whose coefficient b is not a multiple of a. Then C is of the form $ax_i + q + bx_j + q' \geq c$, where $a | q$.

   C2a) If $M_{i-1}(x_j) = 1$, do a **Upper 1 Combine** getting $ax_i + q + (b-r)x_j + q' \geq c - r$.
   C2b) If $M_{i-1}(x_j) = 0$, do a **Lower 0 Combine** getting $ax_i + q + (b+r)x_j + q' \geq c$.

   In both cases a constraint D is obtained that is in S_i, and such that $M \models D$ iff $M \models C$. By doing this for every variable in p whose coefficient is not a multiple of a, we obtain a constraint as in case C1), proving $M_i \models C$. ⊣

5 Examples and Counterexamples

In the previous sections we have given completeness proofs attempting to minimize the applicability of **Combine**, by limiting it to the cases where one of the premises has coefficient 1, where moreover the factors n, m of **Combine** are fully determined, and where a specific variable is either eliminated or increased by a precise amount. Moreover **Divide** is applied only to divide by a, the coefficient of the maximal variable, and only if the polynomial is exactly divisible by a.

A natural question arises: can non-variable eliminating applications of **Combine** be avoided altogether? The answer is negative even in the 0-1 case and even with the more general **Combine** and **Divide**:

EXAMPLE 5.1. *Let S consist of $3x+2y \geq 1$, $-3x+2y \geq -2$, $2x-3y \geq -2$ and $-2x-3y \geq -4$ and the four bounds for $0 \leq x, y \leq 1$, inspired by (and equivalent to) the well-known unsatisfiable example of four clauses that cannot be refuted by resolution without factoring: $x \vee y$, $\overline{x} \vee y$, $x \vee \overline{y}$ and $\overline{x} \vee \overline{y}$. Their standard representation as constraints $x + y \geq 1$, $-x + y \geq 0$, $x - y \geq 0$, and $-x - y \geq -1$ is easily refuted, but all refutations are effectively blocked by representing the clauses in a non-standard way with a clever choice of coefficients $\pm 2, \pm 3$ (adjusting the rhs accordingly). For example, from $3x + 2y \geq 1$ and $-3x + 2y \geq -2$ a variable-eliminating* **Combine** *infers $4y \geq -1$, which can only be* **Divided** *into the completely uninformative $y \geq 0$. One easily verifies that no contradiction can be derived by variable-eliminating* **Combine** *and* **Divide**. ⊣

EXAMPLE 5.2. *Let S be as in Example 5.1. It is instructive to see how the inference rules of Theorem 4.1 refute S without doing much redundant work. Define the ordering by $x > y$, so $x_2 = x$ and $x_1 = y$.* **Ordered VE Combine** *can only be used to infer the tautology $0 \geq 1$ from the bounds for x of y, or with $x \geq 0$ and either $-3x + 2y \geq -2$ or $-2x - 3y \geq -4$, yielding the uninformative $2y \geq -2$ and $-3y \geq -4$, respectively.* **Lower 0 Combine** *and* **Upper 1 Combine** *can only be applied with $3x + 2y \geq 1$ and a bound for y in S, and with $2x - 3y \geq -2$ and a bound for y in S. With $-y \geq -1$ this yields the uninformative $3x \geq -1$ and the informative $2x - 4y \geq -3$, or $x - 2y \geq -1$ after an application of* **Ordered Exact Divide**. *With $y \geq 0$ this yields the informative $3x + 3y \geq 1$, or $x + y \geq 1$ after an application of* **Ordered Exact Divide**, *and the uninformative $2x \geq -2$. We have now inferred two new constraints, $x - 2y \geq -1$ and $x + y \geq 1$ in which the variable x has coefficient 1, which opens up for new applications of* **Ordered VE Combine**. *Indeed, an* **Ordered VE Combine** *of $-3x + 2y \geq -2$ and $x + y \geq 1$ yields $5y \geq 1$, or $y \geq 1$ after an application of* **Ordered Exact Divide**. *Moreover, an* **Ordered VE Combine** *of $-2x - 3y \geq -4$ and $x - 2y \geq -1$ yields $-7y \geq -6$, which contradicts $y \geq 1$, formally by another application of* **Ordered VE Combine**. ⊣

The next question that arises is whether our rules for the 0-1 case are sufficient for the integer case, when *suitably generalized*, i.e., the right premises in **Upper 1 Combine** and **Lower 0 Combine** are the initial bounds for x_j in S, say $-x_j \geq -u_j$ and $x_j \geq l_j$, and the right-hand sides of the conclusion of **Upper 1 Combine** and **Lower 0 Combine** are $c - ru_j$ and $c + rl_j$, respectively. The answer is no, as we will see in the next example.

EXAMPLE 5.3. *Let S be as in Example 5.1 with the only difference that the upper bound $-y \geq -1$ is replaced by $-y \geq -2$. Therefore, the analysis of the previous example largely applies here. Hence we focus on the differences. The upper bound $-y \geq -2$ can only be applied by* **Upper 1 Combine**. *With $3x + 2y \geq 1$ this gives the uninformative $3x \geq -3$. With $2x - 3y \geq -2$ it gives $2x - 4y \geq -4$, and after* **Ordered Exact Divide**, *$x - 2y \geq -2$. From the latter we can infer $-2y \geq -3$ by an application of* **Ordered VE Combine** *with $-x \geq -1$. By applying* **Ordered Exact Divide**, *we can indeed infer a new derived upper bound $-y \geq -1$. However, as remarked after the definition of* **Upper 1 Combine** *and* **Lower 0 Combine**, *derived bounds cannot be used in these two rules, thus blocking the refutation of the previous example. Indeed, with some effort, doing an exhaustive case analysis the reader can now verify that no contradiction can be derived.* ⊣

6 Related Work

It is well-known that from any run of a CDCL-based SAT solver on an unsatisfiable instance one can in fact extract a resolution refutation: at each conflict, by resolution a new clause is derived that enables a backjump, which finally leads to a conflict at decision level zero, giving the empty clause. Proving this formally requires a termination argument based on a well-founded ordering on CDCL states (see e.g., [15]).

We start by the observation that, in a similar way, it is possible to extract a completeness proof of Combine and Divide for bounded ILP from Jovanovic and De Moura's Cutsat procedure [12] (whose termination is based on [15]). But it gives a far longer and more complicated proof than ours, less insight-providing, and it does not lead to our restrictive rules nor to ordered rules. Note however that the Cutsat procedure was not devised with the purpose of providing a completeness proof; it is an elegant and highly non-trivial procedure, the first complete practical CDCL cutting-planes-based solver for ILP.

A similar situation exists for several procedures for Pseudo-Boolean constraint solving [2–4, 6–8, 20], see also the handbook chapter [18]. These procedures perform variable-eliminating Combine steps, and in order to always be able to backjump, they sometimes need to weaken a constraint, e.g., into a cardinality constraint (i.e., a constraint having only unit coefficients) or into a clause, known to be a logical consequence for semantic reasons. These various versions of semantic weakening can be rephrased as sequences of Combine steps and Divide steps. Some papers make these steps explicit; it also follows from the consequence-finding completeness results [13, Theorem 2.14] and [19, Corollary 23.2b]. An independent syntactic proof of this semantic weakening and a formal termination argument together constitute a completeness proof obtained from these procedures.

Hooker has done pioneering work, first proving consequence-finding completeness of a form of cutting planes for cardinality constraints [9]. Later he extended it to arbitrary 0-1 linear constraints, although assuming a subprocedure for finding certain entailed constraints (in particular this is possible for cardinality constraints) [10, 11].

Work on proof complexity of cutting planes, such as [17], hardly relates to ours, since it essentially concentrates on showing that (no) short (in various senses) proofs exist for certain problem families, rather than on completeness of cutting planes. E.g., Cook et al. [5] rely on Schrijver [19] for completeness.

Last but not least, let us again mention the immense body of work on the theory of (I)LP in e.g., Schrijver [19], already described in Section 1.

7 Conclusions and Further Work

We have given short and –we think– insight-providing completeness proofs for general bounded ILPs as well as 0-1 ones, including new (to the best of our knowledge) results for less prolific and hence more easily automatizable systems including ordered ones.

Given the kind of induction used in our proofs it would not be difficult to extend these results to include *abstract redundancy* notions as in, e.g., [1, 16] for (first-order) resolution and paramodulation. This would allow one to remove constraints that follow from (finitely many) *smaller* constraints w.r.t. a given well-founded ordering $\succ$ on constraints. The proof is then by contradiction from the existence of a $\succ$-minimal constraint that is false in the constructed model. Such an abstract redundancy notion covers many concrete criteria. To give just one simple example, a constraint $2x + 4y + z \geq 3$

can be removed (or needs not be generated) if, say, $x + 2y \geq 2$ and $z \geq 0$ are already present.

Some questions arise from the proof complexity point of view. We have proved completeness for inference systems that are particular cases of **Combine** and **Divide** but that are far more "restrictive" than these. But there is usually a trade-off: such restrictive systems tend to become less "efficient" in terms of minimal proof length. Which problem families (if any) do have short (polynomial) proofs by **Combine** and **Divide** but not by our rules? Does this have any practical consequences for CDCL-based ILP provers? If so, are there any more "controllable" appropriate intermediate systems?

Acknowledgments

Both authors have been supported by the Spanish MEC/MICINN project SweetLogics under the grant TIN2010-21062-C02-01 and by the project DAMAS, Spanish MINECO under the grant TIN2013-45732-C4-3-P.

BIBLIOGRAPHY

[1] Leo Bachmair and Harald Ganzinger. Resolution. In J.A. Robinson and A. Voronkov, editors, *Handbook of Automated Reasoning*. Elsevier Science Publishers and MIT Press, 2001.
[2] Peter Barth. A Davis-Putnam based enumeration algorithm for linear pseudo-Boolean optimization. Technical report, Max-Plank-Institut fur Informatik, Saarbrucken, 1995. Technical Report MPII952003.
[3] Daniel Le Berre and Anne Parrain. The sat4j library, release 2.2. *JSAT*, 7(2-3):59–6, 2010.
[4] Donald Chai and Andreas Kuehlmann. A fast pseudo-boolean constraint solver. *IEEE Trans. on CAD of Integrated Circuits and Systems*, 24(3):305–317, 2005.
[5] W. Cook, C. Coullard, and Gy. Turan. On the complexity of cutting-plane proofs. *Discrete Applied Mathematics*, 18:25–38, 1987.
[6] Heidi E. Dixon. *Automating Psuedo-Boolean Inference within a DPLL Framework*. PhD thesis, University of Oregon, 2004.
[7] Heidi E. Dixon and Matthew L. Ginsberg. Inference methods for a pseudo-boolean satisfiability solver. In *Procs 18th National Conf on Artificial Intelligence*, pages 635–640, 2002.
[8] Heidi E. Dixon, Matthew L. Ginsberg, David K. Hofer, Eugene M. Luks, and Andrew J. Parkes. Implementing a generalized version of resolution. In *Procs 19th National Conf on Artificial Intelligence*, pages 55–60, 2004.
[9] John N Hooker. Generalized resolution and cutting planes. *Annals of Operations Research*, 12:217–239, 1988.
[10] John N Hooker. Generalized resolution for 0-1 linear inequalities. *Annals of Mathematics and Artificial Intelligence*, 6:271–286, 1992.
[11] John N Hooker. *Logic-Based Methods for Optimization: Combining Optimization and Constraint Satisfaction*. John Wiley & Sons, 2000.
[12] Dejan Jovanovic and Leonardo Mendonça de Moura. Cutting to the chase - solving linear integer arithmetic. *J. Autom. Reasoning*, 51(1):79–108, 2013.
[13] G. Nemhauser and L. Wolsey. *Integer and Combinatorial Optimization*. John Wiley and Sons, 1999.
[14] Robert Nieuwenhuis. The IntSat method for integer linear programming. In *Principles and Practice of Constraint Programming, 20th International Conference, CP 2014, LNCS 8656*, pages 574–589, 2014.
[15] Robert Nieuwenhuis, Albert Oliveras, and Cesare Tinelli. Solving SAT and SAT Modulo Theories: From an abstract Davis–Putnam–Logemann–Loveland procedure to DPLL(T). *Journal of the ACM, JACM*, 53(6):937–977, 2006.
[16] Robert Nieuwenhuis and Albert Rubio. Paramodulation-based theorem proving. In J.A. Robinson and A. Voronkov, editors, *Handbook of Automated Reasoning*, pages 371–444. Elsevier Science and MIT Press, 2001.
[17] P. Pudlák. Lower bounds for resolution and cutting planes proofs and monotone computations. *Journal of Symbolic Logic*, 62(3):981–998, 1997.
[18] Olivier Roussel and Vasco M. Manquinho. Pseudo-boolean and cardinality constraints. In A. Biere, M. Heule, H. van Maaren, and T. Walsh, editors, *Handbook of Satisfiability*, volume 185 of *Frontiers in AI and Applications*, pages 695–733. IOS Press, 2009.
[19] A. Schrijver. *Theory of Linear and Integer Programming*. John Wiley and Sons, 1998.
[20] Hossein M. Sheini and Karem A. Sakallah. Pueblo: A hybrid pseudo-boolean SAT solver. *JSAT*, 2(1-4):165–189, 2006.

An Algebraic Approach to Filtrations for Superintuitionistic Logics

GURAM BEZHANISHVILI AND NICK BEZHANISHVILI

Dedicated to Albert Visser on the occasion of his 65th birthday.[1]

Abstract

There are two standard model-theoretic methods for proving the finite model property for modal and superintuitionistic logics, the standard filtration and the selective filtration. While the corresponding algebraic descriptions are better understood in modal logic, it is our aim to give similar algebraic descriptions of filtrations for superintuitionistic logics via locally finite reducts of Heyting algebras. We show that the algebraic description of the standard filtration is based on the $\to$-free reduct of Heyting algebras, while that of selective filtration on the $\vee$-free reduct.

1 Introduction

The main tools for establishing the finite model property for modal and superintuitionistic logics are the methods of standard and selective filtrations. If a model $\mathfrak{M}$ refutes a formula φ, then we wish to filter it out so that the resulting model $\mathfrak{N}$ is finite and still refutes φ. The model $\mathfrak{N}$ can be constructed as a factor-model of $\mathfrak{M}$ (standard filtration) or as a submodel of $\mathfrak{M}$ (selective filtration).

The standard filtration (or simply filtration) was originally developed algebraically [24, 25], and later model-theoretically [23, 26]. The model-theoretic approach became a standard tool for proving the finite model property in modal logic. We refer to [12, 14] for a systematic exposition of the method, and its numerous applications. The algebraic and model-theoretic methods are closely related. For modal logics this was first discussed in [21, 22]. For a modern account of the connection see [5, 15, 20].

The method of selective filtration in modal logic was first discussed in [19], and further developed in [18, 29]. In [28] the method was applied to superintuitionistic logics. See [14] for an overview. An algebraic analogue of this technique for superintuionistic logics was developed in [7], and for modal logics in [8] (although [7, 8] do not discuss explicitly the connection of their algebraic method to selective filtration).

In this short note we would like to revisit the methods of standard and selective filtrations for superintuionistic logics from an algebraic point of view. We show that in both cases one has to work

---

[1] Albert Visser has been our senior colleague and friend for many years now. The second author has especially pleasant memories of the time when he shared an office with Albert (and Rosalie Iemhoff) during his postdoc at Utrecht University in 2012–2013. Albert would always give his help and support when needed, and cheer you up by telling a nice joke or anecdote, often his personal encounters with famous logicians. But above all, he has always been a limitless source of inspirational ideas about many different aspects of logic. We are happy to dedicate this short paper to Albert's 65th birthday and wish him many more productive years to come.

with appropriate locally finite reducts of Heyting algebras. In the case of filtrations, the $\rightarrow$-free reduct, and in the case of selective filtrations, the $\vee$-free reduct.

In order to define standard filtrations algebraically we need to work with free Boolean extensions of distributive lattices. This enables us to define the least and greatest filtrations algebraically. We show, via duality for Heyting algebras, that algebraically described standard and selective filtrations exactly correspond to their model-theoretic analogues. In [2, 4] the $\vee$-free and $\rightarrow$-free reducts of Heyting algebras were used for introducing $(\wedge, \rightarrow)$ and $(\wedge, \vee)$-canonical formulas which axiomatize all superintuionistic logics. The proofs essentially use the algebraic versions of standard and selective filtrations.

We assume the reader's familiarity with the algebraic and relational semantics of intuitionistic logic. For basic definitions and facts we refer to [9, 14].

2 Standard Filtration

A *Kripke frame* is a partially ordered set $\mathfrak{F} = (W, \leq)$. We call $U \subseteq W$ an *up-set* (upward closed set) if $w \in U$ and $w \leq v$ imply $v \in U$. A *valuation* ν on $\mathfrak{F}$ assigns to each propositional letter p an up-set $\nu(p)$ of $\mathfrak{F}$. A *Kripke model* is a pair $\mathfrak{M} = (\mathfrak{F}, \nu)$, where $\mathfrak{F}$ is a Kripke frame and ν is a valuation on $\mathfrak{F}$. We recall the definition of filtration (see, e.g., [14, Sec. 5.3]).

DEFINITION 2.1 (Standard filtration model theoretically). *Let Σ be a finite set of formulas closed under subformulas, and let $\mathfrak{M} = (\mathfrak{F}, \nu)$ be a model. Define an equivalence relation $\sim$ on W by*

$$w \sim v \text{ if } (\forall \varphi \in \Sigma)(w \in \nu(\varphi) \text{ iff } v \in \nu(\varphi)). \tag{1}$$

Let $W' = W/\sim$. Since Σ is finite, so is W'. In fact, $|W'| \leq 2^{|\Sigma|}$. Let $\leq'$ be a partial order on W' satisfying the following two conditions for all $w, v \in W$ and $\varphi \in \Sigma$:

$$w \leq v \text{ implies } [w] \leq' [v], \tag{2}$$

$$[w] \leq' [v] \text{ and } w \in \nu(\varphi) \text{ imply } v \in \nu(\varphi). \tag{3}$$

Let ν' be a valuation on $\mathfrak{F}'$ such that

$$\nu'(p) = \{[w] : w \in \nu(p)\} \tag{4}$$

for each $p \in \Sigma$. Then $\mathfrak{M}' = (\mathfrak{F}', \nu')$ is a finite model called a standard filtration *(or simply a* filtration*) of the model $\mathfrak{M}$ through Σ.*

REMARK 2.2. It is a consequence of (3) that $\{[w] : w \in \nu(\varphi)\}$ is an up-set of $\mathfrak{F}'$ for each $\varphi \in \Sigma$. Therefore, there always exists a valuation ν' on $\mathfrak{F}'$ satisfying (4). Thus, $\mathfrak{M}'$ is well defined.

The next lemma is well known (see, e.g., [14, Thm. 5.23]).

LEMMA 2.3 (Filtration Lemma). *Let $\mathfrak{M} = (\mathfrak{F}, \nu)$ be a model and let $\mathfrak{M}' = (\mathfrak{F}', \nu')$ be a filtration of $\mathfrak{M}$ through a finite set Σ of formulas closed under subformulas. Then for each $\varphi \in \Sigma$ and $w \in W$,*

$$w \in \nu(\varphi) \text{ iff } [w] \in \nu'(\varphi). \tag{5}$$

Consequently, if $\varphi \in \Sigma$ is refuted on $\mathfrak{M}$, then it is refuted on $\mathfrak{M}'$ (see, e.g., [14, Cor. 5.25]).

REMARK 2.4. As follows from Remark 2.2, Condition (3) is crucial for $\mathfrak{M}'$ to be well defined. On the other hand, Condition (2) is used in proving Condition (5). But Condition (2) itself is not necessary for refuting $\varphi \in \Sigma$ on $\mathfrak{M}'$ when it is refuted on $\mathfrak{M}$. For this purpose it is sufficient to have Condition (5). In some situations it is even disadvantageous to assume Condition (2); for example, when proving the finite model property for modal logics **GL** and **Grz** via standard filtration [11, 13]. However, it is common to require (2), so we will assume (2) throughout. This condition plays an important role in the study of stable logics [4].

As follows from [14, Sec. 5.3], among the filtrations of $\mathfrak{M}$ through Σ, there always exist the least and greatest filtrations. In other words, among the partial orders on W' that satisfy (2) and (3), there always exist the least and greatest ones. The least filtration is defined as follows. Let

$$[w] \preceq [v] \text{ iff there exist } w' \sim w \text{ and } v' \sim v \text{ such that } w' \leq v', \tag{6}$$

and let $\leq^l$ be the transitive closure of $\preceq$. On the other hand, the greatest filtration is given by

$$[w] \leq^g [v] \text{ iff } (\forall \varphi \in \Sigma)(w \in \nu(\varphi) \Rightarrow v \in \nu(\varphi)). \tag{7}$$

Next we give an algebraic description of filtrations. For a Heyting algebra A, let $\mathfrak{F}_A = (W_A, \subseteq)$ be the frame of prime filters of A ordered by inclusion. We call $\mathfrak{F}_A$ the *spectrum* of A. It is well known that A embeds into the Heyting algebra $\mathrm{Up}(\mathfrak{F}_A)$ of up-sets of $\mathfrak{F}_A$ by $\alpha(a) = \{w \in W_A : a \in w\}$. Each valuation ν on A gives rise to a valuation $\mu = \alpha \circ \nu$ on $\mathfrak{F}_A$.

Let Σ be a finite set of formulas closed under subformulas. Since Σ is finite, $\nu[\Sigma]$ is a finite subset of A. Let S be the bounded sublattice of A generated by $\nu[\Sigma]$. As bounded distributive lattices are locally finite, we see that S is finite. Therefore, S is a Heyting algebra, where

$$a \to_S b = \bigvee \{s \in S : a \wedge s \leq b\} \tag{8}$$

for each $a, b \in S$. Clearly $a \to_S b \leq a \to b$, and

$$a \to_S b = a \to b \text{ provided } a \to b \in S. \tag{9}$$

LEMMA 2.5. *S gives rise to a filtration $\mathfrak{M}'_A = (\mathfrak{F}'_A, \mu')$ of $\mathfrak{M}_A = (\mathfrak{F}_A, \mu)$ through Σ.*

Proof. Define $\sim$ on W_A by $w \sim v$ iff $w \cap S = v \cap S$. We show that $w \sim v$ iff $w \cap \nu[\Sigma] = v \cap \nu[\Sigma]$. The left to right implication is clear. For the right to left implication, let $w \cap \nu[\Sigma] = v \cap \nu[\Sigma]$. If $a \in w \cap S$, then since $\nu[\Sigma]$ generates S, we have $a = \bigvee_{i=1}^{n} \bigwedge_{j=1}^{n_i} b_{ij}$ for some $b_{ij} \in \nu[\Sigma]$. As w is a prime filter of A, there is i with $\bigwedge_{j=1}^{n_i} b_{ij} \in w$, so $b_{ij} \in w$ for all j. Since $w \cap \nu[\Sigma] = v \cap \nu[\Sigma]$, we have $b_{ij} \in v$ for all j. Therefore, $\bigwedge_{j=1}^{n_i} b_{ij} \in v$, and so $\bigvee_{i=1}^{n} \bigwedge_{j=1}^{n_i} b_{ij} \in v$. Thus, $a \in v \cap S$, yielding $w \cap S \subseteq v \cap S$. The other inclusion is proved similarly, hence $w \sim v$.

Now $w \cap \nu[\Sigma] = v \cap \nu[\Sigma]$ is clearly equivalent to $(\forall \varphi \in \Sigma)(\nu(\varphi) \in w \text{ iff } \nu(\varphi) \in v)$, which in turn is equivalent to $(\forall \varphi \in \Sigma)(w \in \mu(\varphi) \text{ iff } v \in \mu(\varphi))$. Let $W'_A = W_A/\sim$, and set $[w] \leq' [v]$ if $w \cap S \subseteq v \cap S$. It is obvious that $\leq'$ is a partial order on W'_A. Let μ' be a valuation on $\mathfrak{F}'_A = (W'_A, \leq')$ such that $\mu'(p) = \{[w] : w \in \mu(p)\}$. To see that such a μ' exists, it is sufficient to show that $\{[w] : w \in \mu(\varphi)\}$ is an up-set for each $\varphi \in \Sigma$. Suppose $w \in \mu(\varphi)$ and $[w] \leq' [v]$. Then $\nu(\varphi) \in w$ and $w \cap S \subseteq v \cap S$. Therefore, $\nu(\varphi) \in v$, so $v \in \mu(\varphi)$, and hence $\{[w] : w \in \mu(\varphi)\}$ is indeed an up-set. Thus, $\mathfrak{M}'_A = (\mathfrak{F}'_A, \mu')$ is a model, and it is straightforward to see that it satisfies Conditions (2) and (3) of Definition 2.1. Thus, $\mathfrak{M}'_A$ is a filtration of $\mathfrak{M} = (\mathfrak{F}_A, \mu)$ through Σ. ⊣

REMARK 2.6. More generally, the construction of Lemma 2.5 can be applied to any finite bounded sublattice L of A that contains $\nu[\Sigma]$, or equivalently has S as a bounded sublattice. In fact, one could define filtration algebraically as a pair (L, ν_L), where L is a sublattice of A that contains $\nu[\Sigma]$ and ν_L is the restriction of ν to L. The filtration lemma for (L, ν_L) holds (see Lemma 2.11 below), so if (A, ν) refutes a formula $\varphi \in \Sigma$, then so does (L, ν_L). But this definition of filtration does not match the model theoretic definition above because the equivalence relation defined from L can be more refined than the one defined from S. To ensure the match, we have to require that $w \cap L = v \cap L$ is equivalent to $w \cap S = v \cap S$. This is achieved by requiring that L and S have isomorphic free Boolean extensions.

LEMMA 2.7. *Let L be a finite bounded sublattice of A that contains S as a bounded sublattice. Then the following two conditions are equivalent.*

1. *$w \cap L = v \cap L$ is equivalent to $w \cap S = v \cap S$.*

2. *$\alpha[L]$ and $\alpha[S]$ generate the same Boolean subalgebra of the powerset of the spectrum of A.*

Proof. (1)$\Rightarrow$(2): Since $\nu[\Sigma]$ generates S, it is obvious that $\mu[\Sigma]$ and $\alpha[S]$ generate the same Boolean subalgebra $\mathfrak{B}$ of the powerset of W_A. Let $\sim$ be the equivalence relation given by $w \sim v$ iff $w \cap S = v \cap S$. Call $U \subseteq W_A$ *saturated* provided $w \in U$ and $w \sim v$ imply $v \in U$. Then $U \subseteq W_A$ is saturated iff U belongs to $\mathfrak{B}$. Therefore, (1) implies that $\alpha[L]$ and $\alpha[S]$ define the same equivalence relation $\sim$ on W_A, and hence both $\alpha[L]$ and $\alpha[S]$ generate the same Boolean algebra $\mathfrak{B}$.

(2)$\Rightarrow$(1): Suppose $\alpha[L]$ and $\alpha[S]$ generate the same Boolean algebra $\mathfrak{B}$. Since S is a bounded sublattice of L, it is clear that $w \cap L = v \cap L$ implies $w \cap S = v \cap S$. Conversely, suppose $w \cap S = v \cap S$. Let $a \in w \cap L$. Then $w \in \alpha(a) \in \alpha[L]$. Therefore, $\alpha(a) \in \mathfrak{B}$. Since $\alpha[S]$ generates $\mathfrak{B}$ and is closed under finite unions and intersections, we can write $\alpha(a) = \bigcup_{i=1}^{n} (\alpha(b_i) \cap \alpha(c_i)^c)$ with $b_i, c_i \in S$. Thus, $w \in \alpha(b_i)$ and $w \notin \alpha(c_i)$ for some i. This yields $b_i \in w$ and $c_i \notin w$ for some i. Since $w \cap S = v \cap S$, we conclude that $b_i \in v$ and $c_i \notin v$ for some i, so $v \in \alpha(a)$. This gives $w \cap L \subseteq v \cap L$, and the reverse inclusion is proved similarly. ⊣

LEMMA 2.8. *Each filtration $\mathfrak{M}'_A$ of $\mathfrak{M}_A$ through Σ gives rise to a finite bounded sublattice L of A that contains S as a bounded sublattice and such that $\alpha[L]$ and $\alpha[S]$ generate the same Boolean subalgebra of the powerset of the spectrum of A.*

Proof. Suppose $\mathfrak{M}'_A = (\mathfrak{F}'_A, \mu')$ is a filtration of $\mathfrak{M}_A = (\mathfrak{F}_A, \mu)$ through Σ. Then saturated subsets of W_A belong to the Boolean subalgebra $\mathfrak{B}$ of the powerset of W_A generated by $\mu(\Sigma)$. Let $\pi : W_A \to W'_A$ be the quotient map given by $\pi(w) = [w]$. Therefore, $\pi^{-1}(U) \in \mathfrak{B}$ for each $U \subseteq W'_A$. Let $\mathrm{Up}(\mathfrak{F}'_A)$ be the Heyting algebra of up-sets of $\mathfrak{F}'_A$. By Condition (2) of Definition 2.1, π is order-preserving. Therefore, $\pi^{-1}(U)$ is an up-set of W_A belonging to $\mathfrak{B}$. Thus, $\pi^{-1}(U) \in \alpha[A]$. This yields that $\pi^{-1}[\mathrm{Up}(\mathfrak{F}'_A)]$ is a bounded sublattice of $\alpha[A]$. It is finite since $\mathrm{Up}(\mathfrak{F}'_A)$ is finite, so $L := \alpha^{-1}\pi^{-1}[\mathrm{Up}(\mathfrak{F}'_A)]$ is a finite bounded sublattice of A. Clearly L contains $\nu[\Sigma]$, hence L contains S as a bounded sublattice. Moreover, for $w, v \in W_A$, we have $w \cap L = v \cap L$ iff $w \sim v$. Indeed, since $\nu[\Sigma] \subseteq L$, the left to right implication is obvious. For the right to left implication, let $w \sim v$. If $a \in w \cap L$, then $a \in w$ and $a = \alpha^{-1}\pi^{-1}(U)$ for some $U \in \mathrm{Up}(\mathfrak{F}'_A)$. This implies that $w \in \alpha(a)$ and $\alpha(a) = \pi^{-1}(U)$. Therefore, $w \in \pi^{-1}(U)$, and so $\pi(w) \in U$. As $w \sim v$, we have $\pi(w) = \pi(v)$. Thus, $\pi(v) \in U$, and hence $v \in \alpha(a)$. This yields that $a \in v \cap L$. Consequently, $w \cap L \subseteq v \cap L$, and the other inclusion is proved similarly. Therefore, we showed that $w \cap L = v \cap L$. Since $w \sim v$ iff $w \cap S = v \cap S$, by Lemma 2.7, $\alpha[L]$ and $\alpha[S]$ generate the same Boolean algebra. ⊣

REMARK 2.9. The Boolean algebra $\mathfrak{B}$ generated by $\alpha[S]$ is nothing more but the free Boolean extension of S. We recall (see, e.g., [1, p. 99, Def. 5]) that the *free Boolean extension* of a bounded distributive lattice L is a pair $(\mathfrak{B}, f)$ such that $\mathfrak{B}$ is a Boolean algebra, $f : L \to \mathfrak{B}$ is a bounded lattice embedding, and for any Boolean algebra $\mathfrak{B}'$ and a bounded lattice homomorphism $g : L \to \mathfrak{B}'$, there is a unique Boolean homomorphism $h : \mathfrak{B} \to \mathfrak{B}'$ such that $h \circ f = g$.

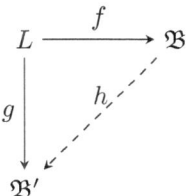

It is well known (see, e.g., [17, p. 39, Constr. 5.7]) that the free Boolean extension of a bounded distributive lattice L can be constructed as follows. Let $\mathfrak{F}_L$ be the spectrum of L and let $\mathfrak{B}$ be the Boolean subalgebra of the powerset of $\mathfrak{F}_L$ generated by $\alpha[L]$. Then $(\mathfrak{B}, \alpha)$ is isomorphic to the free Boolean extension of L. Therefore, up to isomorphism, the free Boolean extension of S is the Boolean subalgebra of the powerset of the spectrum $\mathfrak{F}_A$ of A generated by $\alpha[S]$.

It follows that there is a 1-1 correspondence between filtrations $\mathfrak{M}'_A$ of $\mathfrak{M}_A$ and finite bounded sublattices L of A that contain S as a bounded sublattice and have the same free Boolean extension as S. This motivates the following definition.

DEFINITION 2.10 (Standard filtration algebraically). *Let A be a Heyting algebra, ν be a valuation on A, and Σ be a finite set of formulas closed under subformulas. Suppose L is a finite bounded sublattice of A such that $\nu[\Sigma] \subseteq L$ and S, L have isomorphic free Boolean extensions. Let ν_L be a valuation on L such that $\nu_L(p) = \nu(p)$ for each $p \in \Sigma$. Then we call the pair (L, ν_L) a filtration of (A, ν) through Σ.*

LEMMA 2.11 (Filtration Lemma). *If (L, ν_L) is a filtration of (A, ν) through Σ, then $\nu_L(\varphi) = \nu(\varphi)$ for each $\varphi \in \Sigma$.*

Proof. Induction on the complexity of $\varphi \in \Sigma$. The case $\varphi = p$ follows from the definition of ν_L. The cases $\varphi = \bot$, $\varphi = \psi \wedge \chi$, and $\varphi = \psi \vee \chi$ follow from the fact that L is a bounded sublattice of A. Finally, let $\varphi = \psi \to \chi$. Then $\nu(\psi) \to \nu(\chi) = \nu(\psi \to \chi) \in \nu(\Sigma) \subseteq L$. Therefore, by (9), $\nu(\psi) \to \nu(\chi) = \nu(\psi) \to_L \nu(\chi)$. Thus, $\nu_L(\psi \to \chi) = \nu_L(\psi) \to_L \nu_L(\chi) = \nu(\psi) \to \nu(\chi) = \nu(\psi \to \chi)$. ⊣

REMARK 2.12. In order to prove the filtration lemma, it is not necessary to require that L is a sublattice of A. All that is needed is that $\nu_L(\varphi) \odot_L \nu_L(\psi) = \nu(\varphi) \odot \nu(\psi)$ for $\varphi \odot \psi \in \Sigma$, where $\odot \in \{\vee, \wedge, \to\}$. This is an algebraic import of the discussion in Remark 2.4.

THEOREM 2.1. *Let Σ be a finite set of formula closed under subformulas, A be a Heyting algebra, and ν be a valuation on A. Define μ on the spectrum $\mathfrak{F}_A$ of A by $\mu = \alpha \circ \nu$. Then there is a 1-1 correspondence between the filtrations (L, ν_L) of (A, ν) through Σ and the filtrations $\mathfrak{M}'_A = (\mathfrak{F}'_A, \mu')$ of $\mathfrak{M}_A = (\mathfrak{F}_A, \mu)$ through Σ. Moreover, if (L, ν_L), (K, ν_K) are two filtrations of (A, ν) through Σ and $\mathfrak{M}'_A$, $\mathfrak{M}''_A$ are the corresponding filtrations of $\mathfrak{M}_A$ through Σ, then L is a bounded sublattice of K iff $[w] \leq'' [v]$ implies $[w] \leq' [v]$.*

Proof. If (L, ν_L) is a filtration of (A, ν) through Σ, then by Lemma 2.5 (see also Remark 2.6), $\mathfrak{M}'_A = (W'_A, \leq', \mu')$ is a filtration of $\mathfrak{M}_A$ through Σ, where $[w] \leq' [v]$ iff $w \cap L \subseteq v \cap L$. Conversely, if $\mathfrak{M}'_A$ is a filtration of $\mathfrak{M}_A$ through Σ, then by Lemma 2.8, (L, ν_L) is a filtration of (A, ν) through Σ, where $L = \alpha^{-1} \pi^{-1} [\mathrm{Up}(\mathfrak{F}'_A)]$. This correspondence between the filtrations (L, ν_L) of (A, ν) through Σ and the filtrations $\mathfrak{M}'_A = (\mathfrak{F}'_A, \mu')$ of $\mathfrak{M}_A = (\mathfrak{F}_A, \mu)$ through Σ is 1-1. To see this, it is sufficient to observe that if (L, ν_L) is a filtration of (A, ν) through Σ and $\mathfrak{M}'_A = (\mathfrak{F}'_A, \mu')$ is the corresponding filtration of $\mathfrak{M}_A$ through Σ, then $L = \alpha^{-1} \pi^{-1} [\mathrm{Up}(\mathfrak{F}'_A)]$; and conversely, if $\mathfrak{M}'_A = (\mathfrak{F}'_A, \mu')$ is a filtration of $\mathfrak{M}_A$ through Σ and (L, ν_L) is the corresponding filtration of (A, ν) through Σ, then $w \sim v$ iff $w \cap L = v \cap L$.

For the last statement of the theorem, if L is a bounded sublattice of K, then $w \cap K \subseteq v \cap K$ implies $w \cap L \subseteq v \cap L$. Therefore, $[w] \leq'' [v]$ implies $[w] \leq' [v]$. Conversely, if $[w] \leq'' [v]$ implies $[w] \leq' [v]$, then $\mathrm{Up}(\mathfrak{F}'_A)$ is a bounded sublattice of $\mathrm{Up}(\mathfrak{F}''_A)$. Thus, $L = \alpha^{-1} \pi^{-1} [\mathfrak{F}'_A]$ is a bounded sublattice of $K = \alpha^{-1} \pi^{-1} [\mathfrak{F}''_A]$. ⊣

Among the filtrations (L, ν_L) of (A, ν), the filtration (S, ν_S) is clearly the least one. By Lemma 2.7, there is also the greatest filtration (T, ν_T) given by $T = \alpha^{-1}[\mathfrak{B}]$. By Theorem 2.1, (S, ν_S) corresponds to the greatest filtration $\mathfrak{M}^g = (W'_A, \leq^g, \mu')$ of $\mathfrak{M}_A$ through Σ, while (T, ν_T) to the least filtration $\mathfrak{M}^l = (W'_A, \leq^l, \mu')$. It is instructive to see a direct proof of this.

LEMMA 2.13. *Let Σ be a finite set of formulas closed under subformulas, A be a Heyting algebra, and ν be a valuation on A. Define μ on the spectrum $\mathfrak{F}_A$ of A by $\mu = \alpha \circ \nu$. Then (S, ν_S) gives rise to the greatest filtration $\mathfrak{M}^g = (W'_A, \leq^g, \mu')$, while (T, ν_T) to the least filtration $\mathfrak{M}^l = (W'_A, \leq^l, \mu')$ of $\mathfrak{M}_A = (\mathfrak{F}_A, \mu)$ through Σ.*

Proof. Let $w, v \in W_A$. We have $w \cap S \subseteq v \cap S$ iff $w \cap \nu[\Sigma] \subseteq v \cap \nu[\Sigma]$, which is equivalent to $(\forall \varphi \in \Sigma)(w \in \mu(\varphi) \Rightarrow v \in \mu(\varphi))$. Therefore, $w \cap S \subseteq v \cap S$ iff $[w] \leq^g [v]$. Thus, by Theorem 2.1, (S, ν_S) gives rise to the greatest filtration $\mathfrak{M}^g = (W'_A, \leq^g, \mu')$ of $\mathfrak{M}_A$ through Σ.

First suppose that $[w] \preceq [v]$. Then there are $w' \sim w$ and $v' \sim v$ with $w' \leq v'$. From $w \sim w'$ it follows that $w \cap \nu[\Sigma] = w' \cap \nu[\Sigma]$. Therefore, $w \cap S = w' \cap S$, and so $w \cap T = w' \cap T$ by Lemma 2.7. A similar argument gives $v \cap T = v' \cap T$, yielding $w \cap T \subseteq v \cap T$. Thus, if $\mathfrak{M}' = (W'_A, \leq', \mu')$ is the filtration corresponding to T, then $[w] \preceq [v]$ implies $[w] \leq' [v']$. Consequently, since $\leq'$ is a partial order, we conclude that $[w] \leq^l [v]$ implies $[w] \leq' [v]$. Conversely, suppose $[w] \not\leq^l [v]$. Then $[w] \cap \downarrow[v] = \varnothing$, where as usual, $\downarrow U = \{w : w \leq u \text{ for some } u \in U\}$. If $\downarrow[v]$ is saturated, then $\downarrow[v] \in \mathfrak{B}$, so $W_A \setminus \downarrow[v]$ is an up-set belonging to $\mathfrak{B}$. Therefore, $W_A \setminus \downarrow[v] \in \alpha(A) \cap \mathfrak{B}$, and hence $W_A \setminus \downarrow[v] = \alpha(a)$ for some $a \in T$. Clearly $a \in w \cap T$, but $a \notin v \cap T$, so $w \cap T \not\subseteq v \cap T$.

If $\downarrow[v]$ is not saturated, then let $\downarrow[\downarrow[v]]$ be the saturation of $\downarrow[v]$. Since $[w] \not\leq^l [v]$, we have $[w] \cap \downarrow[\downarrow[v]] = \varnothing$. If $\downarrow[\downarrow[v]]$ is saturated, then the same argument as above yields $w \cap T \not\subseteq v \cap T$. Otherwise we take the saturation of $\downarrow[\downarrow[v]]$. Since there are only finitely many saturated sets, it follows from the definition of $\leq^l$ that after finitely many steps we obtain a saturated down-set missing $[w]$. Thus, repeating the argument above, we conclude that $w \cap T \not\subseteq v \cap T$. Consequently, by Theorem 2.1, (T, ν_T) gives rise to the least filtration $\mathfrak{M}^l = (W'_A, \leq^l, \mu')$ of $\mathfrak{M}_A$ through Σ. ⊣

3 Selective Filtration

The idea of selective filtration is to work with submodels instead of quotient models. Given a model $\mathfrak{M} = (\mathfrak{F}, \nu)$ and a finite set Σ of formulas closed under subformulas, one wants to select a finite submodel $\mathfrak{M}' = (\mathfrak{F}', \nu')$ so that for each $w \in W'$ and $\varphi \in \Sigma$, we have $w \in \nu(\varphi)$ iff $w \in \nu'(\varphi)$. The method was introduced in [19]; see [14, Sec. 5.5] for details.

A more elaborate version of selective filtration was employed in [18] to prove that all transitive subframe logics have the finite model property. This was achieved by finding a subreduction (a p-morphism from a submodel) of $\mathfrak{M}$ onto a finite model $\mathfrak{M}'$. This method was further refined in [29], where it was shown that all transitive cofinal subframe logics have the finite model property. For an algebraic account of the method, as well as for its generalization to all weakly transitive subframe and cofinal subframe logics, consult [8]. Similar results for superintuitionistic logics are obtained in [28], and an algebraic account is given in [7].

We recall that a *p-morphism* between two frames $\mathfrak{F} = (W, \leq)$ and $\mathfrak{F}' = (W', \leq')$ is a map $f : W \to W'$ such that $w \leq v$ implies $f(w) \leq' f(v)$, and $f(w) \leq' u$ implies that there is $v \in W$ with $w \leq v$ and $f(v) = u$. A *subreduction* from a frame $\mathfrak{F}$ to a frame $\mathfrak{F}'$ is a p-morphism from a subframe $\mathfrak{G}$ of $\mathfrak{F}$ onto the frame $\mathfrak{F}'$. We denote the domain of a subreduction f by $\text{dom}(f)$, and call the subreduction f *cofinal* if for each $w \in W$ there is $v \in \text{dom}(f)$ with $w \leq v$.

It is well known (see, e.g., [14, p. 292, Thm. 9.7]) that if f is a subreduction from $\mathfrak{F}$ to $\mathfrak{F}'$, then $f^* : \text{Up}(\mathfrak{F}') \to \text{Up}(\mathfrak{F})$, given by $f^*(U) = W \setminus f^{-1}(W' \setminus U)$, is a $(\wedge, \to)$-homomorphism (meaning that $f^*(U \cap V) = f^*(U) \cap f^*(V)$ and $f^*(U \to V) = f^*(U) \to f^*(V)$). Moreover, f is cofinal iff f^* is a $(\wedge, \to, 0)$-homomorphism (meaning that in addition $f^*(\varnothing) = \varnothing$); see, e.g., [2, Lem. 3.22]. Note that $w \in f^*(U)$ iff $f[\uparrow w] \subseteq U$, where as usual, $\uparrow w = \{v : w \leq v\}$.

A *subreduction* from a model $\mathfrak{M} = (\mathfrak{F}, \nu)$ to a model $\mathfrak{M}' = (\mathfrak{F}', \nu')$ is a subreduction f from the frame $\mathfrak{F}$ to the frame $\mathfrak{F}'$ satisfying $f^*(\nu'(\varphi)) = \nu(\varphi)$.

DEFINITION 3.1 (Selective filtration model theoretically). *Let Σ be a finite set of formulas closed under subformulas and let $\mathfrak{M} = (\mathfrak{F}, \nu)$ be a model. We call a finite model $\mathfrak{M}' = (\mathfrak{F}', \nu')$ a selective filtration of $\mathfrak{M}$ through Σ if there is a subreduction f of $\mathfrak{M}$ to $\mathfrak{M}'$ such that*

$$(\forall w \in W)(\forall \varphi \in \Sigma)(\exists v \in \text{dom}(f) : w \leq v \ \& \ w \in \nu(\varphi) \Leftrightarrow v \in \nu(\varphi)). \tag{10}$$

REMARK 3.2. It follows from Condition (10) that if $\mathfrak{M}'$ is a selective filtration of $\mathfrak{M}$ through Σ, then the subreduction f is cofinal.

LEMMA 3.3 (Filtration Lemma). *Let $\mathfrak{M} = (\mathfrak{F}, \nu)$ be a model and let $\mathfrak{M}' = (\mathfrak{F}', \nu')$ be a selective filtration of $\mathfrak{M}$ through a finite set Σ of formulas closed under subformulas. Then for each $\varphi \in \Sigma$ and $w \in \text{dom}(f)$, we have*

$$w \in \nu(\varphi) \text{ iff } f(w) \in \nu'(\varphi). \tag{11}$$

Proof. We have $w \in \nu(\varphi)$ iff $w \in f^*(\nu'(\varphi))$, which happens iff $f[\uparrow w] \subseteq \nu'(\varphi)$. Since $w \in \text{dom}(f)$, we have $f[\uparrow w] = \uparrow f(w)$, so the last condition is equivalent to $\uparrow f(w) \subseteq \nu'(\varphi)$. Because $\nu'(\varphi)$ is an up-set, this is equivalent to $f(w) \in \nu'(\varphi)$. ⊣

As an immediate consequence of Lemma 3.3, we obtain:

LEMMA 3.4. *Let $\mathfrak{M} = (\mathfrak{F}, \nu)$ be a model and let $\mathfrak{M}' = (\mathfrak{F}', \nu')$ be a selective filtration of $\mathfrak{M}$ through a finite set Σ of formulas closed under subformulas. If $\mathfrak{M}$ refutes $\varphi \in \Sigma$, then so does $\mathfrak{M}'$.*

Proof. If $\mathfrak{M}$ refutes $\varphi \in \Sigma$, then there is $w \in W$ such that $w \notin \nu(\varphi)$. Let f be the subreduction of $\mathfrak{M}$ to $\mathfrak{M}'$. By (10), there is $v \in \text{dom}(f)$ such that $w \leq v$ and $v \notin \nu(\varphi)$. Therefore, by (11), $f(v) \notin \nu'(\varphi)$. Thus, $\mathfrak{M}'$ refutes φ. ⊣

We next give an algebraic description of selective filtrations. Let A be a Heyting algebra, ν be a valuation on A, and Σ be a finite set of formulas closed under subformulas. Then $\nu[\Sigma]$ is a finite subset of A. Let S be the bounded implicative subsemilattice of A generated by $\nu[\Sigma]$ (so S is closed under $\wedge, \to, 0$, but not necessarily under $\vee$). By Diego's Theorem [16], S is finite. Therefore, S is a Heyting algebra, where

$$a \vee_S b = \bigwedge \{s \in S : a, b \leq s\} \tag{12}$$

for each $a, b \in S$. It follows from the definition that $a \vee b \leq a \vee_S b$, and that

$$a \vee b = a \vee_S b \text{ provided } a \vee b \in S. \tag{13}$$

DEFINITION 3.5 (Selective filtration algebraically). *Let A be a Heyting algebra, ν be a valuation on A, and Σ be a finite set of formulas closed under subformulas. Suppose L is a finite bounded implicative subsemilattice of A containing $\nu[\Sigma]$. Let ν_L be a valuation on L such that $\nu_L(p) = \nu(p)$ for each $p \in \Sigma$. Then we call (L, ν_L) a selective filtration of (A, ν) through Σ.*

LEMMA 3.6 (Filtration Lemma). *If (L, ν_L) is a selective filtration of (A, ν) through Σ, then $\nu_L(\varphi) = \nu(\varphi)$ for each $\varphi \in \Sigma$.*

Proof. Induction on the complexity of $\varphi \in \Sigma$. The case $\varphi = p$ follows from the definition of ν_L. The cases $\varphi = \bot$, $\varphi = \psi \wedge \chi$, and $\varphi = \psi \to \chi$ follow from the fact that L is a bounded implicative subsemilattice of A. Finally, let $\varphi = \psi \vee \chi$. Then $\nu(\psi) \vee \nu(\chi) = \nu(\psi \vee \chi) \in \nu(\Sigma) \subseteq L$. Therefore, by (13), $\nu(\psi) \vee \nu(\chi) = \nu(\psi) \vee_L \nu(\chi)$. Thus, $\nu_L(\psi \vee \chi) = \nu_L(\psi) \vee_L \nu_L(\chi) = \nu(\psi) \vee \nu(\chi) = \nu(\psi \vee \chi)$. ⊣

REMARK 3.7. Like with standard filtrations (see Remark 2.12), in order to prove the filtration lemma for selective filtrations, it is not necessary to require that L is an implicative subsemilattice of A. All that is needed is that $\nu_L(\varphi) \odot_L \nu_L(\psi) = \nu(\varphi) \odot \nu(\psi)$ for $\varphi \odot \psi \in \Sigma$, where $\odot \in \{\vee, \wedge, \to\}$.

REMARK 3.8. While the construction of the selective filtration (S, ν_S) of (A, ν) through Σ is relatively straightforward (see also [7, Sec. 7]), it is rather involved in modal logic, both model-theoretically (see [18] and [14, Thm. 9.34]) and algebraically (see [8]).

THEOREM 3.1. *Let Σ be a finite set of formula closed under subformulas. If $\mathfrak{M}' = (\mathfrak{F}', \nu')$ is a selective filtration of $\mathfrak{M} = (\mathfrak{F}, \nu)$ through Σ, then $(\mathrm{Up}(\mathfrak{F}'), \nu')$ is a selective filtration of $(\mathrm{Up}(\mathfrak{F}), \nu)$ through Σ. Conversely, suppose (L, ν_L) is a selective filtration of (A, ν) through Σ. Let $\mathfrak{F}_A$ be the spectrum of A and $\mathfrak{F}_L$ be the spectrum of L. Define μ on $\mathfrak{F}_A$ by $\mu = \alpha \circ \nu$ and μ' on $\mathfrak{F}_L$ by $\mu' = \alpha \circ \nu_L$. Then $(\mathfrak{F}_L, \mu')$ is a selective filtration of $(\mathfrak{F}_A, \mu)$ through Σ.*

Proof. First suppose that $\mathfrak{M}'$ is a selective filtration of $\mathfrak{M}$ through Σ. Let f be the subreduction. Then $f^*[\mathrm{Up}(\mathfrak{F}')]$ is a finite bounded implicative subsemilattice of $\mathrm{Up}(\mathfrak{F})$. Since $\nu(\varphi) = f^*(\nu'(\varphi))$ for each $\varphi \in \Sigma$, we see that $\nu[\Sigma]$ is contained in $f^*[\mathrm{Up}(\mathfrak{F}')]$. Therefore, $(\mathrm{Up}(\mathfrak{F}'), \nu')$ is a selective filtration of $(\mathrm{Up}(\mathfrak{F}), \nu)$ through Σ.

Next suppose that (L, ν_L) is a selective filtration of (A, ν) through Σ. Then L is a finite bounded implicative subsemilattice of A. Therefore, there is a cofinal subreduction f from $\mathfrak{F}_A$ to $\mathfrak{F}_L$ such that $f^*(\mu'(\varphi)) = \mu(\varphi)$ for each $\varphi \in \Sigma$ (see [2, Thm. 3.14, Lem. 3.29, and Lem. 3.31]). To see that $(\mathfrak{F}_L, \mu')$ satisfies Condition (10), let $w \in W_A$ and $\varphi \in \Sigma$. Since f is cofinal, there is $v \in \mathrm{dom}(f)$ with $w \leq v$. Clearly $w \in \mu(\varphi)$ implies $v \in \mu(\varphi)$. On the other hand, if $w \notin \mu(\varphi)$, then $w \notin f^*(\mu'(\varphi))$. Thus, $f[\uparrow w] \not\subseteq \mu'(\varphi)$. So there is $u \in \mathrm{dom}(f)$ with $w \leq u$ and $f(u) \notin \mu'(\varphi)$. Since $u \in \mathrm{dom}(f)$, by Lemma 3.3, $u \notin \mu(\varphi)$. Therefore, replacing v with u, we see that Condition (10) is satisfied. Thus, $(\mathfrak{F}_L, \mu')$ is a selective filtration of $(\mathfrak{F}_A, \mu)$ through Σ. ⊣

REMARK 3.9. The 1-1 correspondence of Theorem 2.1 between filtrations (L, ν_L) of (A, ν) and filtrations $\mathfrak{M}'_A = (\mathfrak{F}'_A, \mu')$ of $\mathfrak{M}_A = (\mathfrak{F}_A, \mu)$ through Σ does not have an immediate analogue for selective filtrations. The reason is that different subreductions of $\mathfrak{M}_A$ may give rise to the same bounded implicative subsemilattice of A (see [2, Sec. 4]). To remedy this, we need to strengthen the notion of a subreduction to that of an onto partial Esakia morphism of [2]. In order to keep the definition of selective filtrations relatively simple, we decided to work with a more familiar concept of subreductions. This discrepancy (roughly speaking, there are more model-theoretic selective filtrations than algebraic ones) goes away if in Definition 3.1, the subreduction f is strengthened to be an onto partial Esakia morphism.

REMARK 3.10. Clearly among selective filtrations of (A, ν) through Σ there is a least one, namely (S, ν_S). However, there is no greatest selective filtration of (A, ν). This means that among selective filtrations of $\mathfrak{M}$ through Σ there is a greatest one, but there is no least selective filtration of $\mathfrak{M}$. This is in contrast with the theory of standard filtrations, where we always have the least and greatest filtrations. However, we would get the same situation for standard filtrations if we defined them as suggested in Remark 2.6.

REMARK 3.11. Let L be a superintuitionistic logic. Following [14, p. 142, Sec. 5.3], we say that L *admits filtration* if for each non-theorem φ of L and some countermodel $\mathfrak{M} = (\mathfrak{F}, \nu)$ of φ, there is a filtration $\mathfrak{M}' = (\mathfrak{F}', \nu')$ of $\mathfrak{M}$ through some finite set Σ closed under subformulas and containing φ such that $\mathfrak{F}' \models L$. Clearly every superintuitionistic logic admitting filtration has the finite model property. This notion depends on at least three different parameters: formulas, models, frames. In [4] *stable superinutionistic logics* are introduced as the logics that are sound and complete with respect to a class of frames closed under order-preserving images. Every stable logic admits filtration, and hence has the finite model property. Thus, stable logics in some sense formalize the notion of admitting filtration by avoiding mentioning models and formulas. However, since a notion of filtration is not unique (there is a whole spectrum of them between the least and greatest ones), not every logic that admits filtration is stable. In [4] it is shown how to axiomatize these logics by special types of formulas, called *stable formulas*. We refer to [4] and [6] for more details.

Similarly, we can define when a superintuitionistic logic admits selective filtration. Fine [18] formalizes this notion by defining subframe transitive modal logics, and Zakharyaschev [28, 29] defines subframe and cofinal subframe modal and superintuitionistic logics. These are logics that are sound and complete with respect to a class of frames closed under subframes or cofinal subframes. An algebraic characterization of these logics can be found in [2, 3, 7, 8]. These logics are axiomatizable by the so-called subframe and cofinal subframe formulas [14, 18, 29]. They are also axiomatizable by the NNIL-formulas of Visser et al. [27], while stable logics are axiomatizable by ONNILLI-formulas (only NNIL to the left of implications) [10]. Thus, we see yet another influence of Albert Visser's ideas!

BIBLIOGRAPHY

[1] R. Balbes and P. Dwinger. *Distributive lattices*. University of Missouri Press, Columbia, Mo., 1974.

[2] G. Bezhanishvili and N. Bezhanishvili. An algebraic approach to canonical formulas: Intuitionistic case. *Rev. Symb. Log.*, 2(3):517–549, 2009.
[3] G. Bezhanishvili and N. Bezhanishvili. Canonical formulas for wK4. *Rev. Symb. Log.*, 5(4):731–762, 2012.
[4] G. Bezhanishvili and N. Bezhanishvili. Locally finite reducts of Heyting algebras and canonical formulas. *Notre Dame J. Form. Log.*, 2015. To appear. Available at: *http://dspace.library.uu.nl/handle/1874/273468*.
[5] G. Bezhanishvili, N. Bezhanishvili, and R. Iemhoff. Stable canonical rules. *J. Symb. Logic*, 2015. To appear. Available as ILLC Prepublications Series Report PP-2014-08.
[6] G. Bezhanishvili, N. Bezhanishvili, and J. Ilin. Cofinal stable logics. Submitted. Available as ILLC Prepublications Series Report PP-2015-08, 2015.
[7] G. Bezhanishvili and S. Ghilardi. An algebraic approach to subframe logics. Intuitionistic case. *Ann. Pure Appl. Logic*, 147(1-2):84–100, 2007.
[8] G. Bezhanishvili, S. Ghilardi, and M. Jibladze. An algebraic approach to subframe logics. Modal case. *Notre Dame J. Form. Log.*, 52(2):187–202, 2011.
[9] N. Bezhanishvili. *Lattices of Intermediate and Cylindric Modal Logics*. PhD thesis, University of Amsterdam, 2006.
[10] N. Bezhanishvili and D. de Jongh. Stable formulas in intuitionistic logic. Submitted. Available as ILLC Prepublication Series Report PP-2014-19, 2014.
[11] N. Bezhanishvili and B. ten Cate. Transfer results for hybrid logic. I. The case without satisfaction operators. *J. Logic Comput.*, 16(2):177–197, 2006.
[12] P. Blackburn, M. de Rijke, and Y. Venema. *Modal logic*. Cambridge University Press, 2001.
[13] G. Boolos. *The logic of provability*. Cambridge University Press, Cambridge, 1993.
[14] A. Chagrov and M. Zakharyaschev. *Modal logic*. The Clarendon Press, New York, 1997.
[15] W. Conradie, W. Morton, and C. van Alten. An algebraic look at filtrations in modal logic. *Log. J. IGPL*, 21(5):788–811, 2013.
[16] A. Diego. *Sur les Algèbres de Hilbert*. Translated from the Spanish by Luisa Iturrioz. Collection de Logique Mathématique, Sér. A, Fasc. XXI. Gauthier-Villars, Paris, 1966.
[17] L. Esakia. *Heyting algebras. I. Duality theory*. "Metsniereba", Tbilisi, 1985. (In Russian).
[18] K. Fine. Logics containing K4. II. *J. Symb. Logic*, 50(3):619–651, 1985.
[19] D. M. Gabbay. Selective filtration in modal logic. I. Semantic tableaux method. *Theoria*, 36:323–330, 1970.
[20] S. Ghilardi. Continuity, freeness, and filtrations. *J. Appl. Non-Classical Logics*, 20(3):193–217, 2010.
[21] E. J. Lemmon. Algebraic semantics for modal logics. I. *J. Symb. Logic*, 31:46–65, 1966.
[22] E. J. Lemmon. Algebraic semantics for modal logics. II. *J. Symb. Logic*, 31:191–218, 1966.
[23] E. J. Lemmon. *An introduction to modal logic*. Basil Blackwell, Oxford, 1977. The "Lemmon notes", In collaboration with Dana Scott, Edited by Krister Segerberg, American Philosophical Quarterly, Monograph Series, No. 11.
[24] J. C. C. McKinsey. A solution of the decision problem for the Lewis systems S2 and S4, with an application to topology. *J. Symb. Logic*, 6:117–134, 1941.
[25] J. C. C. McKinsey and A. Tarski. The algebra of topology. *Ann. of Math.*, 45:141–191, 1944.
[26] K. Segerberg. *An essay in classical modal logic. Vols. 1, 2, 3*. Filosofiska Föreningen och Filosofiska Institutionen vid Uppsala Universitet, Uppsala, 1971.
[27] A. Visser, D. de Jongh, J. van Benthem, and G. Renardel de Lavalette. NNIL, a study in intuitionistic propositional logic. In A. Ponse, M. de Rijke, and Y. Venema, editors, *Modal logic and process algebra*, pages 289–326, 1995.
[28] M. Zakharyaschev. Syntax and semantics of superintuitionistic logics. *Algebra and Logic*, 28(4):262–282, 1989.
[29] M. Zakharyaschev. Canonical formulas for K4. II. Cofinal subframe logics. *J. Symb. Logic*, 61(2):421–449, 1996.

Uniform Interpolation in Provability Logics

MARTA BÍLKOVÁ

To Albert Visser

Abstract

We prove the uniform interpolation theorem in modal provability logics **GL** and **Grz** by a proof-theoretical method, using analytical and terminating sequent calculi for the logics. The calculus for Gödel-Löb's logic **GL** is a variant of the standard sequent calculus of [20], in the case of Grzegorczyk's logic **Grz**, the calculus implements an explicit loop-preventing mechanism inspired by work of Heuerding [12, 13].

1 Introduction

Uniform Interpolation

Uniform Interpolation Property for a logic L is a strong interpolation property, stating that, for any formula α and any propositional variable p, there is a post-interpolant $\exists p(\alpha)$ not containing p, such that $\vdash_L \alpha \to \exists p(\alpha)$, and $\vdash_L \alpha \to \beta$ implies $\vdash_L \exists p(\alpha) \to \beta$ for each β not containing p. Similarly, for any β and p there is a pre-interpolant $\forall p(\beta)$ not containing p, such that $\vdash_L \forall p(\beta) \to \beta$ and $\vdash_L \alpha \to \beta$ implies $\vdash_L \alpha \to \forall p(\beta)$ for each α not containing p. Uniform interpolation property entails Craig interpolation property, and uniform interpolants are unique up to the provable equivalence, they are the the minimal and the maximal interpolants of a given implication w.r.t. the provability ordering.

While for classical propositional logic, and also for other locally tabular logics like modal logic **S5**, uniform interpolation property is easily obtained, in other logics it is not the case. Interest in the topic arose with a seminal work by Pitts [16], who proved uniform interpolation for intuitionistic propositional logic using a terminating sequent calculus. For modal logic **K** uniform interpolation was first proved by Visser [27] and Ghilardi [6], for provability logic **GL** by Shavrukov [22]. The failure of uniform interpolation in modal logic **S4**, which applies to **K4** as well, was proved by Ghilardi and Zawadowski [7]. More recent is a proof for monotone modal logic by Venema and Santocanale [21], using a coalgebraic perspective: uniform interpolants are constructed via erasing variable in a disjunctive normal form. This relates to the way the problem of computing uniform interpolants is understood in Artificial intelligence, which is variable forgetting. Similar motivation, but different approach based on resolution calculi and conjunctive normal forms, is applied to modal logic **K** by Herzig et al. [11].

As the notation suggests, uniform interpolants relate to a certain type of propositional quantifiers: if propositional quantifiers satisfying at least the usual quantifier axioms and rules are expressible in the language, they are the uniform interpolants. On the other hand, if we construct uniform interpolants so that the construction commutes with substitutions, we can use them to interpret the propositional quantifiers, precisely as was done by Pitts' in [16]. Visser [26] proved uniform interpolation for various modal logics, provability logics **GL** and **Grz** among them, via a semantical argument which yields a semantic characterization of the resulting quantifiers as *bisimulation* quantifiers: from the semantic point of view, quantifying over p, they quantify over possible valuations of p in models bisimilar to the current one up to p. A complexity bound of uniform interpolants in terms of $\Box$-depth is obtained in the proof, however, the proof does not provide us with a direct construction of the interpolants.

Using a method similar to Pitts' and using sequent calculi for modal logics, the author proved effective uniform interpolation for modal logics **K** and **T** in [2, 3]. The thesis [2] also contains proofs of uniform interpolation for provability logics **GL** and **Grz** which are reconsidered in this paper. The reason we came back to the topic is a recent interest in uniform interpolation in modal and modal intuitionistic logic by Iemhoff [14].

Provability modal logics

In this paper, we concentrate solely on the Gödel-Löb's provability logic **GL**, and Grzegorczyk!s logic **Grz**, also known as **S4Grz**. The main reference for provability modal logics and their properties is Boolos' book [4], for a history of provability logic see also [19].

The logic **GL** is a normal modal logic, extending the basic modal logic **K** with the Löb's axiom

$$L : \Box(\Box p \to p) \to \Box p.$$

It is known to be complete with respect to transitive and conversely well-founded Kripke frames. The logic **Grz** is a normal modal logic, extending the basic modal logic **K** with the axiom $T : \Box p \to p$, and the Grzegorczyk's axiom

$$Grz : \Box(\Box(p \to \Box p) \to p) \to \Box p.$$

It is known to be complete with respect to transitive, reflexive and conversely well-founded Kripke frames. Both logics have the finite model property as well and are therefore decidable, as was shown e.g. in [1].

The $\Box$ modality of Gödel-Löb's logic **GL** can be interpreted as formalized provability in an arithmetical recursively axiomatizable theory T: assume an axiomatization of T is expressed by a sentence τ and consider a standard proof predicate $Pr_\tau(\overline{\varphi})$ for T. An arithmetical interpretation of modal formulas is a function from propositional variables to arithmetical sentences such that it commutes with logical connectives, and $e(\bot) = (0 = S(0))$, and $e(\Box\alpha) = Pr_\tau(\overline{e(\alpha)})$. Arithmetical completeness is established as the following statement:

$$\vdash_{GL} \alpha \quad \text{iff} \quad \forall e(T \vdash e(\alpha)).$$

Gödel-Löb's logic **GL** was proved to be arithmetically complete for Peano arithmetic by Solovay [24]. Later it was shown that it is the logic of provability of a large family of reasonable formal theories.

Using the above interpretation of **GL**, we obtain the following arithmetical interpretation of Grzegorczyk's logic: an arithmetical interpretation of modal formulas is as before, only now $e(\Box\alpha) = Pr_\tau(\overline{e(\alpha)}) \wedge \overline{e(\alpha)}$.

2 Calculi

To prove the uniform interpolation theorem we use sequent calculi with good structural properties. The particular form of sequent calculi has been chosen for proof-search related manipulations. In particular, we use finite multisets of formulas to formulate a sequent, a notation which does not hide contractions (contraction rules are not part of the definition and are to be proved admissible rules), we use a definition without the cut rule (which is to be proved admissible), and structural rules of contraction and weakening are built in logical rules and axioms. Since the proof of the uniform interpolation theorem contained in the next section is closely related to termination of a proof-search in the calculi, we will devote some space in this section to explain proof-search in provability logics and its termination. Namely, we employ simple implicit loop-preventing mechanisms provided naturally by diagonal formulas, and in the case of Grzegorczyk's logics also an explicit syntactic loop-preventing mechanism to avoid reflexive loops due to the presence of the T axiom.

We assume the reader is familiar with basics on sequent calculi as contained e.g. in the Schwichtenberg's and Troelstra's book [25]. For sequent calculi of modal logics having arithmetical interpretation we refer to Sambin and Valentini's paper [20], or Avron's paper [1].

Preliminaries

Formulas are given by the following grammar of the basic modal language, where atoms p are taken from a fixed countable set of propositional variables:

$$\alpha := \bot \mid p \mid \neg\alpha \mid \alpha \wedge \alpha \mid \alpha \vee \alpha \mid \Box\alpha,$$

the notion of subformulas is standard, and with the term atomic formula we refer to atoms as well as the constant $\bot$. We moreover define $\top = \neg\bot$ and $\alpha \to \beta = \neg\alpha \vee \beta$, $\Diamond\alpha = \neg\Box\neg\alpha$, and we use $\bigvee \emptyset = \bot$ and $\bigwedge \emptyset = \top$. The *weight* $w(\alpha)$ of a formula α is the number of symbols it contains, and the *box-depth* $d(\alpha)$ of a formula α is defined as the maximum number of boxes along a branch in the corresponding formula tree. By capital Greeks we denote finite multisets of formulas. Formally, Γ is a function from the set of formulas to natural numbers with finite support (finitely many non-zero values), but we mostly use a relaxed notation and treat multisets as sets with multiple occurrences, in particular by $\alpha \in \Gamma$ we mean that $\Gamma(\alpha) > 0$. For a multiset Γ, we denote the underlying set by Γ°. $\Box\Gamma$ denotes the multiset resulting from prefixing elements of Γ with box while keeping the multiplicities intact. A *sequent* is a syntactic object of the form $\Gamma \Rightarrow \Delta$, or, in the case of Grzegorczyk's logic, of the form $\Box\Sigma|\Gamma \Rightarrow \Delta$. The weight $w(\Gamma)$ of a multiset Γ is the sum of the weights of elements in Γ, the weight $w(\Gamma, \Delta)$ of a sequent $\Gamma \Rightarrow \Delta$ is the sum $w(\Gamma) + w(\Delta)$.

A rule consists of a finite set of sequents called premises and a single sequent called the conclusion, rules with zero premises are called axioms. A calculus is given by a set of rule-schemes, a proof in the calculus is then a finite rooted tree labeled with sequents in such a way that leaves are labeled with axioms and labels of parent-children nodes respect correct instances of the rules of the calculus. The height of a proof is the height of the tree. A sequent is provable if there is a proof whose root is labeled with the sequent.

We call a rule *invertible* if whenever the conclusion is provable, then all its premises are provable as well, we call a rule height-preserving invertible if moreover the premises have proofs of at most the height of the proof of the conclusion. We call a rule *admissible* if whenever its premises are provable, so is the conclusion, and height-preserving admissible if moreover the conclusion has a proof of at most the maximum of the heights of the proofs of the premises.

By a *proof-search* in a calculus we mean a procedure based on applying rules of the calculus backwards to a sequent in such a way that, for a provable sequent, the resulting tree contains a proof of the sequent. We call a proof-search *terminating* if it results in a finite tree. Particular instances of proof search will be defined later.

Sequent calculus for GL

The following calculus is a variant of the sequent calculus introduced in [20], and reconsidered in [9, 10] using multisets in place of sets.

DEFINITION 2.1. *Sequent calculus G_{GL}*:

$$\Gamma, p \Rightarrow p, \Delta \qquad \Gamma, \bot \Rightarrow \Delta$$

$$\frac{\Gamma, \alpha, \beta \Rightarrow \Delta}{\Gamma, \alpha \wedge \beta \Rightarrow \Delta} \wedge\text{-}l \qquad \frac{\Gamma \Rightarrow \alpha, \beta, \Delta}{\Gamma \Rightarrow \alpha \vee \beta, \Delta} \vee\text{-}r$$

$$\frac{\Gamma, \alpha \Rightarrow \Delta}{\Gamma \Rightarrow \neg \alpha, \Delta} \neg\text{-}r \qquad \frac{\Gamma \Rightarrow \alpha, \Delta}{\Gamma, \neg \alpha \Rightarrow \Delta} \neg\text{-}l$$

$$\frac{\Gamma \Rightarrow \alpha, \Delta \quad \Gamma \Rightarrow \beta, \Delta}{\Gamma \Rightarrow \alpha \wedge \beta, \Delta} \wedge\text{-}r \qquad \frac{\Gamma, \alpha \Rightarrow \Delta \quad \Gamma, \beta \Rightarrow \Delta}{\Gamma, \alpha \vee \beta \Rightarrow \Delta} \vee\text{-}l$$

$$\frac{\Box\Gamma, \Gamma, \Box\alpha \Rightarrow \alpha}{\Box\Gamma, \Pi \Rightarrow \Box\alpha, \Delta} \Box_{GL}$$

In the $\Box_{GL}$ rule, Π contains only propositional variables and Δ contains only propositional variables and boxed formulas, and we call formulas $\Box\alpha$ as well as $\Box\Gamma$ principal formulas (formula occurrences). In the case of axioms, p (resp. $\bot$) are principal formulas, and in the remaining rules, the principal formula is the one to which a connective is introduced.

The propositional (non-modal) part of the calculus is a slight variant of the propositional part of the calculus **G3c** from [25]. The propositional rules of the calculus are height-preserving invertible, for a proof of this fact we refer to [25]. It is also not hard to prove that sequents of the form $\Gamma, \alpha \Rightarrow \alpha, \Delta$ are provable for arbitrary α.

LEMMA 2.2. *Weakening and contraction rules are height-preserving admissible in G_{GL}.*

Proof. Weakening and contraction rules are:

$$\frac{\Gamma \Rightarrow \Delta}{\alpha, \Gamma \Rightarrow \Delta} \text{w-l} \qquad \frac{\Gamma \Rightarrow \Delta}{\Gamma \Rightarrow \Delta, \alpha} \text{w-r} \qquad \frac{\alpha, \alpha, \Gamma \Rightarrow \Delta}{\alpha, \Gamma \Rightarrow \Delta} \text{c-l} \qquad \frac{\Gamma \Rightarrow \Delta, \alpha, \alpha}{\Gamma \Rightarrow \Delta, \alpha} \text{c-r}$$

Proof is by induction on the weight of the principal formula α of the weakening (resp. contraction) inference, and for each weight on the height of the proof of the premise of the rule. We prove the admissibility of both the left and right weakening rules simultaneously, and the same applies to the two contraction rules.

Weakening: For an atomic weakening formula the proof is obvious - note that atomic weakening is built in axioms as well as in the $\Box_{GL}$-rule. (The case of weakening-r by $\Box\alpha$ when the last inference is a $\Box_{GL}$-inference is then also obvious since it is built-in the rule as well.) For a non atomic and not boxed formula we consider its main connective and use height-preserving invertibility of the corresponding propositional rule, then weakening by subformula(s) of lower weight admissible by the induction hypothesis, and finally apply the propositional rule.

Let us therefore only spell out the step for a formula of the form $\alpha = \Box\beta$. If it is not principal in the last step of the proof of the premise of the weakening, we simply permute the weakening upwards and use the induction hypothesis. So consider $\Box\beta$ being the principle formula of a $\Box_{GL}$ inference. Notice that weakening-right by boxed formulas is built in the $\Box_{GL}$ rule, therefore it is enough to consider weakening-left rule.

- Weakening-l by $\Box\beta$, the last step is a $\Box_{GL}$ inference — we permute the proof as follows:

$$\dfrac{\dfrac{\Box\Gamma, \Gamma, \Box\gamma \Rightarrow \gamma}{\Box\Gamma, \Pi \Rightarrow \Box\gamma, \Lambda}\Box_{GL}}{\Box\beta, \Box\Gamma, \Pi \Rightarrow \Box\gamma, \Lambda}\text{w-l} \quad \Longrightarrow \quad \dfrac{\dfrac{\Box\Gamma, \Gamma, \Box\gamma \Rightarrow \gamma}{\Box\beta, \beta, \Box\Gamma, \Gamma, \Box\gamma \Rightarrow \gamma}\text{w-l, i.h.}}{\Box\beta, \Box\Gamma, \Pi \Rightarrow \Box\gamma, \Lambda}\Box_{GL}$$

Contraction: For α atomic, if the premise is an axiom, the conclusion is an axiom as well. If not, α is not principal and we use the induction hypothesis and permute contraction one step above or, in the case of $\Box_{GL}$ rule, we apply the rule so that the conclusion is weakened by only one occurrence of α. For α not atomic and not boxed we consider its main connective and use the height preserving invertibility of the appropriate rule and by i.h. we apply contraction on formula(s) of lower weight and then the rule again.

Let us therefore only spell out the step for a formula of the form $\alpha = \Box\beta$. If neither of its two occurrences is principal in the last step of the proof of the premise of the contraction, we simply permute the weakening upwards and use the induction hypothesis. So consider one occurrence of $\Box\beta$ is the principle formula of a $\Box_{GL}$ inference.

- Contraction-right on $\Box\beta$, with one of the occurrences of $\Box\beta$ principle of a $\Box_{GL}$ inference: we use the $\Box_{GL}$ rule so that we do not weaken by the other occurrence of $\Box\beta$ in the conclusion.

- Contraction-left on $\Box\beta$, with the occurrences of $\Box\beta$ principle of a $\Box_{GL}$ inference: we permute the proof as follows and use a contraction on a simpler formula, h.p. admissible by the induction hypothesis:

$$\dfrac{\dfrac{\beta, \beta, \Gamma, \Box\gamma \Rightarrow \gamma}{\Box\beta, \Box\beta, \Box\Gamma, \Pi \Rightarrow \Box\gamma, \Sigma}\Box_{GL}}{\Box\beta, \Box\Gamma, \Pi \Rightarrow \Box\gamma, \Sigma}\text{c-l} \quad \Longrightarrow \quad \dfrac{\dfrac{\beta, \beta, \Gamma \Rightarrow \gamma}{\beta, \Gamma, \Box\gamma \Rightarrow \gamma}\text{c-l, i.h.}}{\Box\beta, \Box\Gamma, \Pi \Rightarrow \Box\gamma, \Sigma}\Box_{GL}$$

All the above permutations are easily seen, using the induction hypothesis, to be height-preserving.

⊣

Terminating Proof-search in G_{GL}

The proof-search strategy we adopt is based on applying the rules of G_{GL} backwards to a given sequent, so that we always first apply the invertible rules and then, when it is no longer possible and if we haven't reached an axiom or a sequent with no boxed formulas on the right, we perform a modal jump — we apply the $\Box_{GL}$ rule backwards. We prefer to pack all the invertible steps into a single step, therefore it is useful to define the following notions of a critical sequent and a closure of a sequent first:

DEFINITION 2.3. *A sequent is called* critical, *if no invertible rule can be applied to it backwards. For a sequent $(\Gamma \Rightarrow \Delta)$, consider the smallest set of sequents containing $(\Gamma \Rightarrow \Delta)$ and closed under backward applications of the invertible rules of G_{GL}. The* closure *of a sequent $(\Gamma \Rightarrow \Delta)$, denoted $Cl(\Gamma; \Delta)$, is then the subset of all* critical *sequents contained in the set.*

Note that the closure of any sequent is finite, and that a critical sequent is of the form $(\Pi, \Box\Gamma \Rightarrow \Box\Delta, \Lambda)$, with Π, Λ multisets of atomic formulas, and its closure is the singleton of the sequent itself.

For a sequent $S = (\Gamma \Rightarrow \Delta)$ and finite multisets Θ, Ω, let $S(\Theta; \Omega)$ denote the sequent $\Gamma, \Theta \Rightarrow \Delta, \Omega$. The closure satisfies the following lemma, proof of which is immediate from the definition of the closure:

LEMMA 2.4. *Let S be a sequent, $Cl(S) = \{S_1, \ldots, S_n\}$, and Θ, Ω arbitrary finite multisets of formulas. Then:*

- $S_1, \ldots, S_n \vdash_{G_{GL}} S$
  if $\vdash_{G_{GL}} S$ then $\vdash_{G_{GL}} S_i$ for each i.

- $S_1(\Theta; \Omega), \ldots, S_n(\Theta; \Omega) \vdash_{G_{GL}} S(\Theta; \Omega)$
  if $\vdash_{G_{GL}} S(\Theta; \Omega)$ then $\vdash_{G_{GL}} S_i(\Theta; \Omega)$ for each i.

The proof-search procedure can now be described by creating a proof-search tree as follows: we start with creating a root and labeling it with the given sequent. For every node we have created, we proceed as follows: if it is labeled with a non-critical sequent, we compute its closure, and create a child-node for each sequent in the closure and label it with the sequent (thus creating a finite conjunctive branching). If a node is labeled with a critical sequent of the form $(\Pi, \Box\Gamma \Rightarrow \Box\Delta, \Lambda)$, we distinguish the following cases: if $\Pi \cap \Lambda \neq \emptyset$ or $\bot \in \Pi$ or $\Gamma \cap \Delta \neq \emptyset$ we mark the node a provable leaf, if it is not the case and $\Delta = \emptyset$ we mark the node an unprovable leaf, and in the remaining case we apply the $\Box_{GL}$ rule backwards: we create $|\Delta|$ children nodes and label them, for each $\Box\alpha \in \Delta$, with the premise of the $\Box_{GL}$ inference with $\Box\alpha$ principal (thus creating a finite disjunctive branching).

Checking whether $\Gamma \cap \Delta \neq \emptyset$ before applying the modal rule backwards works as a simple loop preventing mechanism — we do not apply the $\Box_{GL}$ rule backwards with $\Box\alpha$ principle if the diagonal formula $\Box\alpha$ is already in the antecedent (in which case the sequent in question is clearly provable). This is crucial since it enables us to bound the number of $\Box_{GL}$ inferences along each branch, and consequently also the weight of sequents occurring in a proof search for a fixed sequent.

LEMMA 2.5. *Proof search in the calculus G_{GL} always terminates.*

Proof. Consider a proof search for a sequent $(\Phi \Rightarrow \Psi)$. Let n be the number of boxed subformulas contained in multisets Φ, Ψ. This is, by the subformula property, by the nature of the $\Box_{GI}$ rule, and by the loop-preventing mechanism described above, an upper bound of the number of $\Box_{GL}$ inferences along a branch of the proof search tree. The reason is that, along a given branch, we never apply the $\Box_{GL}$ rule with the same principal formula twice. In general, the weight of sequents increases whenever the $\Box_{GL}$ rule is applied backwards. But the bound on the number of such steps along a single branch enables us to give an upper bound on the weight of sequents occurring in the fixed proof search: namely, $c = 2^n w(\Phi, \Psi)$ is an upper bound of the weight of a sequent occurring in the proof search for a sequent $(\Phi \Rightarrow \Psi)$.

For any multiset Γ occurring in the proof-search tree, let now $b(\Gamma)$ denote the number of boxed formulas in Γ counted as a set. For a sequent $(\Gamma \Rightarrow \Delta)$ occurring during the proof-search, consider an ordered pair $\langle c - b(\Gamma), w(\Gamma, \Delta) \rangle$. This measure strictly decreases in every backward application of a rule in terms of the lexicographical ordering: c is certainly greater or equal to the maximal number of boxed formulas in the antecedent which can occur during the proof search, so the first number does not decrease below zero. When an invertible rule is applied backwards, the weight of a sequent strictly decreases, therefore for a non-critical sequent, all sequents from its closure are of strictly smaller weight. When the $\Box_{GL}$ rule is applied backwards, b increases, and so $c - b$ decreases, therefore the measure decreases. [1]   ⊣

Extracting a Proof

From a proof-search tree for a provable sequent we are expected to be able to extract an actual proof of the sequent. The tree is finite, and all the leaves are marked either provable, or not provable. We can extend the marking in an obvious way to all the nodes: if a node is a parent node of a conjunctive branching, we mark it provable if and only if all its children are marked provable, if a node is a parent node of a disjunctive branching, we mark it provable if and only if at least one of its children is marked provable. If the root is marked provable, the sequent we started with has a proof presented by a tree of nodes marked provable, generated by the root. It is a routine induction to see it is indeed a proof.

Sequent calculi for Grz

We present a calculus for Grzegorczyk's logic with a loop-preventing mechanism built into the syntax of sequents. Namely we include a third multiset of boxed formulas in a sequent, thus sequents are now of the form $\Box\Sigma|\Gamma \Rightarrow \Delta$. The third multiset is used to store the boxed formulas of the $\Box_T^+$ inferences and the diagonal formulas of the $\Box_{Grz2}^+$ inferences when the rules are applied backwards to prevent unnecessary looping. This strategy was inspired by work of Heuerding [12, 13].

To improve readability, we denote the diagonal formula $\Box(\alpha \to \Box\alpha)$ of the Grzegorczyk's axiom by $D(\alpha)$ in the following text.

DEFINITION 2.6. *Sequent calculus G_{Grz}^+:*

$$\Box\Sigma|\Gamma, p \Rightarrow p, \Delta \qquad \Box\Sigma|\Gamma, \bot \Rightarrow \Delta$$

---

[1] Another way (closer to the approach of [13] or [12]) how to formulate a measure is the following: for a sequent $(\Gamma; \Delta)$ consider the function $f(\Gamma; \Delta) = c^2 - cb(\Gamma) + w(\Gamma, \Delta)$. The function (values of which are non-negative integers) decreases in every backward application of a rule in a proof search for $(\Phi \Rightarrow \Psi)$. (c^2 is included to ensure that f doesn't decrease below zero, and $cb(\Gamma)$ balances the possible increase of $w(\Gamma \Rightarrow \Delta)$ in the case of a backward application of the $\Box_{GL}$-rule.)

$$\frac{\Box\Sigma|\Gamma,\alpha,\beta \Rightarrow \Delta}{\Box\Sigma|\Gamma,\alpha \wedge \beta \Rightarrow \Delta} \wedge\text{-}l \qquad \frac{\Box\Sigma|\Gamma \Rightarrow \alpha,\beta,\Delta}{\Box\Sigma|\Gamma \Rightarrow \alpha \vee \beta,\Delta} \vee\text{-}r$$

$$\frac{\Box\Sigma|\Gamma,\alpha \Rightarrow \Delta}{\Box\Sigma|\Gamma \Rightarrow \neg\alpha,\Delta} \neg\text{-}r \qquad \frac{\Box\Sigma|\Gamma \Rightarrow \alpha,\Delta}{\Box\Sigma|\Gamma,\neg\alpha \Rightarrow \Delta} \neg\text{-}l$$

$$\frac{\Box\Sigma|\Gamma \Rightarrow \alpha,\Delta \quad \Box\Sigma|\Gamma \Rightarrow \beta,\Delta}{\Box\Sigma|\Gamma \Rightarrow \alpha \wedge \beta,\Delta} \wedge\text{-}r \qquad \frac{\Box\Sigma|\Gamma,\alpha \Rightarrow \Delta \quad \Box\Sigma|\Gamma,\beta \Rightarrow \Delta}{\Box\Sigma|\Gamma,\alpha \vee \beta \Rightarrow \Delta} \vee\text{-}l$$

$$\frac{\Box\alpha,\Box\Sigma|\Gamma,\alpha \Rightarrow \Delta}{\Box\Sigma|\Gamma,\Box\alpha \Rightarrow \Delta} \Box_T^+$$

$$\frac{\Box\Gamma|\emptyset \Rightarrow \alpha}{\Box\Gamma|\Pi \Rightarrow \Box\alpha,\Delta} \Box_{Grz1}^+, D(\alpha) \in \Box\Gamma \qquad \frac{\Box\Gamma,D(\alpha)|\Gamma \Rightarrow \alpha}{\Box\Gamma|\Pi \Rightarrow \Box\alpha,\Delta} \Box_{Grz2}^+, D(\alpha) \notin \Box\Gamma$$

In the $\Box_{Grz}^+$ rules, Π contains only propositional variables and Δ contains only propositional variables and boxed formulas. The notion of a principal formula is similar to the previous case. All the propositional rules are easily seen to be height-preserving invertible, there is one additional invertible rule here:

LEMMA 2.7. *$\Box_T^+$ is height-preserving invertible.*

Proof. Proof is a routine induction on the height of the proof of the premise and we leave it to the reader. ⊣

LEMMA 2.8. *Weakening rules are admissible in G_{Grz}^+.*

Proof. The weakening rules we consider in this paper are:

$$\frac{\Box\Sigma|\Gamma \Rightarrow \Delta}{\Box\Sigma|\Gamma,\alpha \Rightarrow \Delta} \text{w-l} \qquad \frac{\Box\Sigma|\Gamma \Rightarrow \Delta}{\Box\Sigma|\Gamma \Rightarrow \Delta,\alpha} \text{w-r} \qquad \frac{\Box\Sigma|\Gamma \Rightarrow \Delta}{\Box\Sigma,\Box\alpha|\Gamma \Rightarrow \Delta} \text{w-l+}$$

The proof is by induction on the weight of the principal weakening formula α and, for each weight, on the height of the proof of the premise. We prove admissibility of the three weakening rules simultaneously.

For an atomic weakening formula the proof is obvious - note that atomic weakening is built in axioms as well as in the $\Box_{Grz}^+$-rules. (The case of weakening-r by $\Box\alpha$ when the last inference is one of the $\Box_{Grz}^+$-rules is then also obvious since it is built-in the rules as well.) For non atomic and not boxed formula we consider its main connective and use height-preserving invertibility of the corresponding propositional rule, weaken by formula(s) of lower weight (admissible by the induction hypothesis), and then apply the rule. We next consider the weakening formula being of the form $\Box\beta$.

Weakening-right: consider the last step of the proof of the premise of the weakening inference. If it is a $\Box_{Grz}^+$ rule, we can use the rule so that the weakening by $\Box\beta$ is built-in its conclusion. If it is an invertible rule, we permute the weakening upwards.

Weakening-left: consider the last step of the proof of the premise of the weakening inference. If it is an invertible rule, we permute the weakening upwards. Let us consider last step is a $\Box^+_{Grz}$ rule, then we permute as follows:

$$\dfrac{\dfrac{D(\gamma), \Box\Sigma|\emptyset \Rightarrow \gamma}{D(\gamma), \Box\Sigma|\Pi \Rightarrow \Box\gamma, \Delta}\Box^+_{Grz1}}{D(\gamma), \Box\Sigma|\Box\beta, \Pi \Rightarrow \Box\gamma, \Delta}\text{w-l} \quad \Longrightarrow \quad \dfrac{\dfrac{\dfrac{\dfrac{D(\gamma), \Box\Sigma|\emptyset \Rightarrow \gamma}{D(\gamma), \Box\Sigma, \Box\beta|\emptyset \Rightarrow \gamma}\text{w-l}^+}{D(\gamma), \Box\Sigma, \Box\beta|\Pi \Rightarrow \Box\gamma, \Delta}\Box^+_{Grz1}}{D(\gamma), \Box\Sigma, \Box\beta|\beta, \Pi \Rightarrow \Box\gamma, \Delta}\text{w-l}}{D(\gamma), \Box\Sigma|\Box\beta, \Pi \Rightarrow \Box\gamma, \Delta}\Box^+_T$$

$$\dfrac{\dfrac{D(\beta), \Box\Sigma|\Sigma \Rightarrow \beta}{\Box\Sigma|\Pi \Rightarrow \Box\beta, \Delta}\Box^+_{Grz2}}{\Box\Sigma|\Box\alpha, \Pi \Rightarrow \Box\beta, \Delta}\text{w-l} \quad \Longrightarrow \quad \dfrac{\dfrac{\dfrac{\dfrac{D(\beta), \Box\Sigma|\Sigma \Rightarrow \beta}{D(\beta), \Box\Sigma, \Box\alpha|\alpha, \Sigma \Rightarrow \beta}\text{w-l}^+,1}{\Box\Sigma, \Box\alpha|\Pi \Rightarrow \Box\beta, \Delta}\Box^+_{Grz2}}{\Box\Sigma, \Box\alpha|\alpha, \Pi \Rightarrow \Box\beta, \Delta}\text{w-l}}{\Box\Sigma|\Box\alpha, \Pi \Rightarrow \Box\beta, \Delta}\Box^+_T$$

Remark: The two transformations above are clearly not height-preserving, therefore weakenings are in general not height-preserving admissible. However, one can show, that weakening rules with α principal are admissible and the height only increases by the box depth $d(\alpha)$.

Weakening-l+: Notice that w-l+ is built in the axioms. If the last inference of the proof of the premise of the weakening is a propositional inference or a $\Box^+_T$ inference, we just use the i.h., a weakening one step above, and use the appropriate rule again.

Let the last inference of the proof of the premise of the weakening be a $\Box^+_{Grz1}$ inference, w-l+ permutes over the inference as follows:

$$\dfrac{\dfrac{D(\gamma), \Box\Sigma|\emptyset \Rightarrow \gamma}{D(\gamma), \Box\Sigma|\Pi \Rightarrow \Box\gamma, \Delta}\Box^+_{Grz1}}{D(\gamma), \Box\Sigma, \Box\beta|\Pi \Rightarrow \Box\gamma, \Delta}\text{w-l}^+ \quad \Longrightarrow \quad \dfrac{\dfrac{D(\gamma), \Box\Sigma|\emptyset \Rightarrow \gamma}{D(\gamma), \Box\Sigma, \Box\beta|\emptyset \Rightarrow \gamma}\text{w-l}^+}{D(\gamma), \Box\Sigma, \Box\beta|\Pi \Rightarrow \Box\gamma, \Delta}\Box^+_{Grz1}$$

Let the last inference be a $\Box^+_{Grz2}$ inference, w-l+ permutes over the inference as follows:

$$\dfrac{\dfrac{D(\gamma), \Box\Sigma|\Sigma \Rightarrow \gamma}{\Box\Sigma|\Pi \Rightarrow \Box\gamma, \Delta}\Box^+_{Grz2}}{\Box\Sigma, \Box\beta|\Pi \Rightarrow \Box\gamma, \Delta}\text{w-l}^+ \quad \Longrightarrow \quad \dfrac{\dfrac{D(\gamma), \Box\Sigma|\Sigma \Rightarrow \gamma}{D(\gamma), \Box\Sigma, \Box\beta|\beta, \Sigma \Rightarrow \gamma}\text{w-l},1^+}{\Box\Sigma, \Box\beta|\Pi \Rightarrow \Box\gamma, \Delta}\Box^+_{Grz2}$$

The two permutations above are in fact height-preserving. ⊣

LEMMA 2.9. *Contraction rules are height-preserving admissible in* G^+_{Grz}.

Proof. The contraction rules are:

$$\dfrac{\Sigma|\Gamma, \alpha, \alpha \Rightarrow \Delta}{\Sigma|\Gamma, \alpha \Rightarrow \Delta}\text{c-l} \qquad \dfrac{\Sigma|\Gamma \Rightarrow \Delta, \alpha, \alpha}{\Sigma|\Gamma \Rightarrow \Delta, \alpha}\text{c-r} \qquad \dfrac{\Sigma, \Box\alpha, \Box\alpha|\Gamma \Rightarrow \Delta}{\Sigma, \Box\alpha|\Gamma \Rightarrow \Delta}\text{c-l+}$$

The proof is by induction on the weight of the contraction formula and, for each weight, on the height of the proof of the premise. The induction runs simultaneously for all the contraction rules. We use the height preserving invertibility of rules. Note that in the contr-l+ rule the contraction formula is always of the form $\Box\alpha$.

For α atomic, if the premise is an axiom, the conclusion is an axiom as well. If not, α is not principal and we use the induction hypothesis and apply contraction one step above or, in the case of $\Box^+_{Grz}$ rules, we apply the rule so that the conclusion is weakened by only one occurrence of α. For α not atomic and not boxed we consider its main connective and use the height preserving invertibility of the corresponding propositional rule and by the induction hypothesis we apply contraction on subformula(s) of lower weight and then the rule again. The third multiset does not make any difference here and it works precisely as in the classical logic. All the steps described so far are height preserving.

Now suppose the contraction formula to be of the form $\Box\beta$. We distinguish the following cases:

(i) Both occurrences of the contraction formula are principal of a $\Box^+_{Grz1}$ inference in the antecedent, in this case the only possibility is c-l+. Then we permute the proof as follows using the the induction hypothesis:

$$\dfrac{\dfrac{\dfrac{\Box\Sigma, \Box\beta, \Box\beta|\emptyset \Rightarrow \gamma}{\Box\Sigma, \Box\beta, \Box\beta|\Pi \Rightarrow \Box\gamma, \Delta}\Box^+_{Grz1}}{\Box\Sigma, \Box\beta|\Pi \Rightarrow \Box\gamma, \Delta}\text{c-l+}} \quad \Longrightarrow \quad \dfrac{\dfrac{\dfrac{\Box\Sigma, \Box\beta, \Box\beta|\emptyset \Rightarrow \gamma}{\Box\Sigma, \Box\beta|\emptyset \Rightarrow \gamma}\text{c-l+}}{\Box\Sigma, \Box\beta|\Pi \Rightarrow \Box\gamma, \Delta}\Box_{Grz1}}$$

In this case, $D(\gamma) \in \Box\Sigma$, or $D(\gamma) = \Box\beta$. The permutation is obviously height preserving.

(ii) Both occurrences of the contraction formula are principal of a $\Box^+_{Grz2}$ inference in the antecedent, again c-l+ is the only possibility. Then we permute the proof as follows:

$$\dfrac{\dfrac{\dfrac{D(\gamma), \Box\Sigma, \Box\beta, \Box\beta|\Sigma, \beta, \beta \Rightarrow \gamma}{\Box\Sigma, \Box\beta, \Box\beta|\Pi \Rightarrow \Box\gamma, \Delta}\Box^+_{Grz2}}{\Box\Sigma, \Box\beta|\Pi \Rightarrow \Box\gamma, \Sigma}\text{c-l+}} \quad \Longrightarrow \quad \dfrac{\dfrac{\dfrac{D(\gamma), \Box\Sigma, \Box\beta, \Box\beta|\Sigma, \beta, \beta \Rightarrow \gamma}{D(\gamma), \Box\Sigma, \Box\beta|\Sigma, \beta \Rightarrow \gamma}\text{c-l,l+}}{\Box\Sigma, \Box\beta|\Pi \Rightarrow \Box\gamma, \Sigma}\Box_{Grz2}}$$

Here, $D(\gamma) \notin \Box\Sigma$ and $D(\gamma) \neq \Box\beta$. The permutation is obviously height preserving.

(iii) One occurrence of the contraction formula is the principal formula of a $\Box^+_T$ inference in the antecedent. Then we permute the proof as follows using the i.h. and the height preserving invertibility of the $\Box^+_T$ rule:

$$\dfrac{\dfrac{\dfrac{\Sigma, \Box\beta|\Box\beta, \beta, \Gamma \Rightarrow \Delta}{\Sigma|\Box\beta, \Box\beta, \Gamma \Rightarrow \Delta}\Box^+_T}{\Sigma|\Box\beta, \Gamma \Rightarrow \Delta}\text{c-l}} \quad \Longrightarrow \quad \dfrac{\dfrac{\dfrac{\dfrac{\Sigma, \Box\beta|\Box\beta, \beta, \Gamma \Rightarrow \Delta}{\Sigma, \Box\beta, \Box\beta|\beta, \beta, \Gamma \Rightarrow \Delta}\text{invert.}}{\Sigma, \Box\beta|\beta, \Gamma \Rightarrow \Delta}\text{c-l,l+}}{\Sigma|\Box\beta, \Gamma \Rightarrow \Delta}\Box^+_T}$$

The permutation is height preserving since the steps c-l, c-l+, and invert. do not increase the height of the proof.

(iv) One occurrence of the contraction formula is the principal formula in the succedent and the last inference is a $\Box^+_{Grz}$ inference. Then we use the $\Box^+_{Grz}$ rule so that the conclusion is not weakened by the other occurrence of $\Box\beta$. This step is obviously height preserving. If the contraction formula is not the principal formula and the last step is a $\Box^+_{Grz}$ inference, $\Box\beta$ is in Δ. Then we use the $\Box^+_{Grz}$ rule so that the conclusion is weakened by only one occurrence of the contraction formula. If the last step is another inference, we use contraction one step above on the proof of lower height. If it is an axiom, the conclusion of the desired contraction is an axiom as well. Again, all the steps are height preserving. ⊣

Next we want to relate the calculus G^+_{Grz} to the standard sequent calculus G_{Grz} of [1]. For this, we consider a multiset variant of the latter. The calculus is known to be complete, and the rules of weakening are easily proved to be admissible by a similar argument that used in Lemma 2.2.

DEFINITION 2.10. *Sequent calculus G_{Grz} results from the non-modal part of the calculus G_{GL} adding the following two modal rules:*

$$\frac{\Gamma, \Box\alpha, \alpha \Rightarrow \Delta}{\Gamma, \Box\alpha \Rightarrow \Delta} \Box_T \qquad \frac{\Box\Gamma, D(\alpha) \Rightarrow \alpha}{\Box\Gamma, \Pi \Rightarrow \Box\alpha, \Delta} \Box_{Grz}$$

LEMMA 2.11. *The calculi G_{Grz} and G^+_{Grz} are equivalent:*

$$\vdash_{G_{Grz}} \Gamma \Rightarrow \Delta \quad \text{iff} \quad \vdash_{G^+_{Grz}} \emptyset | \Gamma \Rightarrow \Delta.$$

Proof.
The right-left implication: deleting the "|" symbol from a G^+_{Grz} proof of $(\emptyset | \Gamma \Rightarrow \Delta)$ yields correct instances of rules of G_{Grz}, except the $\Box^+_{Grz1}$ rule. It has to be simulated as follows:

$$\frac{\dfrac{\Box\Gamma, D(\alpha) \Rightarrow \alpha}{\Box\Gamma, \Pi \Rightarrow \Box\alpha, \Delta} \Box_{Grz}}{\Box\Gamma, D(\alpha), \Pi \Rightarrow \Box\alpha, \Delta} \text{ admiss. w-l}$$

We end up with a G_{Grz} proof of $\Gamma \Rightarrow \Delta$. (Lemma 2.14 below states that the calculus G_{Grz} is complete, and soundness of weakening entails that weakening is indeed admissible in G_{Grz}.)

The left-right implication: the idea is to add a third, empty multiset to all the sequents in a proof. This yields correct instances of the axioms as well as the propositional rules. The $\Box_T$ rule has to be simulated as follows, using invertibility of $\Box^+_T$ rule and admissibility of contraction:

$$\dfrac{\dfrac{\dfrac{\emptyset | \Box\alpha, \alpha\Gamma \Rightarrow \Delta}{\Box\alpha | \alpha, \alpha, \Gamma \Rightarrow \Delta} \text{ inv. of } \Box^+_T}{\Box\alpha | \alpha, \Gamma \Rightarrow \Delta} \text{ c-l}}{\emptyset | \Box\alpha, \Gamma \Rightarrow \Delta} \Box^+_T$$

The $\Box_{Grz}$ rule is simulated as follows $(D(\alpha) \notin \Box\Gamma)$:

$$\frac{\displaystyle \frac{\displaystyle \frac{\displaystyle \frac{\displaystyle \frac{\emptyset|\Box\Gamma,\Box(\neg\alpha\vee\Box\alpha)\Rightarrow\alpha}{\Box\Gamma,\Box(\neg\alpha\vee\Box\alpha)|\Gamma,\neg\alpha\vee\Box\alpha\Rightarrow\alpha}\text{ inv. of }\Box_T^+}{\Box\Gamma,\Box(\neg\alpha\vee\Box\alpha)|\Gamma\Rightarrow\alpha,\alpha}\text{ inv. of }\vee\text{-l and }\neg\text{-l}}{\Box\Gamma,\Box(\neg\alpha\vee\Box\alpha)|\Gamma\Rightarrow\alpha}\text{ admiss. c-r}}{\displaystyle \frac{\Box\Gamma|\Pi\Rightarrow\Box\alpha,\Delta}{\Box\Gamma|\Gamma,\Pi\Rightarrow\Box\alpha,\Delta}\Box_{Grz2}^+}\text{ admiss. w-l inferences}}{\emptyset|\Box\Gamma,\Pi\Rightarrow\Box\alpha,\Delta}\Box_T^+\text{ inferences}$$

If $D(\alpha)\in\Gamma$, we use some admissible c-l+ inferences before the $\Box_{Grz1}^+$ inference is used. ⊣

Terminating proof-search in G_{Grz}^+

We will restrict ourselves to proof-search for sequents of the form $(\emptyset|\Phi\Rightarrow\Psi)$ with the third multiset empty and Φ and Ψ arbitrary finite multisets of formulas (Lemma 2.11 justifies this restriction).

The notion of critical sequent and the closure of a sequent for G_{Grz}^+ is the same as given in Definition 2.3, only with a third multiset added. Recall that now also the $\Box_T^+$ rule is invertible, so critical sequents are of the form: $\Box\Gamma|\Pi\Rightarrow\Box\Delta,\Lambda$ with Π,Λ atomic.

For a sequent $S=(\Box\Sigma|\Gamma\Rightarrow\Delta)$ and finite multisets Λ,Θ,Ω, let $S(\Lambda|\Theta;\Omega)$ denote the sequent $(\Box\Sigma,\Box\Lambda|\Gamma,\Theta\Rightarrow\Delta,\Omega)$. The closure satisfies the following lemma, essentially the same as Lemma 2.4:

LEMMA 2.12. *Let S be a sequent, $Cl(S)=\{S_1,\ldots,S_n\}$, and $\Lambda|\Theta,\Omega$ arbitrary finite multisets of formulas. Then:*

- $S_1,\ldots,S_n\vdash_{G_{GL}} S$
  if $\vdash_{G_{GL}} S$ then $\vdash_{G_{GL}} S_i$ for each i.

- $S_1(\Lambda|\Theta;\Omega),\ldots,S_n(\Lambda|\Theta;\Omega)\vdash_{G_{GL}} S(\Lambda|\Theta;\Omega)$
  if $\vdash_{G_{GL}} S(\Lambda|\Theta;\Omega)$ then $\vdash_{G_{GL}} S_i(\Lambda|\Theta;\Omega)$ for each i.

Before we continue to describe proof-search and prove its termination, we briefly discuss forms of looping we prevent by using the specific form of the calculus.

Reflexive looping

This simple looping occurs when one searches for proofs in the calculus G_{Grz} and applies the $\Box_T$ rule backwards repeatedly with the same principal formula. Such looping is prevented by the presence of the third storage multiset in sequents and by the particular form of $\Box_T^+$ rule we use — when this rule is applied backwards, it remembers that the principle formula has already been treated.

Transitive looping

Another looping phenomenon arises when one tries to search for a proof of the sequent $\Box\neg\Box p\Rightarrow\Box p$ in the calculus G_{Grz} — it loops on the sequent

$$\Box\neg\Box p, D(p)\Rightarrow p,\Box p.$$

Such looping can be avoided and the diagonal formula plays a crucial role here as a natural loop-preventing mechanism again. We have made this mechanism explicit by splitting the $\Box_{Grz}$ rule into two cases distinguishing if the diagonal formula is present in the antecedent or not. Consider the $\Box_{Grz1}^+$ rule bottom up. When the diagonal formula is already in the third multiset, we apply the rule so that we neither add the diagonal formula to the third multiset, nor we add Γ to the antecedent.

The proof-search procedure for Grzegorczyk logic is fully analogous to that for logic **GL**: we create a proof-search tree, using the strategy of alternating the closure step for non-critical sequents and a modal jump step for critical sequents. Also the labeling and extraction of an actual proof is carried out similarly.

LEMMA 2.13. *Proof search in Gm_{Grz}^+ for sequents of the form $(\emptyset|\Phi \Rightarrow \Psi)$ always terminates.*

Proof. Consider a proof search for a sequent $(\emptyset|\Phi \Rightarrow \Psi)$. Let n be the number of boxed subformulas occurring in the sequent $(\emptyset|\Phi \Rightarrow \Psi)$. This number, as in the case of **GL**, is an upper bound on the number of the $\Box_{Grz}^+$ rules applied backwards along one branch of the proof search tree. Each backward application of the $\Box_{Grz2}^+$ rule adds a new boxed formula in the storage multiset, but also a $\Box_T^+$ rule does so during the closure steps. Therefore n^2 is an upper bound of the number of formulas stored in Σ if we do not duplicate them and count them as a set. (If we allowed duplicate formulas in Σ, we would need an exponential function of n.)

With each sequent $(\Box\Sigma|\Gamma \Rightarrow \Delta)$ occurring during the proof search, we associate an ordered pair $\langle n^2 - |\Sigma^\circ|, w(\Gamma, \Delta)\rangle$. Therefore the first number does not decrease below zero. The measure obviously decreases in every backward application of a rule of the calculus. For the $\Box_{Grz2}^+$ rule, $|\Sigma^\circ|$ increases and so $n^2 - |\Sigma^\circ|$ decreases, while for other rules the weight $w(\Gamma, \Delta)$ decreases. ⊣

Cut admissibility via completeness

We do not give a constructive proof of completeness of the two calculi without the cut rule in this paper. Such a proof can be established using a proof-search method described in the previous subsections. One can argue that, for any given sequent, the proof-search tree either yields a proof, or can be used to construct a (finite) counterexample. Instead, we state the completeness without the cut rule, and refer for a proof to Avron [1] who proved that the calculi G_{GL} and G_{Grz} are complete without the cut rule w.r.t. their respective Kripke semantics. Then an easy semantical argument of soundness of the cut rule entails its admissibility. Lemma 2.11 then yields admissibility in G_{Grz}^+ of the cut rule we will use later in the proof of Theorem 3.3.

Proofs of completeness can be found in [1] for **GL** and Grzegorczyk's logic, and [28] or [20] for **GL**, where redundancy of the cut rule is established through a decision procedure which either creates a cut-free proof, or a Kripke counterexample to a given sequent. Although both authors use a formulation via sets of formulas, observe, that a cut-free proof with sets can be equivalently formulated using multisets and contraction rules, which are, as we have proved, admissible in our cut-free calculi. Equivalently, if a sequent does not have a cut-free proof in the system based on multisets, its set-based counterpart sequent does not have a cut-free proof in the system based on sets.

LEMMA 2.14. *(Avron [1]:) There are a canonical Kripke model $(W, <)$ and a canonical valuation V such that:*

- $<$ *is irreflexive and transitive*

- *for every $w \in W$, the set $\{v|v < w\}$ is finite*

- *if $(\Gamma \Rightarrow \Delta)$ has no cut-free proof in G_{GL}, then there is a $w \in W$ such that $w \Vdash_V \alpha$ for every $\alpha \in \Gamma$ and $w \nVdash_V \beta$ for every $\beta \in \Delta$.*

There are a canonical Kripke model $(W, \leq)$ and a canonical valuation V such that:

- $\leq$ *partially orders W*

- *for every $w \in W$, the set $\{v|v \leq w\}$ is finite*

- *if $(\Gamma \Rightarrow \Delta)$ has no cut-free proof in G_{Grz}, then there is a $w \in W$ such that $w \Vdash_V \alpha$ for every $\alpha \in \Gamma$ and $w \nVdash_V \beta$ for every $\beta \in \Delta$.*

Proof. See [1]. The canonical model is built from all saturated sequents (sequents closed under subformulas) that have no cut-free proof in appropriate calculi. The lemma entails completeness of G_{GL} w.r.t. transitive, conversely well-founded Kripke models; and completeness of G_{Grz} w.r.t. transitive, reflexive and conversely well-founded Kripke models. ⊣

COROLLARY 2.15. *The cut rule*

$$\frac{\Gamma \Rightarrow \Delta, \gamma \quad \gamma, \Pi \Rightarrow \Lambda}{\Gamma, \Pi \Rightarrow \Delta, \Lambda}$$

is admissible in G_{GL} and G_{Grz}.

Proof. It is easy to give a semantic argument of soundness of the cut rule. Given a counterexample of the conclusion $(\Gamma, \Pi \Rightarrow \Delta, \Lambda)$ of a cut inference, there is a counterexample to one of its premises: consider the counterexample (W,R) and a world $w \in W$ in it such that $w \Vdash_V A$ for every $\alpha \in \Gamma \cup \Pi$ and $w \nVdash_V \beta$ for some $\beta \in \Delta \cup \Lambda$. For any formula γ it is either the case that $w \Vdash_V \gamma$, and then w refutes $(\gamma, \Pi \Rightarrow \Lambda)$, or $w \nVdash_V \gamma$, and then w refutes $(\Gamma \Rightarrow \Delta, \gamma)$.

Now Lemma 2.14 (completeness of G_{GL} and G_{Grz}) entails admissibility of the cut rule in the calculi. ⊣

COROLLARY 2.16. *The cut rule*

$$\frac{\emptyset|\Gamma \Rightarrow \Delta, \gamma \quad \emptyset|\gamma, \Pi \Rightarrow \Lambda}{\emptyset|\Gamma, \Pi \Rightarrow \Delta, \Lambda}$$

is admissible in G_{Grz}^+.

Proof. Follows from Corollary 2.15 and Lemma 2.11. ⊣

REMARK 2.17. *The above cut rule cannot be replaced by the expected form of cut:*

$$\frac{\Box\Sigma|\Gamma \Rightarrow \Delta, \alpha \quad \Box\Theta|\alpha, \Pi \Rightarrow \Lambda}{\Box\Sigma, \Box\Theta|\Gamma, \Pi \Rightarrow \Delta, \Lambda} \; cut',$$

since it is not admissible. The counterexample is the following instance of cut':

$$\dfrac{\dfrac{\Box p, D(p)|p \Rightarrow p}{\Box p|\emptyset \Rightarrow \Box p}\,\Box^+_{Grz2} \quad \dfrac{\Box p|p \Rightarrow p}{\emptyset|\Box p \Rightarrow p}\,\Box^+_T}{\Box p|\emptyset \Rightarrow p}\,cut,$$

which results in the sequent $(\Box p|\emptyset \Rightarrow p)$ unprovable in G^+_{Grz}.

A fixed-point trick: Before we proceed to the proof of uniform interpolation, we state a lemma which later will play a crucial role in a termination argument for the definition of the interpolants. The lemma is a simple example of a particular fixed point existence in modal logic **K4**. Namely, a recursive equivalence $\Diamond((\alpha \vee x) \wedge \beta) \equiv x$ has a solution $x = \Diamond(\alpha \wedge \beta)$. We will however need a bit more complicated form of the statement, which is the following lemma:

LEMMA 2.18. *Let $\delta = \bigwedge_i \alpha_i \wedge \beta$, $i = 1 \ldots k$. Then the following sequents are provable:*

$$\vdash_{G_{GL}} \Diamond(\bigwedge_i(\alpha_i \vee \Diamond\delta) \wedge \beta) \Leftrightarrow \Diamond(\bigwedge_i \alpha_i \wedge \beta), \quad \vdash_{G^+_{Grz}} \emptyset|\Diamond(\underbrace{\bigwedge_i(\alpha_i \vee \Diamond\delta) \wedge \beta}_{\delta}) \Leftrightarrow \Diamond(\underbrace{\bigwedge_i \alpha_i \wedge \beta}_{\delta}).$$

Proof. The corresponding two implications (same in both cases) are easily seen to hold on transitive models. The claim now follows from the completeness results of Lemma 2.14. It is however also not hard to write down proofs in G_{GL} and G^+_{Grz}, which we omit for space reasons. ⊣

3 Uniform Interpolation

Uniform Interpolation in GL

We will prove the uniform interpolation by constructing, for each formula α, the pre-interpolant $\forall p(\alpha)$. The post-interpolant can be defined by $\exists(\alpha) = \neg\forall p(\neg\alpha)$. The construction is based on a proof-search for the sequent $\emptyset \Rightarrow \alpha$. To make it work we need to define interpolants for sequents instead of formulas. The uniform interpolation is then obtained via

$$\forall p(\alpha) = \forall p(\emptyset; \alpha).$$

THEOREM 3.1. *Let Γ, Δ be finite multisets of formulas. For every propositional variable p there exists a formula $\forall p(\Gamma; \Delta)$ such that:*

(i)
$$Var(\forall p(\Gamma; \Delta)) \subseteq Var(\Gamma, \Delta) \setminus \{p\}$$

(ii)
$$\vdash_{G_{GL}} \Gamma, \forall p(\Gamma; \Delta) \Rightarrow \Delta$$

(iii) *moreover let Φ, Ψ be multisets of formulas not containing p and*

$$\vdash_{G_{GL}} \Phi, \Gamma \Rightarrow \Delta, \Psi.$$

Then

$$\vdash_{G_{GL}} \Phi \Rightarrow \forall p(\Gamma; \Delta), \Psi.$$

Proof. In the following construction of the interpolant, it is instructive to imagine that with the formula $\forall p(\Gamma; \Delta)$ we are describing (a relevant part of) the proof-search tree for the sequent $(\Phi, \Gamma \Rightarrow \Delta, \Psi)$ for any context Φ, Ψ not containing p, namely the part only depending on Γ, Δ. This description has to be finite. The interpolant is defined recursively, closely following the proof-search strategy: for a non-critical sequent we simply use its closure, while for a critical sequent we apply a matching argument, similar to the strategy used by Pitts [16]. We start with a definition of the formula $\forall p(\Gamma; \Delta)$, then we prove that the definition terminates, and proceed with proving it satisfies items (i)-(iii) of the Theorem 3.1.

Definition of the interpolant. We describe the construction of the interpolant recursively. The formula $\forall p(\Gamma; \Delta)$ is for a noncritical $(\Gamma; \Delta)$ defined by

$$\forall p(\Gamma; \Delta) = \bigwedge_{(\Gamma_i \Rightarrow \Delta_i) \in Cl(\Gamma; \Delta)} \forall p(\Gamma_i; \Delta_i) \tag{1}$$

The recursive steps for $\Gamma \Rightarrow \Delta$ being a critical sequent of the form $(\Box\Gamma', \Pi; \Box\Delta', \Lambda)$, with Π, Λ atomic, are given below in Table 1. The first line of Table 1 corresponds to some of the cases when the critical sequent is provable - it is either an axiom or the diagonal formula is already in the antecedent (here we are using the loop preventing mechanism from the termination argument in Lemma 2.5). The line 2 of Table 1 corresponds to a critical step in a proof-search, the corresponding disjunction covering:

- propositional variables from multisets Π, Λ, where all $q, r \neq p$,
- all the possibilities of a $\Box_{GL}$ inference with the principal formula from $\Box\Delta'$,
- and, by the diamond formula $\Diamond \bigwedge N(\Box\Gamma', \Gamma'; \emptyset)$, which we will define below, also the possibility of a $\Box_{GL}$ inference with the principal formula not from $\Box\Delta'$ (i.e. from a context not containing p). Morally, we should include $\Diamond\forall p(\Box\Gamma', \Gamma'; \emptyset)$ instead, but then the definition would not terminate. This is the trick we describe below in Remark 3.1.

	$\Box\Gamma', \Pi; \Box\Delta', \Lambda$ matches	$\forall p(\Box\Gamma', \Pi; \Box\Delta', \Lambda)$ equals
1	if $p \in \Pi \cap \Lambda$ or $\bot \in \Pi$ or $\Gamma' \cap \Delta' \neq \emptyset$	$\top$
2	otherwise (here all $q, r \neq p$)	$\bigvee_{q \in \Lambda} q \vee \bigvee_{r \in \Pi} \neg r$ $\bigvee_{\beta \in \Delta'} \Box\forall p(\Box\Gamma', \Gamma', \Box\beta; \beta) \vee$ $\Diamond \bigwedge N(\Box\Gamma', \Gamma'; \emptyset)$

Table 1.

For a sequent of the form $(\Box\Gamma', \Gamma'; \emptyset)$, a set of formulas $N(\Box\Gamma', \Gamma'; \emptyset)$ is defined as the smallest set given by Table 2:

	$(\Box\Sigma, \Upsilon; \Box\Omega, \Theta) \in Cl(\Box\Gamma', \Gamma'; \emptyset)$ matches	$N(\Box\Gamma', \Gamma'; \emptyset)$ contains
1	$\Sigma^\circ \supset \Gamma'^\circ$ or $p \in \Upsilon \cap \Theta$ or $\Sigma \cap \Omega \neq \emptyset$ or $\bot \in \Upsilon$	$\forall p(\Box\Sigma, \Upsilon; \Box\Omega, \Theta)$
2	otherwise (here all $q, r \neq p$)	$\bigvee_{q \in \Theta} q \vee \bigvee_{r \in \Upsilon} \neg r$ $\vee \bigvee_{\beta \in \Omega} \Box\forall p(\Box\Sigma, \Sigma, \Box\beta; \beta)$

Table 2.

In the first line of Table 2, we use the fact that, given $\Sigma^\circ \supset \Gamma'^\circ$, the sequent $(\Box\Sigma, \Upsilon; \Box\Omega, \Theta)$ is strictly simpler then $(\Box\Gamma', \Gamma'; \emptyset)$ in terms of the measure we will use below to prove termination of the definition, and therefore it is safe to recursively call the procedure. In the remaining cases the sequent is provable and the value in the second column therefore equals $\top$. The second line of Table 2 covers the case when $\Sigma^\circ = \Gamma'^\circ$, and resembles the line 2 of Table 1, only to ensure termination, we have omitted the diamond-part of the disjunction.

Termination. Let us see that the definition terminates. The argument is similar to that we have used to prove termination of the calculus G_{GL} in 2.5. Consider a run of the procedure for $\forall p(\Phi; \Psi)$ and let n be the number of boxed subformulas occurring in $(\Phi; \Psi)$, which bounds the maximal number of critical steps occurring along a branch in the tree corresponding to the run of the procedure. This is crucial since it enables us to consider an upper bound of the weight of an argument of $\forall p$ occurring during this run of the procedure. Put $c = 4^n w(\Phi; \Psi)$, i.e. an upper bound of the weight of an argument of $\forall p$ occurring during the run of the procedure for $(\Phi; \Psi)$. Here, in contrast to the termination argument for the calculus G_{GL}, we need 4^n since the weight of a recursively called argument of $\forall p$ can increase more. This is because in the construction of $N(\Box\Gamma', \Gamma'; \emptyset)$ we look one level deeper. Let, for any multiset Γ, $b(\Gamma)$ be the number of boxed formulas in Γ counted as a set. For a $\forall p$ argument $(\Gamma; \Delta)$ occurring during the construction, consider an ordered pair $\langle c - b(\Gamma), w(\Gamma, \Delta) \rangle$. This measure decreases in every recursive step of the procedure in terms of the lexicographical ordering:

- It is obvious that, for each noncritical sequent $(\Gamma' \Rightarrow \Delta') \in Cl(\Gamma; \Delta)$, $w(\Gamma', \Delta') < w(\Gamma, \Delta)$ and that b does not decrease.

- for a critical argument $(\Box\Gamma', \Pi; \Box\Delta', \Lambda)$, i.e., line 2 of Table 1, and whole Table 2 constructing $N(\Box\Gamma', \Gamma'; \emptyset)$. For all the three recursively called arguments b increases, thus $c - b$, and therefore the whole measure, decreases.

REMARK 3.1. *[termination trick] To retain termination of the definition, we cannot replace*

$$\Diamond \bigwedge N(\Box\Gamma', \Gamma'; \emptyset)$$

in line 2 of Table 1 with $\Diamond \forall p(\Box\Gamma', \Gamma'; \emptyset)$, *which in fact seems to be needed to prove the part (iii) of the theorem. The reason is that its recursively called argument need not be simpler then the sequent in question. However, in the prove of (iii)* $\Diamond \forall p(\Box\Gamma', \Gamma'; \emptyset)$ *can and will be used, because, as we show next, it is the case that*

$$\vdash_{GGL} \Diamond \bigwedge N(\Box\Gamma', \Gamma'; \emptyset) \Leftrightarrow \Diamond \forall p(\Box\Gamma', \Gamma'; \emptyset). \tag{2}$$

To see this, we use the fixed point observation we made earlier in Lemma 2.18. Consider sequents $(\Box\Sigma, \Upsilon; \Box\Omega, \Theta)$ in the closure of $(\Box\Gamma', \Gamma'; \emptyset)$, and refer by S to sequents with $\Sigma^\circ = \Gamma'^\circ$, and by S' to sequents in the closure with $\Sigma^\circ \supset \Gamma'^\circ$, i.e. strictly simpler then $(\Box\Gamma', \Gamma'; \emptyset)$. Since

$$\forall p(\Box\Gamma', \Gamma'; \emptyset) = \bigwedge_S \forall p(S) \wedge \bigwedge_{S'} \forall p(S'),$$

and for each $S = (\Box\Sigma, \Upsilon; \Box\Omega, \Theta)$ with $\Sigma^\circ = \Gamma'^\circ$ we obtain by the line 2 of Table 1

$$\forall p(S) = \bigvee_{q\in\Theta} q \vee \bigvee_{r\in\Upsilon} \neg r \vee \bigvee_{\beta\in\Omega} \Box\forall p(\Box\Sigma, \Sigma, \Box\beta; \beta) \vee \Diamond \bigwedge N(\Box\Sigma, \Sigma; \emptyset),$$

and using the definition of $N(\Box\Gamma', \Gamma'; \emptyset)$ in the Table 2, the above equivalence (2) becomes the following:

$$\vdash_{GGL} \Diamond(\bigwedge_S (\bigvee_{q\in\Theta} q \vee \bigvee_{r\in\Upsilon} \neg r \vee \bigvee_{\beta\in\Omega} \Box\forall p(\Box\Sigma, \Sigma, \Box\beta; \beta))) \wedge \bigwedge_{S'} \forall p(S')$$

$$\Leftrightarrow \Diamond(\bigwedge_S (\bigvee_{q\in\Theta} q \vee \bigvee_{r\in\Upsilon} \neg r \vee \bigvee_{\beta\in\Omega} \Box\forall p(\Box\Sigma, \Sigma, \Box\beta; \beta) \vee \Diamond \bigwedge N(\Box\Sigma, \Sigma; \emptyset)) \wedge \bigwedge_{S'} \forall p(S')).$$

Observe, that $N(\Box\Sigma, \Sigma; \emptyset)$ is equivalent to $N(\Box\Gamma', \Gamma'; \emptyset)$ by $\Sigma^\circ = \Gamma'^\circ$. Therefore the result now follows by Lemma 2.18, instantiated with

$$\alpha_S = \bigvee_{q\in\Theta} q \vee \bigvee_{r\in\Upsilon} \neg r \vee \bigvee_{\beta\in\Omega} \Box\forall p(\Box\Sigma, \Sigma, \Box\beta; \beta)$$

$$\beta = \bigwedge_{S'} \forall p(S')$$

$$\delta = \bigwedge N(\Box\Gamma', \Gamma'; \emptyset)$$

We have thus established the termination of the definition of the uniform interpolants. Now we proceed in proving the three items of Theorem 3.1.

(i). The item (i) follows easily by induction on Γ, Δ, because we never use p during the definition of the formula $\forall p(\Gamma; \Delta)$.

(ii). We proceed by induction on the complexity of Γ, Δ given by the measure function defined above, and prove that $\vdash_{G_{GL}} \Gamma, \forall p(\Gamma; \Delta) \Rightarrow \Delta$.

First, let $(\Gamma \Rightarrow \Delta)$ be a noncritical sequent. Then sequents $(\Gamma_i \Rightarrow \Delta_i) \in Cl(\Gamma; \Delta)$ are of lower complexity and by the induction hypotheses $\vdash_{G_{GL}} \Gamma_i, \forall p(\Gamma_i; \Delta_i) \Rightarrow \Delta_i$ for each i. Then by admissibility of weakening and by Lemma 2.4

$$\vdash_{G_{GL}} \Gamma, \forall p(\Gamma_1; \Delta_1), \ldots, \forall p(\Gamma_k; \Delta_k) \Rightarrow \Delta,$$

therefore by a $\wedge$-l inference

$$\vdash_{G_{GL}} \Gamma, \bigwedge_{(\Gamma_i \Rightarrow \Delta_i) \in Cl(\Gamma; \Delta)} \forall p(\Gamma_i; \Delta_i) \Rightarrow \Delta,$$

which is by (1)

$$\vdash_{G_{GL}} \Gamma, \forall p(\Gamma; \Delta) \Rightarrow \Delta.$$

Second, let $(\Gamma \Rightarrow \Delta)$ be a critical sequent. If $(\Gamma \Rightarrow \Delta)$ is a critical sequent matching the line 1 of Table 1, then (ii) is an axiom or a provable sequent. Let $(\Gamma \Rightarrow \Delta)$ be a critical sequent matching the line 2 of Table 1. We prove

$$\vdash_{G_{GL}} \Pi, \delta, \Box\Gamma' \Rightarrow \Box\Delta', \Lambda$$

for each disjunct δ used in the line 2 of Table 1 to define the interpolant:

- For each $r \in \Pi$ obviously $\vdash_{G_{GL}} \Pi, \neg r, \Box\Gamma' \Rightarrow \Box\Delta', \Lambda$, therefore
  $\vdash_{G_{GL}} \Pi, \bigvee_{r \in \Pi} \neg r, \Box\Gamma' \Rightarrow \Box\Delta', \Lambda$.

- for each $q \in \Lambda$ obviously $\vdash_{G_{GL}} \Pi, q, \Box\Gamma' \Rightarrow \Box\Delta', \Lambda$, therefore
  $\vdash_{G_{GL}} \Pi, \bigvee_{q \in \Lambda} q, \Box\Gamma' \Rightarrow \Box\Delta', \Lambda$

- For each $\beta \in \Delta'$, $\vdash_{G_{GL}} \Box\Gamma', \Gamma', \Box\beta, \forall p(\Box\Gamma', \Gamma', \Box\beta; \beta) \Rightarrow \beta$ by the induction hypothesis, which gives $\vdash_{G_{GL}} \Box\Gamma', \Pi, \Box\forall p(\Box\Gamma', \Gamma', \Box\beta; \beta) \Rightarrow \Box\Delta', \Lambda$ by a $\Box_L$ inference.

It remains to be proved that

$$\vdash_{G_{GL}} \Box\Gamma', \Pi, \Diamond\bigwedge N(\Box\Gamma', \Gamma'; \emptyset) \Rightarrow \Box\Delta', \Lambda.$$

For each $(\Box\Sigma, \Upsilon \Rightarrow \Box\Omega, \Theta) \in Cl(\Box\Gamma', \Gamma'; \emptyset)$ from the first line of Table 2, we know by the induction hypotheses that

$$\vdash_{G_{GL}} \Box\Sigma, \Upsilon, \forall p(\Box\Sigma, \Upsilon; \Box\Omega, \Theta) \Rightarrow \Box\Omega, \Theta.$$

For each $(\Box\Sigma, \Upsilon \Rightarrow \Box\Omega, \Theta) \in Cl(\Box\Gamma', \Gamma'; \emptyset)$ from the second line of Table 2 we have the following:

- $\vdash_{G_{GL}} \Box\Sigma, \Upsilon, \bigvee_{q \in \Theta} q \Rightarrow \Box\Omega, \Theta$.

- $\vdash_{G_{GL}} \Box\Sigma, \Upsilon, \bigvee_{\neg r \in \Upsilon} \neg r \Rightarrow \Box\Omega, \Theta$.

- for each $\beta \in \Omega$ by the induction hypotheses

$$\vdash_{G_{GL}} \Box\Sigma, \Sigma, \Box\beta, \forall p(\Box\Sigma, \Sigma, \Box\beta; \beta) \Rightarrow \beta$$

and by weakening and a $\Box_{GL}$ inference

$$\vdash_{G_{GL}} \Box\Sigma, \Box\forall p(\Box\Sigma, \Sigma, \Box\beta; \beta) \Rightarrow \Box\beta.$$

Together this yields, using $\vee$-l inferences,

$$\vdash_{G_{GL}} \Box\Sigma, \Upsilon, \bigvee_{q \in \Theta} \vee \bigvee_{\neg r \in \Upsilon} \vee \bigvee_{\beta \in \Omega} \Box\forall p(\Box\Sigma, \Sigma, \Box\beta; \beta) \Rightarrow \Box\Omega, \Theta.$$

Therefore, for each $(\Box\Sigma, \Upsilon \Rightarrow \Box\Omega, \Theta) \in Cl(\Box\Gamma', \Gamma'; \emptyset)$, we obtain, using weakening and $\wedge$-r inferences,

$$\vdash_{G_{GL}} \Box\Sigma, \Upsilon, \bigwedge N(\Box\Gamma', \Gamma'; \emptyset) \Rightarrow \Box\Omega, \Theta.$$

Now, by Lemma 2.4,

$$\vdash_{G_{GL}} \Box\Gamma', \Gamma', \bigwedge N(\Box\Gamma', \Gamma'; \emptyset) \Rightarrow \emptyset.$$

By negation and weakening inferences

$$\vdash_{G_{GL}} \Box\Gamma', \Gamma', \Box\neg \bigwedge N(\Box\Gamma', \Gamma'; \emptyset) \Rightarrow \neg \bigwedge N(\Box\Gamma', \Gamma'; \emptyset)$$

and by a $\Box_{GL}$ inference

$$\vdash_{G_{GL}} \Box\Gamma', \Pi \Rightarrow \Box\neg \bigwedge N(\Box\Gamma', \Gamma'; \emptyset), \Box\Delta', \Lambda.$$

Now, using a negation inference again, we obtain

$$\vdash_{G_{GL}} \Box\Gamma', \Pi, \Diamond \bigwedge N(\Box\Gamma', \Gamma'; \emptyset) \Rightarrow \Box\Delta', \Lambda.$$

Putting finally all the above disjuncts together then yields, using $\vee$-l inferences,

$$\vdash_{G_{GL}} \Pi, \Box\Gamma', \bigvee_{q \in \Lambda} q \bigvee_{r \in \Pi} \neg r \bigvee_{\beta \in \Delta'} \Box\forall p(\Box\Gamma', \Gamma', \Box\beta; \beta) \vee \Diamond \bigwedge N(\Box\Gamma', \Gamma'; \emptyset) \Rightarrow \Box\Delta', \Lambda,$$

that is, by the line 2 of Table 1,

$$\vdash_{G_{GL}} \Pi, \Box\Gamma', \forall p(\Pi, \Box\Gamma'; \Box\Delta', \Lambda) \Rightarrow \Box\Delta', \Lambda.$$

(iii). We proceed by induction on the height of a proof of $(\Phi, \Gamma \Rightarrow \Delta, \Psi)$, and by sub-induction on the measure of the sequent $(\Gamma; \Delta)$ used to show termination of the definition. We show that $\vdash_{G_{GL}} \Phi \Rightarrow \forall p(\Gamma; \Delta), \Psi$.

First, consider $(\Phi, \Gamma \Rightarrow \Delta, \Psi)$ is an axiom. The following cases apply:

- $\bot$ is principal and $\bot \in \Phi$, then (iii) is an axiom.

- ⊥ is principal and ⊥ ∈ Γ, then $\forall p(\Gamma; \Delta) = \top$ and $\Phi \Rightarrow \top, \Psi$ is provable.

- p is principal, i.e. $p \in \Gamma \cap \Delta$ and $\forall p(\Gamma; \Delta) = \top$ and $\Phi \Rightarrow \top, \Psi$ is provable.

- $q \neq p$ is principal, and $q \in \Phi \cap \Psi$. Then $\Phi \Rightarrow \forall p(\Gamma; \Delta), \Psi$ is an axiom.

- $q \neq p$ is principal, and $q \in \Phi \cap \Delta$. Then $\vdash_{G_{GL}} q \Rightarrow \forall p(\Gamma; \Delta)$ by the line 1 of Table 1, and we obtain the result by weakening.

- $q \neq p$ is principal, and $q \in \Gamma \cap \Psi$. Then $\vdash_{G_{GL}} \neg q \Rightarrow \forall p(\Gamma; \Delta)$ by the line 1 of Table 1, and $\vdash_{G_{GL}} \emptyset \Rightarrow \forall p(\Gamma; \Delta), q$ by ¬-l invertibility, and we obtain the result by weakening.

- $q \neq p$ is principal, and $q \in \Gamma \cap \Delta$. Then $\vdash_{G_{GL}} q \vee \neg q \Rightarrow \forall p(\Gamma; \Delta)$ by the line 1 of the table, and therefore $\vdash_{G_{GL}} \emptyset \Rightarrow \forall p(\Gamma; \Delta)$, and we obtain the result by weakening.

Consider then $(\Phi, \Gamma \Rightarrow \Delta, \Psi)$ is not an axiom. We distinguish two main cases: Consider first $(\Gamma; \Delta)$ is a noncritical sequent. Then all $(\Gamma' \Rightarrow \Delta') \in Cl(\Gamma; \Delta)$ are strictly simpler in terms of the measure, and for all of them we have $\vdash_{G_{GL}} \Phi, \Gamma' \Rightarrow \Delta', \Psi$ by Lemma 2.4. Then, using the induction hypothesis and (1), the following are equivalent:

$$\vdash_{G_{GL}} \Phi \Rightarrow \forall p(\Gamma'; \Delta'), \Psi \quad \text{for all} \quad (\Gamma' \Rightarrow \Delta') \in Cl(\Gamma; \Delta)$$

$$\vdash_{G_{GL}} \Phi \Rightarrow \bigwedge_{(\Gamma' \Rightarrow \Delta') \in Cl(\Gamma;\Delta)} \forall p(\Gamma'; \Delta'), \Psi$$

$$\vdash_{G_{GL}} \Phi \Rightarrow \forall p(\Gamma; \Delta), \Psi.$$

Consider $(\Gamma; \Delta)$ is a critical sequent and the last inference is an instance of an invertible rule. Then the principal formula of the inference is in Φ, Ψ. We apply the induction hypothesis to the premise of the last inference, and then the invertible rule in question again.

Finally assume that $(\Gamma; \Delta)$ is a critical sequent and the last inference is a $\Box_L$ inference:

Consider first that the principal formula $\Box \alpha \in \Psi$, in particular, α doesn't contain p. Then the proof ends with the step:

$$\frac{\Box \Phi', \Box \Gamma', \Phi', \Gamma', \Box \alpha \Rightarrow \alpha}{\Box \Pi', \Box \Gamma', \Pi'', \Gamma'' \Rightarrow \Box \alpha, \Psi', \Delta} \Box_L$$

where $\Box \Phi', \Phi''$ is Φ; $\Box \Gamma', \Gamma''$ is Γ; and $\Box \alpha, \Psi'$ is Ψ. Consider $\Box \Gamma' \cap \Delta = \emptyset$ (otherwise the line 1 of Table 1 applies and $\forall p(\Gamma; \Delta) = \top$, and therefore (iii) holds). So we can use the line 2 of Table 1. Then the induction hypothesis gives

$$\vdash_{G_{GL}} \Phi', \Box \alpha \Rightarrow \forall p(\Box \Gamma', \Gamma'; \emptyset), \alpha$$

and by a ¬-l inference we obtain

$$\vdash_{G_{GL}} \Phi', \Box \alpha, \neg \forall p(\Box \Gamma', \Gamma'; \emptyset) \Rightarrow \alpha.$$

Now, by a $\Box_L$ and a negation inferences, we obtain

$$\vdash_{G_{GL}} \Box\Phi', \Phi'' \Rightarrow \Diamond\forall p(\Box\Gamma', \Gamma'; \emptyset), \Box\alpha, \Psi'.$$

By the line 2 of Table 1, invertibility of the $\vee$-l rule, and by (2) we have

$$\vdash_{G_{GL}} \Diamond\forall p(\Box\Gamma', \Gamma'; \emptyset) \Rightarrow \forall p(\Box\Gamma', \Gamma''; \Delta).$$

The two sequents above yield (iii) by cut admissibility.

Consider the principal formula $\Box\alpha \in \Delta$. Again, consider $\Box\Gamma' \cap \Delta = \emptyset$ so we can use the line 2 of Table 1. Then the proof ends with:

$$\frac{\Box\Phi', \Box\Gamma', \Phi', \Gamma', \Box\alpha \Rightarrow \alpha}{\Box\Phi', \Box\Gamma', \Phi'', \Gamma'' \Rightarrow \Box\alpha, \Delta', \Psi} \Box_L$$

where $\Box\Phi', \Phi''$ is Φ; $\Box\Gamma', \Gamma''$ is Γ; and $\Box\alpha, \Delta'$ is Δ. Now the induction hypothesis gives

$$\vdash_{G_{GL}} \Box\Phi', \Phi' \Rightarrow \forall p(\Box\Gamma', \Gamma', \Box\alpha; \alpha),$$

and by weakening and a $\Box_L$ inference we obtain

$$\vdash_{G_{GL}} \Box\Phi', \Phi'' \Rightarrow \Box\forall p(\Box\Gamma', \Gamma', \Box\alpha; \alpha), \Psi.$$

The line 2 of Table 1 and invertibility of the $\vee$-l rule yields

$$\vdash_{G_{GL}} \Box\forall p(\Box\Gamma', \Gamma', \Box\alpha; \alpha) \Rightarrow \forall p(\Box\Gamma', \Gamma''; \Box\alpha, \Delta').$$

Finally, we obtain (iii) by cut admissibility. ⊣

REMARK 3.2 (**Constructivity of the proof**). *We have given a construction of uniform interpolant which is effective and implementable. However, since we argued semantically to claim cut-free completeness of the calculus, reader might object that our proof of the uniform interpolation theorem is not fully constructive. To this point we say the following: one can look at the cut-elimination proof in [9, 10] and prove constructively that the two calculi are equivalent. Another way is to use the proof-search procedure described in subsection 2 and prove completeness via decidability. The point is that an unsuccessful proof-search tree can be used to construct a counterexample to a given sequent, in spirit of the proof contained in [28]. We have not included such an argument here mainly for space reasons and because it is not essential to understand the proof of uniform interpolation.*

Fixed points

Uniform interpolation theorem for **GL** entails Sambin's and de Jongh's fixed point theorem. Our proof then presents an alternative constructive proof of the fixed point theorem:

THEOREM 3.2. *Fixed point theorem: Suppose p is modalized in β (i.e., any occurrence of p is in the scope of a $\Box$). Then we can find a formula γ in the variables of β without p such that*

$$\vdash_{GL} \gamma \leftrightarrow \beta(\gamma).$$

Already Craig interpolation entails fixed point theorem: a fixed point of a formula β is an interpolant of a sequent expressing the uniqueness of the fixed point

$$\boxdot(p \leftrightarrow \beta(p)) \wedge \boxdot(q \leftrightarrow \beta(q)) \Rightarrow p \leftrightarrow q,$$

which is provable in **GL** - proofs of this fact in [20] and [4] are easily adaptable to our variant of the calculus. However, to construct the fixed point using this method requires to have an actual proof of the sequent expressing the uniqueness.

Direct proofs of fixed point theorem were given by Sambin [18], Sambin and Valentini in [20] (a construction of explicit fixed points which is effective and implementable), Smoryński [23] from Beth's definability property, Reidhar-Olson [17], Gleit and Goldfarb [8]. A proof from Beth's property can be found also in Kracht's book [15], for three different proofs see Boolos' book [4]. A different and effective constructive proof of fixed point theorem is the one by Sambin and Valentini in [20]. We present a proof of fixed point theorem based on uniform interpolation, it is an effective proof alternative to those above. We learnt this simple argument from Albert Visser, and we found it an interesting application of the uniform interpolation theorem.

Proof. (*of Theorem 3.2.*) Let us consider a formula $\beta(p,\bar{q})$ with p modalized in β. The fixed point of β then would be the simulation of

$$\exists \bar{p}(\Box(p \leftrightarrow \beta(p)) \wedge \beta(p))$$

or, equivalently, of

$$\forall r(\Box(r \leftrightarrow \beta(r)) \to \beta(r)).$$

Let us denote them γ_1 and γ_2 and observe they are both interpolants of the sequent

$$(\Box(p \leftrightarrow \beta(p)) \wedge \beta(p) \Rightarrow \Box(r \leftrightarrow \beta(r)) \to \beta(r))$$

and that neither of them contains p, r. We show that any of them is the fixed point of $\beta(p)$ and that they are indeed equivalent. To keep readability we just sketch the proofs in G_{GL} below.

First we show that $(\Box(p \leftrightarrow \beta(p)) \wedge \beta(p) \Rightarrow \Box(r \leftrightarrow \beta(r)) \to \beta(r))$ is provable from the uniqueness statement:

$$\cfrac{\cfrac{\boxdot(p \leftrightarrow \beta(p)) \wedge \boxdot(r \leftrightarrow \beta(r)) \Rightarrow p \leftrightarrow r \quad p \leftrightarrow r, p \leftrightarrow \beta(p), r \leftrightarrow \beta(r), \beta(p) \Rightarrow \beta(r)}{\cfrac{\Box(p \leftrightarrow \beta(p)), \Box(r \leftrightarrow \beta(r)), \beta(p) \Rightarrow \beta(r)}{\cfrac{(\Box(p \leftrightarrow \beta(p)), \beta(p) \Rightarrow \Box(r \leftrightarrow \beta(r)) \to \beta(r))}{(\Box(p \leftrightarrow \beta(p)) \wedge \beta(p) \Rightarrow \Box(r \leftrightarrow \beta(r)) \to \beta(r))}}}}{} \text{cut}$$

Now let us see that any of γ_i is a fixed point and thus, by the uniqueness, $\gamma_1 \leftrightarrow \gamma_2$. First observe, that whenever $(\Gamma(p) \Rightarrow \Delta(p))$ is provable, $(\Gamma[p/\alpha] \Rightarrow \Delta[p/\alpha])$ where we substitute α for p is provable as well (we substitute everywhere in the proof, to treat $\Box_{GL}$ inferences can require some admissible weakenings, and we add proofs of sequents $(\Gamma, \alpha \Rightarrow \Delta, \alpha)$ in place of axioms with p principal). The label "subst." in the following proof-tree refers to such a substitution, the label "inv." refers to invertibility of a rule:

$$\dfrac{\dfrac{\Box(p\leftrightarrow\beta(p))\wedge\beta(p)\Rightarrow\gamma_i}{\dfrac{\Box(\gamma_i\leftrightarrow\beta(\gamma_i))\wedge\beta(\gamma_i)\Rightarrow\gamma_i}{\dfrac{\Box(\gamma_i\leftrightarrow\beta(\gamma_i)),\beta(\gamma_i)\Rightarrow\gamma_i}{\dfrac{\Box(\gamma_i\leftrightarrow\beta(\gamma_i))\Rightarrow\neg\beta(\gamma_i),\gamma_i}{\Box(\gamma_i\leftrightarrow\beta(\gamma_i))\Rightarrow\beta(\gamma_i)\to\gamma_i}}\text{inv.}}\text{subst.}} \quad \dfrac{\dfrac{\gamma_i\Rightarrow\Box(r\leftrightarrow\beta(r))\to\beta(r)}{\dfrac{\gamma_i\Rightarrow\Box(\gamma_i\leftrightarrow\beta(\gamma_i))\to\beta(\gamma_i)}{\dfrac{\gamma_i\Rightarrow\neg\Box(\gamma_i\leftrightarrow\beta(\gamma_i)),\beta(\gamma_i)}{\dfrac{\gamma_i,\Box(\gamma_i\leftrightarrow\beta(\gamma_i))\Rightarrow\beta(\gamma_i)}{\dfrac{\Box(\gamma_i\leftrightarrow\beta(\gamma_i))\Rightarrow\neg\gamma_i,\beta(\gamma_i)}{\Box(\gamma_i\leftrightarrow\beta(\gamma_i))\Rightarrow\gamma_i\to\beta(\gamma_i)}}}\text{inv.}}\text{subst.}}}{\dfrac{\Box(\gamma_i\leftrightarrow\beta(\gamma_i))\Rightarrow\gamma_i\leftrightarrow\beta(\gamma_i)}{\emptyset\Rightarrow\Box(\gamma_i\leftrightarrow\beta(\gamma_i))}\Box_{GL}}$$

Now by a cut

$$\dfrac{\emptyset\Rightarrow\Box(\gamma_i\leftrightarrow\beta(\gamma_i))\qquad \Box(\gamma_i\leftrightarrow\beta(\gamma_i))\Rightarrow\gamma_i\leftrightarrow\beta(\gamma_i)}{\emptyset\Rightarrow\gamma_i\leftrightarrow\beta(\gamma_i)}\text{ cut}$$

From this proof one can see that already ordinary interpolation does the job. The point of using uniform interpolation here is that we do not need to have an actual proof of $(\Box(p\leftrightarrow\beta(p))\wedge\beta(p)\Rightarrow \Box(r\leftrightarrow\beta(r))\to\beta(r))$ to construct a fixed point - we just need to know that the sequent is provable to show that we have indeed constructed a fixed point. ⊣

Uniform interpolation in Grz

The proof of uniform interpolation in **Grz** follows the same ideas and is very similar to the previous one, only syntactically a bit more more complicated.

THEOREM 3.3. *Let Γ, Δ, Σ be finite multisets of formulas. For every propositional variable p there exists a formula $\forall p(\Box\Sigma|\Gamma; \Delta)$ such that:*

(i)
$$Var(\forall p(\Box\Sigma|\Gamma;\Delta)) \subseteq Var(\Sigma,\Gamma,\Delta)\setminus\{p\}$$

(ii)
$$\vdash_{G^+_{Grz}} \Box\Sigma|\Gamma, \forall p(\Box\Sigma|\Gamma;\Delta) \Rightarrow \Delta$$

(iii) *moreover let Φ, Ψ, Θ be multisets of formulas not containing p and*

$$\vdash_{G^+_{Grz}} \Box\Theta, \Box\Sigma|\Phi, \Gamma \Rightarrow \Psi, \Delta.$$

Then

$$\vdash_{G^+_{Grz}} \emptyset|\Box\Theta, \Phi \Rightarrow \forall p(\Box\Sigma|\Gamma;\Delta), \Psi.$$

Proof. We start with a definition of the formula $\forall p(\Box\Sigma|\Gamma; \Delta)$, then we prove that the definition terminates, and proceed with proving it satisfies items (i)-(iii) of the Theorem. We remark that the item (iii) is formulated with the third multiset empty because we only have a particular form of cut admissible, see Remark 2.17.

Definition of the interpolant. We describe the construction of the interpolant recursively. The formula $\forall p(\Box\Sigma|\Gamma; \Delta)$ is defined by

$$\forall p(\Box\Sigma|\Gamma;\Delta) = \bigwedge_{(\Box\Sigma_i|\Gamma_I\Rightarrow\Delta_i)\in Cl(\Box\Sigma|\Gamma;\Delta)} \forall p(\Box\Sigma_i|\Gamma_i;\Delta_i) \qquad (3)$$

The recursive steps for $(\Box\Sigma|\Gamma \Rightarrow \Delta)$ being a critical sequent of the form $(\Box\Gamma'|\Pi; \Box\Delta', \Lambda)$, with Π, Λ atomic, are given by the following table:

	$\Box\Gamma'\|\Pi; \Box\Delta', \Lambda$ matches	$\forall p(\Box\Gamma'\|\Pi; \Box\Delta', \Lambda)$ equals
1	if $p \in \Pi \cap \Lambda$ or $\bot \in \Pi$ or $\Gamma' \cap \Delta' \neq \emptyset$	$\top$
2	otherwise (here all $q, r \neq p$)	$\bigvee_{q\in\Lambda} q \vee \bigvee_{r\in\Pi} \neg r$ $\bigvee_{\beta\in\Delta', D(\beta)\notin\Box\Gamma'} \Box\forall p(\Box\Gamma', D(\beta)\|\Gamma'; \beta)$ $\bigvee_{\beta\in\Delta', D(\beta)\in\Box\Gamma'} \Box\forall p(\Box\Gamma'\|\emptyset; \beta))$ $\vee \Diamond \bigwedge N(\Box\Gamma'\|\Gamma'; \emptyset)$

Table 3.

As in Table 1 before, the first line corresponds to some of the cases when the critical sequent is provable, and the line 2 corresponds to a critical step, the corresponding disjunction covering

- propositional variables from multisets Π, Λ,
- all the possibilities of $\Box^+_{Grz1}$ and $\Box^+_{Grz2}$ inferences with the principal formula from $\Box\Delta'$,
- and, by the diamond formula $\Diamond \bigwedge N(\Box\Gamma'|\Gamma'; \emptyset)$ defined below in Table 4, also the possibility of a $\Box^+_{Grz1}$ or a $\Box^+_{Grz2}$ inference with the principal formula not from $\Box\Delta'$ (i.e. from a context not containing p). For a sequent of the form $(\Box\Gamma'|\Gamma'; \emptyset)$, a set of formulas $N(\Box\Gamma'|\Gamma'; \emptyset)$ is defined as the smallest set given by the Table 4.

Termination. We adopt the same simplification as we have used proving termination of the calculus G_{Grz+} — we treat the third multiset as a set (i.e., we remove duplicate formulas stored in the set). Consider a run of the procedure for $\forall p(\emptyset|\Phi; \Psi)$. Let n be the number of boxed subformulas occurring in $\Phi; \Psi$, which is, as in the case of **GL**, maximal number of critical steps along one branch of the corresponding tree. With each $\forall p$ argument $(\Box\Sigma|\Gamma; \Delta)$ occurring during the run of the procedure, we associate an ordered pair $\langle n^2 - |\Sigma^\circ|, w(\Gamma, \Delta)\rangle$, where n^2 is an upper bound of the number of formulas stored in $\Box\Sigma$ if we do not duplicate them. The measure strictly decreases in each step of the run of the procedure in terms of the lexicographical ordering:

- For a noncritical argument $(\Box\Sigma|\Gamma; \Delta)$ and for each $(\Box\Sigma'|\Gamma'; \Delta') \in Cl(\Box\Sigma|\Gamma; \Delta), w(\Gamma', \Delta') < w(\Gamma, \Delta)$.

	$(\Box\Sigma\|\Upsilon; \Box\Omega, \Theta) \in Cl(\Box\Gamma'\|\Gamma'; \emptyset)$ matches	$N(\Box\Gamma', \Gamma'; \emptyset)$ contains
1	$\Sigma^\circ \supset \Gamma'^\circ$ or $p \in \Upsilon \cap \Theta$, or $\Sigma \cap \Omega \neq \emptyset$ or $\bot \in \Upsilon$	$\forall p(\Box\Sigma\|\Upsilon; \Box\Omega, \Theta)$
2	otherwise (here all $q, r \neq p$)	$\bigvee_{q \in \Theta} q \vee \bigvee_{r \in \Upsilon} \neg r \vee$ $\bigvee_{\beta \in \Omega, D(\beta) \notin \Box\Sigma} \Box\forall p(\Box\Sigma, D(\beta)\|\Sigma; \beta) \vee$ $\bigvee_{\beta \in \Omega, D(\beta) \in \Box\Sigma} \Box\forall p(\Box\Sigma\|\emptyset; \beta))$

Table 4.

- For a critical argument $(\Box\Gamma'\|\Pi; \Box\Delta', \Lambda)$ let us see that, in Table 3 and Table 4, for each of the five recursively called arguments the measure decreases.

    - the line 2 in Table 3, $(\Box\Gamma', D(\beta)\|\Gamma'; \beta)$ where $\beta \in \Delta'$ and $D(\beta) \notin \Box\Gamma'$: here obviously $|(\Box\Gamma' \cup D(\beta))^\circ| > |\Box\Gamma'^\circ|$.
    - the line 2 in Table 3, $(\Box\Gamma'\|\emptyset; \beta)$ where $\beta \in \Delta'$ and $D(\beta) \in \Box\Gamma'$: in this case, $w(\emptyset, \beta) < w(\Pi, \Box\Delta', \Lambda)$.
    - the first line in Table 4, $(\Box\Sigma\|\Upsilon; \Box\Omega, \Theta)$ where $(\Box\Sigma\|\Upsilon \Rightarrow \Box\Omega, \Theta) \in Cl(\Box\Gamma'\|\Gamma'\emptyset)$ and $\Sigma^\circ \supset \Gamma'^\circ$: Since $\Sigma^\circ \supset \Gamma'^\circ$, $|\Sigma^\circ| > |\Gamma'^\circ|$.
    - the second line in Table 4, $(\Box\Sigma, D(\beta)\|\Sigma; \beta)$ where $\beta \in \Omega$, $D(\beta) \notin \Box\Sigma$, and $(\Box\Sigma\|\Upsilon \Rightarrow \Box\Omega, \Theta) \in Cl(\Box\Gamma'\|\Gamma'; \emptyset)$ with $\Sigma^\circ = \Gamma'^\circ$:
    Since $\Sigma^\circ = \Gamma'^\circ$, also $|\Box\Sigma^\circ| = |\Box\Gamma'^\circ|$. Hence $|(\Box\Sigma \cup D(\beta))^\circ| > |\Box\Gamma'^\circ|$.
    - the second line in Table 4, $(\Box\Sigma\|\emptyset; \beta)$ where $\beta \in \Omega$, $D(\beta) \in \Sigma$, $(\Box\Sigma\|\Upsilon \Rightarrow \Box\Omega, \Theta) \in Cl(\Box\Gamma'\|\Gamma'; \emptyset)$ and $\Sigma^\circ = \Gamma'^\circ$:
    here $w(\emptyset, \beta) < w(\Pi, \Box\Delta', \Lambda)$.

REMARK 3.3 (termination trick). *Analogously to (2), we want to use $\forall p(\Box\Gamma'\|\Gamma'; \emptyset)$ in place of $\Diamond \bigwedge N(\Box\Gamma'\|\Gamma'; \emptyset)$ while proving the item (iii) of the theorem. We show next that it is indeed the case that*

$$\vdash_{G^+_{Grz}} \Diamond \bigwedge N(\Box\Gamma'\|\Gamma'; \emptyset) \Leftrightarrow \Diamond\forall p(\Box\Gamma'\|\Gamma'; \emptyset). \qquad (4)$$

Consider sequents $(\Box\Sigma\|\Upsilon; \Box\Omega, \Theta)$ in the closure of $(\Box\Gamma'\|\Gamma'; \emptyset)$, and refer by S to sequents with $\Sigma^\circ = \Gamma'^\circ$, and by S' to sequents in the closure with $\Sigma^\circ \supset \Gamma'^\circ$, i.e. strictly simpler then $(\Box\Gamma'\|\Gamma'; \emptyset)$. Since

$$\forall p(\Box\Gamma'\|\Gamma'; \emptyset) \equiv \bigwedge_S \forall p(S) \wedge \bigwedge_{S'} \forall p(S'),$$

and for each $S = (\Box\Sigma\|\Upsilon; \Box\Omega, \Theta)$ with $\Sigma^\circ = \Gamma'^\circ$ we obtain $\forall p(S)$ by the line 2 of Table 3 to be the *following formula:*

$$\bigvee_{q\in\Theta} q \vee \bigvee_{r\in\Upsilon} \neg r \vee \bigvee_{\substack{\beta\in\Omega \\ D(\beta)\notin\Box\Sigma}} \Box\forall p(\Box\Sigma, D(\beta)|\Sigma;\beta) \vee \bigvee_{\substack{\beta\in\Omega \\ D(\beta)\in\Box\Sigma}} \Box\forall p(\Box\Sigma|\emptyset;\beta)) \vee \Diamond\bigwedge N(\Box\Sigma,\Sigma;\emptyset),$$

which we can shorten as

$$\alpha_S \vee \Diamond\bigwedge N(\Box\Sigma,\Sigma;\emptyset).$$

Now using the definition of $N(\Box\Gamma'|\Gamma';\emptyset)$ in Table 4, the left-hand side of the above sequent (4) becomes the following:

$$\Diamond(\bigwedge_S \alpha_S \wedge \bigwedge_{S'} \forall p(S'))$$

and the right-hand side becomes the following:

$$\Diamond(\bigwedge_S(\alpha_S \vee \Diamond\bigwedge N(\Box\Sigma,\Sigma;\emptyset)) \wedge \bigwedge_{S'} \forall p(S')).$$

Observe, that $N(\Box\Sigma|\Sigma;\emptyset)$ is equivalent to $N(\Box\Gamma'|\Gamma';\emptyset)$ by $\Sigma^\circ = \Gamma'^\circ$. The result now follows by Lemma 2.18, putting $\beta = \bigwedge_{S'}\forall p(S')$ and $\delta = \bigwedge N(\Box\Gamma'|\Gamma';\emptyset)$.

(i). The item (i) follows easily by induction on $(\Box\Sigma|\Gamma;\Delta)$ just because we never add p during the definition of the formula $\forall p(\Box\Sigma|\Gamma;\Delta)$.

(ii). we proceed by induction on the complexity of $(\Box\Sigma|\Gamma;\Delta)$ given by the measure function used above to prove termination, and prove that

$$\vdash_{G^+_{Grz}} \Box\Sigma|\Gamma, \forall p(\Box\Sigma|\Gamma;\Delta) \Rightarrow \Delta.$$

First let $(\Box\Sigma|\Gamma \Rightarrow \Delta)$ be a noncritical sequent. Then sequents $(\Box\Sigma_i|\Gamma_i \Rightarrow \Delta_i) \in Cl(\Box\Sigma|\Gamma;\Delta)$ are of lower complexity and by the induction hypotheses

$$\vdash_{G^+_{Grz}} \Box\Sigma_i|\Gamma_i, \forall p(\Box\Sigma_i|\Gamma_i;\Delta_i) \Rightarrow \Delta_i$$

for each i. Then by admissibility of weakening and by Lemma 2.12

$$\vdash_{G^+_{Grz}} \Sigma|\Gamma, \forall p(\Box\Sigma_1|\Gamma_1;\Delta_1), \ldots, \forall p(\Box\Sigma_k|\Gamma_k;\Delta_k) \Rightarrow \Delta,$$

therefore by a $\wedge$-l inference

$$\vdash_{G^+_{Grz}} \Box\Sigma|\Gamma, \bigwedge_{(\Box\Sigma_i|\Gamma_i\Rightarrow\Delta_i)\in Cl(\Box\Sigma|\Gamma;\Delta)} \forall p(\Box\Sigma_i|\Gamma_i;\Delta_i) \Rightarrow \Delta,$$

which is by (3)

$$\vdash_{G^+_{Grz}} \Gamma, \forall p(\Box\Sigma|\Gamma;\Delta) \Rightarrow \Delta.$$

Let $(\Box\Sigma|\Gamma \Rightarrow \Delta)$ be a critical sequent matching the line 1 of Table 3. Then either (ii) is an axiom in the case that $p \in \Pi \cap \Lambda$ or $\bot \in \Pi$, or (ii) is provable in the case that $\Gamma' \cap \Delta' \neq \emptyset$.

Let $(\Box\Sigma|\Gamma \Rightarrow \Delta)$ be a critical sequent matching the line 2 of Table 3. We prove

$$\vdash_{G^+_{Grz}} \Box\Gamma'|\Pi, \delta \Rightarrow \Box\Delta', \Lambda$$

for each disjunct δ used in the line 2 of Table 3 to define the interpolant.

- For each $r \in \Pi$ obviously $\vdash_{G^+_{Grz}} \Box\Gamma'|\Pi, \neg r, \Box\Gamma' \Rightarrow \Box\Delta', \Lambda$, therefore
$\vdash_{G^+_{Grz}} \Box\Gamma'|\Pi, \bigvee_{r\in\Pi} \neg r, \Box\Gamma' \Rightarrow \Box\Delta', \Lambda$.

- for each $q \in \Lambda$ obviously $\vdash_{G^+_{Grz}} \Box\Gamma'|\Pi, q, \Box\Gamma' \Rightarrow \Box\Delta', \Lambda$, therefore
$\vdash_{G^+_{Grz}} \Box\Gamma'|\Pi, \bigvee_{q\in\Lambda} q, \Box\Gamma' \Rightarrow \Box\Delta', \Lambda$

- For each $\beta \in \Delta'$ with $D(\beta) \notin \Box\Gamma'$ we have

$$\vdash_{G^+_{Grz}} \Box\Gamma', D(\beta)|\Gamma', \forall p(\Box\Gamma', D(\beta)|\Gamma'; \beta) \Rightarrow \beta$$

by the induction hypothesis, which gives

$$\vdash_{G^+_{Grz}} \Box\Gamma', D(\beta), \Box\forall p(\Box\Gamma', D(\beta)|\Gamma'; \beta)|\Gamma', \forall p(\Box\Gamma', D(\beta)|\Gamma'; \beta) \Rightarrow \beta$$

by admissible weakening inferences. This yields

$$\vdash_{G^+_{Grz}} \Box\Gamma', \Box\forall p(\Box\Gamma', D(\beta)|\Gamma'; \beta)|\Pi \Rightarrow \Box\Delta', \Lambda$$

by a $\Box^+_{Grz2}$ inference. Then by weakening and $\Box^+_T$ inferences

$$\vdash_{G^+_{Grz}} \Box\Gamma'|\Box\forall p(\Box\Gamma', D(\beta)|\Gamma'; \beta), \Pi \Rightarrow \Box\Delta', \Lambda.$$

- For each $\beta \in \Delta'$ with $D(\beta) \in \Box\Gamma'$ we have

$$\vdash_{G^+_{Grz}} \Box\Gamma'|\forall p(\Box\Gamma'|\emptyset; \beta) \Rightarrow \beta$$

by the induction hypothesis, which gives

$$\vdash_{G^+_{Grz}} \Box\Gamma', \Box\forall p(\Box\Gamma'|\emptyset; \beta)|\forall p(\Box\Gamma'|\emptyset; \beta), \Gamma' \Rightarrow \beta$$

by admissible weakening inferences. This yields

$$\vdash_{G^+_{Grz}} \Box\Gamma', \Box\forall p(\Box\Gamma'|\emptyset; \beta)|\Pi \Rightarrow \Box\Delta', \Lambda$$

by a $\Box^+_{Grz2}$ inference and a weakening (notice there is an occurrence of $D(\beta)$ missing in Γ'). Then by weakening and $\Box^+_T$ inferences

$$\vdash_{G^+_{Grz}} \Box\Gamma'|\Box\forall p(\Box\Gamma'|\emptyset; \beta), \Pi \Rightarrow \Box\Delta', \Lambda.$$

It remains to be proved that

$$\vdash_{G^+_{Grz}} \Box\Gamma'|\Pi, \Diamond\bigwedge N(\Box\Gamma'|\Gamma';\emptyset) \Rightarrow \Box\Delta', \Lambda.$$

For each $(\Box\Sigma|\Upsilon \Rightarrow \Box\Omega, \Theta) \in Cl(\Box\Gamma'|\Gamma';\emptyset)$ of the first line of Table 4, we know by the induction hypotheses that

$$\vdash_{G^+_{Grz}} \Box\Sigma|\Upsilon, \forall p(\Box\Sigma|\Upsilon; \Box\Omega, \Theta) \Rightarrow \Box\Omega, \Theta.$$

For each $(\Box\Sigma|\Upsilon \Rightarrow \Box\Omega, \Theta) \in Cl(\Box\Gamma'|\Gamma';\emptyset)$ of the second line of Table 4 we have the following:

- $\vdash_{G^+_{Grz}} \Box\Sigma|\Upsilon, \bigvee_{q\in\Theta} q \Rightarrow \Box\Omega, \Theta.$

- $\vdash_{G^+_{Grz}} \Box\Sigma|\Upsilon, \bigvee_{\neg r\in\Upsilon} \neg r \Rightarrow \Box\Omega, \Theta.$

- for each $\beta \in \Omega$ with $D(\beta) \notin \Box\Sigma$ by the induction hypotheses

$$\vdash_{G^+_{Grz}} \Box\Sigma, D(\beta)|\Sigma, \forall p(\Box\Sigma, D(\beta)|\Sigma; \beta) \Rightarrow \beta$$

and by weakening and a $\Box^+_{Grz2}$ inference

$$\vdash_{G^+_{Grz}} \Box\Sigma, \Box\forall p(\Box\Sigma, D(\beta)|\Sigma; \beta)|\emptyset \Rightarrow \Box\beta.$$

and by weakening and $\Box^+_T$

$$\vdash_{G^+_{Grz}} \Box\Sigma|\Box\forall p(\Box\Sigma, D(\beta)|\Sigma; \beta) \Rightarrow \Box\beta.$$

- for each $\beta \in \Omega$ with $D(\beta) \in \Box\Sigma$ by the induction hypotheses

$$\vdash_{G^+_{Grz}} \Box\Sigma|\emptyset, \forall p(\Box\Sigma|\emptyset; \beta) \Rightarrow \beta$$

and by weakening and a $\Box^+_{Grz2}$ inference

$$\vdash_{G^-_{Grz}} \Box\Sigma, \Box\forall p(\Box\Sigma|\emptyset; \beta)|\emptyset \Rightarrow \Box\beta.$$

and by weakening and $\Box^+_T$

$$\vdash_{G^+_{Grz}} \Box\Sigma|\Box\forall p(\Box\Sigma|\emptyset; \beta) \Rightarrow \Box\beta.$$

Together this yields, using ∨-l inferences,

$$\vdash_{G^+_{Grz}} \Box\Sigma|\Upsilon, \bigvee_{q\in\Theta} \vee \bigvee_{\neg r\in\Upsilon} \vee \bigvee_{D(\beta)\notin\Box\Sigma} \Box\forall p(\Box\Sigma, D(\beta)|\Sigma; \beta) \vee \bigvee_{D(\beta)\in\Box\Sigma} \Box\forall p(\Box\Sigma|\emptyset; \beta) \Rightarrow \Box\Omega, \Theta.$$

Therefore finally, putting things together for each $(\Box\Sigma|\Upsilon \Rightarrow \Box\Omega, \Theta) \in Cl(\Box\Gamma'|\Gamma';\emptyset)$, we obtain, using weakening and ∧-r inferences,

$$\vdash_{G^+_{Grz}} \Box\Sigma, \Upsilon, \bigwedge N(\Box\Gamma'|\Gamma';\emptyset) \Rightarrow \Box\Omega, \Theta.$$

Now, by closure properties in Lemma 2.4,

$$\vdash_{G^+_{Grz}} \Box\Gamma'|\Gamma', \bigwedge N(\Box\Gamma'|\Gamma'; \emptyset) \Rightarrow \emptyset.$$

By negation and weakening inferences

$$\vdash_{G^+_{Grz}} \Box\Gamma', D(\neg \bigwedge N(\Box\Gamma', \Gamma'; \emptyset))|\Gamma' \Rightarrow \neg \bigwedge N(\Box\Gamma', \Gamma'; \emptyset)$$

and by a $\Box^+_{Grz2}$ inference

$$\vdash_{G^+_{Grz}} \Box\Gamma'|\Pi \Rightarrow \Box\neg \bigwedge N(\Box\Gamma', \Gamma'; \emptyset), \Box\Delta', \Lambda.$$

Now, using a negation inference again, we obtain

$$\vdash_{G^+_{Grz}} \Box\Gamma'|\Pi, \Diamond \bigwedge N(\Box\Gamma', \Gamma'; \emptyset) \Rightarrow \Box\Delta', \Lambda.$$

Putting finally all the above disjuncts together for a critical sequent $(\Box\Gamma'|\Pi; \Box\Delta', \Lambda)$ yields, using $\vee$-l inferences and the line 2 of Table 3,

$$\vdash_{G^+_{Grz}} \Box\Gamma'|\Pi, \forall p(\Box\Gamma'|\Pi; \Box\Delta', \Lambda) \Rightarrow \Box\Delta', \Lambda.$$

(iii) We proceed by induction on the height of the proof of the sequent $(\Box\Theta, \Box\Sigma|\Phi, \Gamma \Rightarrow \Psi, \Delta)$ in G^+_{Grz}, and sub-induction on the measure of the sequent $(\Box\Sigma|\Gamma; \Delta)$. We show that

$$\vdash_{G^+_{Grz}} \Box\Theta|\Phi \Rightarrow \forall p(\Box\Sigma|\Gamma; \Delta), \Psi.$$

Let us first consider the last step of the proof of $(\Box\Theta, \Box\Sigma|\Phi, \Gamma \Rightarrow \Psi, \Delta)$ is an axiom, or, if it is not an axiom, then $(\Box\Sigma|\Gamma; \Delta)$ is a noncritical sequent. In this case we proceed similarly as in Theorem 3.1 (iii), the third multiset makes no difference here.

Let us then consider that the last inference of the proof of $(\Box\Theta, \Box\Sigma|\Phi, \Gamma \Rightarrow \Psi, \Delta)$ is a $\Box^+_{Grz2}$ inference. There are two cases to distinguish:

- Consider first the case when the principal formula $\Box\alpha \in \Delta$. Then the proof ends with:

$$\frac{\Box\Theta, \Box\Sigma, D(\alpha)|\Theta, \Sigma \Rightarrow \alpha}{\Box\Theta, \Box\Sigma|\Gamma, \Phi \Rightarrow \Box\alpha, \Delta', \Psi} \Box^+_{Grz2}$$

where $\Box\alpha, \Delta'$ is Δ. Consider $\Box\Sigma \cap \Delta = \emptyset$ (otherwise $\forall p(\Box\Sigma|\Gamma; \Delta) \equiv \top$ and (iii) holds). Then by the induction hypotheses

$$\vdash_{G^+_{Grz}} \emptyset|\Box\Theta, \Theta \Rightarrow \forall p(\Box\Sigma, D(\alpha)|\Sigma; \alpha).$$

By invertibility of $\Box^+_T$ inferences, by contraction inferences, and weakening

$$\vdash_{G^+_{Grz}} \Box\Theta, D(\forall p(\Box\Sigma, D(\alpha)|\Sigma; \alpha))|\Theta \Rightarrow \forall p(\Box\Sigma, D(\alpha)|\Sigma; \alpha),$$

Now, by a $\Box^+_{Grz2}$ inference, we obtain

$$\vdash_{G^+_{Grz}} \exists\Theta|\Phi \Rightarrow \Box\forall p(\Box\Sigma, D(\alpha)|\Sigma;\alpha), \Psi.$$

By weakening inferences

$$\vdash_{G^+_{Grz}} \Box\Theta|\Theta, \Phi \Rightarrow \Box\forall p(\Box\Sigma, D(\alpha)|\Sigma;\alpha), \Psi.$$

By $\Box^+_T$ inferences we obtain

$$\vdash_{G^+_{Grz}} \emptyset\;\Box\Theta, \Phi \Rightarrow \Box\forall p(\Box\Sigma, D(\alpha)|\Sigma;\alpha), \Psi.$$

By the line 2 of Table 3 and invertibility of the ∨-l rule

$$\vdash_{G^+_{Grz}} \emptyset|\Box\forall p(\Box\Sigma, D(\alpha)|\Sigma;\alpha) \Rightarrow \forall p(\Box\Sigma|\Gamma; \Box\alpha, \Delta').$$

The two sequents above yield (iii) by admissibility of the cut rule in G^+_{Grz}.

- Consider next the case when the principal formula $\Box\alpha \in \Psi$, i.e., α doesn't contain p. Then the proof ends with:

$$\frac{\Box\Theta, \Box\Sigma, D(\alpha)|\Theta, \Sigma \Rightarrow \alpha}{\Box\Theta, \Box\Sigma|\Gamma, \Phi \Rightarrow \Delta, \Box\alpha, \Psi'}\;\Box^+_{Grz2}$$

where $\Box\alpha, \Psi'$ is Ψ. Then by the induction hypotheses

$$\vdash_{G^+_{Grz}} \emptyset|D(\alpha), \Box\Theta, \Theta \Rightarrow \forall p(\Box\Sigma|\Sigma; \emptyset), \alpha.$$

By invertibility of $\Box^+_T$ inferences and by contraction inferences we obtain

$$\vdash_{G^+_{Grz}} D(\alpha), \Box\Theta|(\alpha \to \Box\alpha), \Theta \Rightarrow \forall p(\Box\Sigma|\Sigma; \emptyset), \alpha.$$

To get rid of $(\alpha \to \Box\alpha)$, which is $(\neg\alpha \vee \Box\alpha)$, we use invertibility of the ∨-l and ¬-l rules, and contraction, to obtain

$$\vdash_{G^+_{Grz}} D(\alpha), \Box\Theta|\Theta \Rightarrow \forall p(\Box\Sigma|\Sigma; \emptyset), \alpha.$$

By a ¬-l inference and weakening

$$\vdash_{G^+_{Grz}} D(\alpha), \Box\Theta, \Box\neg\forall p(\Box\Sigma|\Sigma; \emptyset)|\Theta, \neg\forall p(\Box\Sigma|\Sigma; \emptyset) \Rightarrow \alpha.$$

By a $\Box^+_{Grz2}$ inference

$$\vdash_{G^+_{Grz}} \Box\Theta, \Box\neg\forall p(\Box\Sigma|\Sigma; \emptyset)|\Phi \Rightarrow \Box\alpha, \Psi'.$$

Since weakening is admissible in Gm^+_{Grz}, we obtain

$$\vdash_{G^+_{Grz}} \Box\Theta, \Box\neg\forall p(\Box\Sigma|\Sigma; \emptyset)|\Theta, \neg\forall p(\Box\Sigma|\Sigma; \emptyset), \Phi \Rightarrow \Box\alpha, \Psi'$$

and now $\Box_T^+$ inferences and a $\neg$-l inference yield

$$\vdash_{G_{Grz}^+} \emptyset|\Box\Theta,\Phi \Rightarrow \Diamond\forall p(\Box\Sigma|\Sigma;\emptyset),\Box\alpha,\Psi'.$$

By weakening inferences

$$\vdash_{G_{Grz}^+} \Box\Theta|\Theta,\Phi \Rightarrow \Diamond\forall p(\Box\Sigma|\Sigma;\emptyset),\Box\alpha,\Psi'.$$

By $\Box_T^+$ inferences

$$\vdash_{G_{Grz}^+} \emptyset|\Box\Theta,\Phi \Rightarrow \Diamond\forall p(\Box\Sigma|\Sigma;\emptyset),\Box\alpha,\Psi'.$$

By the line 2 of Table 3, invertibility of the $\lor$-l rule, and by (4) we have

$$\vdash_{G_{Grz}^+} \emptyset|\Diamond\forall p(\Box\Sigma|\Sigma;\emptyset) \Rightarrow \forall p(\Box\Sigma|\Sigma;\emptyset).$$

The two sequents above yield (iii) by admissibility of the cut rule in G_{Grz}^+.

Let us consider that the last inference of the proof of $(\Box\Theta,\Box\Sigma|\Phi,\Gamma \Rightarrow \Psi,\Delta)$ is a $\Box_{Grz1}^+$ inference. Again, we distinguish two cases:

- Consider first the case when the principal formula $\Box\alpha \in \Delta$. Then the proof ends with:

$$\frac{\Box\Theta,\Box\Sigma|\emptyset \Rightarrow \alpha}{\Box\Theta,\Box\Sigma|\Gamma,\Phi \Rightarrow \Box\alpha,\Delta',\Psi} \Box_{Grz1}^+$$

where $\Box\alpha,\Delta'$ is Δ. Consider $\Box\Sigma \cap \Delta = \emptyset$ (otherwise $\forall p(\Box\Sigma|\Gamma;\Delta) \equiv \top$ and (iii) holds). Then by the induction hypotheses

$$\vdash_{G_{Grz}^+} \emptyset|\Box\Theta \Rightarrow \forall p(\Box\Sigma|\emptyset;\alpha).$$

By invertibility of $\Box_T^+$ inferences, by contraction inferences, and weakening

$$\vdash_{G_{Grz}^+} \Box\Theta, D(\forall p(\Box\Sigma|\emptyset;\alpha))|\Theta \Rightarrow \forall p(\Box\Sigma|\emptyset;\alpha),$$

Now, by a $\Box_{Grz2}^+$ inference, we obtain

$$\vdash_{G_{Grz}^+} \Box\Theta|\Phi \Rightarrow \Box\forall p(\Box\Sigma|\emptyset;\alpha),\Psi.$$

By weakening inferences and $\Box_T^+$ inferences we obtain

$$\vdash_{G_{Grz}^+} \emptyset|\Box\Theta,\Phi \Rightarrow \Box\forall p(\Box\Sigma|\emptyset;\alpha)\Psi.$$

By the line 2 of Table 3 and invertibility of the $\lor$-l rule we have

$$\vdash_{G_{Grz}^+} \emptyset|\Box\forall p(\Box\Sigma|\emptyset;\alpha) \Rightarrow \forall p(\Box\Sigma|\Gamma;\Box\alpha,\Delta').$$

The two sequents above yield (iii) by admissibility of the cut rule in G_{Grz}^+.

- Consider next the case that the principal formula $\Box\alpha \in \Psi$, i.e., α doesn't contain p. Then the proof ends with:

$$\dfrac{\Box\Theta, \Box\Sigma | \emptyset \Rightarrow \alpha}{\Box\Theta, \Box\Sigma | \Gamma, \Phi \Rightarrow \Delta, \Box\alpha, \Psi'} \Box^+_{Grz1}$$

where $\Box\alpha, \Psi'$ is Ψ. Then by the induction hypotheses

$$\vdash_{G^+_{Grz}} \emptyset | \Box\Theta \Rightarrow \forall p(\Box\Sigma | \emptyset; \emptyset), \alpha.$$

Notice that $(\Box\Sigma|\emptyset;\emptyset)$ is a critical sequent with all but one multisets empty, and by the table $\forall p(\Box\Sigma|\emptyset;\emptyset) \equiv \bot$. Thus we have in fact

$$\vdash_{G^+_{Grz}} \emptyset | \Box\Theta \Rightarrow \alpha.$$

By invertibility of $\Box^+_T$ inferences we obtain

$$\vdash_{G^+_{Grz}} \Box\Theta | \Theta \Rightarrow \alpha$$

and by weakening

$$\vdash_{G^+_{Grz}} \Box\Theta, D(\alpha) | \Theta \Rightarrow \alpha.$$

By a $\Box^+_{Grz2}$ inference

$$\vdash_{G^+_{Grz}} \Box\Theta | \Phi \Rightarrow \Box\alpha, \Psi'.$$

By admissibility of weakening we obtain

$$\vdash_{G^+_{Grz}} \Box\Theta | \Phi \Rightarrow \forall p(\Box\Sigma | \Gamma; \Delta), \Box\alpha, \Psi'.$$

⊣

Concluding remarks

We have provided an effective construction of uniform interpolants in provability logics. We would like to point out, that even if the proofs as presented are not fully constructive, the only part that is not constructive is the completeness of the two calculi without the cut rule. This can be, in both cases, repaired by completing the proof search argument and make it into a decision procedure.

What we also left open in this paper is to investigate which distribution laws the quantifiers satisfy. For example, it is the case that, in the basic modal logic **K**, the universal bisimulation quantifier commutes with the diamond modality [3], [5]. In fact, it commutes with (the dual of) the cover modality, which is a principle that, besides the usual axioms and rules for quantification, axiomatizes bisimulation quantifiers over **K**. Whether a similar insight can be obtained for **GL** is not clear at the moment.

Acknowledgement The work of this paper has been supported by the project No. P202/11/1632 of the Czech Science Foundation. I would like to thank Nick Bezhanishvilli for pointing up the topic years ago, Albert Visser for inspiration and discussions on the topic, and Rosalie Iemhoff and Tadeusz Litak for encouraging me to write this up once again.

BIBLIOGRAPHY

[1] A. Avron, *On modal systems having arithmetical interpretations*, The Journal of Symbolic Logic **49** (1984), 935–942.
[2] M. Bílková, *Interpolation in modal logic*, Ph.D. thesis, Charles University in Prague, 2006.
[3] _____, *Uniform interpolation and propositional quantifiers in modal logics*, Studia Logica **85** (2007), 1–31.
[4] G. Boolos, *The logic of provability*, Cambridge University Press, New York and Cambridge, 1993.
[5] T. French, *Bisimulation quantifiers for modal logics*, Ph.D. thesis, School of Computer Science and Software Engineering, University of Western Australia, 2006.
[6] S. Ghilardi, *An algebraic theory of normal forms*, Annals of Pure and Applied Logic **71** (1995), 189–245.
[7] S. Ghilardi and M. Zawadovski, *Undefinability of propositional quantifiers in the modal system S4*, Studia Logica **55** (1995), 259–271.
[8] Z. Gleit and W. Goldfarb, *Characters and fixed points in provability logic*, Notre Dame Journal of Formal Logic **31** (1990), no. 1, 26–36.
[9] R. Gore and R. Ramanayake, *Valentini's cut-elimination for provability logic resolved*, Advances in Modal logic **7** (2008), 67–86.
[10] _____, *Valentini's cut-elimination for provability logic resolved*, The Review of Symbolic Logic **5** (2012), 212–238.
[11] A. Herzig and J. Mengin, *Uniform interpolation by resolution in modal logic*, Logics in Artificial Intelligence (S. Hlldobler, C. Lutz, and H. Wansing, eds.), Lecture Notes in Computer Science, vol. 5293, Springer, 2008, pp. 219–231.
[12] A. Heuerding, *Sequent calculi for proof search in some modal logics*, Ph.D. thesis, University of Bern, Switzerland, 1998.
[13] A. Heuerding, M. Seyfried, and H. Zimmermann, *Efficient loop-check for backward proof search in some non-classical propositional logics*, Tableaux (1996), 210–225.
[14] R. Iemhoff, *Uniform interpolation and sequent calculi in modal logic*, submitted.
[15] M. Kracht, *Tools and techniques in modal logic*, Elsevier, 1999.
[16] A. Pitts, *On an interpretation of second order quantification in first order intuitionistic propositional logic*, The Journal of Symbolic Logic **57** (1992), 33–52.
[17] L. Reidhaar-Olson, *A new proof of the fixed point theorem of provability logic*, Notre Dame Journal of Formal Logic **31** (1990), no. 1, 37–43.
[18] G. Sambin, *An effective fixed-point theorem in intuitionistic diagonalizable algebras*, Studia Logics **35** (1976), 345–361.
[19] G. Sambin and G. Boolos, *Provability: The emergence of a mathematical modality*, Studia Logica **50** (1991), 1–23.
[20] G. Sambin and S. Valentini, *The modal logic of provability. the sequential approach*, Journal of Philosophical Logic **11** (1982), 311–342.
[21] L. Santocanale and Y. Venema, *Uniform interpolation for monotone modal logic*, Advances in Modal Logic (L. Beklemishev, V. Goranko, and V. Shehtman, eds.), vol. 8, College Publications, 2010, pp. 350–370.
[22] V. Yu. Shavrukov, *Subalgebras of diagonalizable algebras of theories containing arithmetic*, Ph.D. thesis, Dissertationes Mathematicae CCCXXIII, Polska Akademia Nauk, Mathematical Institute, Warszawa, 1993.
[23] C. Smoryński, *Beth's theorem and self-referential sentences*, Logic Colloquium 77 (1978), 253–261.
[24] R. Solovay, *Provability interpretations of modal logics*, Israel Journal of Mathematics **25** (1976), 287–304.
[25] A. S. Troelstra and H. Schwichtenberg, *Basic proof theory*, Cambridge University Press, 1996.
[26] A. Visser, *Bisimulations, model descriptions and propositional quantifiers*, Logic Group Preprint Series, Utrecht University **161** (1996).
[27] _____, *Uniform interpolation and layered bisimulation*, Gödel 96 (Brno, 1996), Lecture Notes Logic, vol. 6, Springer, 2002.
[28] V. Švejdar, *On provability logic*, Nordic Journal of Philosophical Logic **4** (2000), no. 2, 95–116.

Gossip in Dynamic Networks

HANS VAN DITMARSCH, JAN VAN EIJCK, PERE PARDO, RAHIM RAMEZANIAN, FRANÇOIS SCHWARZENTRUBER

Abstract

A gossip protocol is a procedure for spreading secrets among a group of agents, using a connection graph. In this paper the problem of designing and analyzing gossip protocols is given a dynamic twist by assuming that when a call is established not only secrets are exchanged but also contact list, i.e., links in the gossip graph. Thus, each call in the gossip graph changes both the graph and the distribution of secrets. This paper gives a full characterization for the class of dynamic gossip graphs where the Learn New Secrets protocol (make a call to an agent if you know the number but not the secret of that agent) is successful.[1]

For Albert Visser

1 How to Spread Secrets

This contribution is offered to Albert Visser in the knowledge that the topic will delight him.[2] Gossip is idle talk about other people, and it typically involves details not confirmed as true. Not something that Albert engages in, but still connected to his interests in various ways. The formal study of how gossip spreads investigates the mechanisms behind the diffusion of information, and information and its growth are at the core of logic, from intuitionism to dynamic semantics for natural language.

Gossip protocols are procedures for spreading secrets among a group of agents, using a connection graph. It is assumed that everyone has a unique secret. The assumption that each agent starts out with a secret only known to that agent will enable us to trace each piece of information back to its unique source.

In the original set-up a totally connected graph was assumed. One of the key questions was to find a minimal sequence of calls to achieve a state where all agents knew all secrets. The assumption was that during a call, all secrets were exchanged. As it turns out, in a totally connected graph with $n > 3$ agents, $2n - 4$ calls are sufficient for this. Consider the totally connected graph with four agents.

---

[1]This research was initiated when Jan van Eijck visited the other authors at LORIA in Nancy, in April 2015.
[2]Jan has many fond memories of interactions with Albert in the past: a joint talk on Montague grammar for the *Vereniging voor Logica*, the writing of *Inzien en Bewijzen*, running the Parallels Project together, and so on. Hans kindly remembers Albert from his early days in Utrecht as a mathematics and philosophy student, and from many other occasions such as the Alice in Wonderland workshop at the *Internationale School voor Wijsbegeerte*. Rahim's PhD-supervisor Mohammad Ardeshir recalls a memorable visit of Albert to Iran and their shared interests in intuitionistic logics.

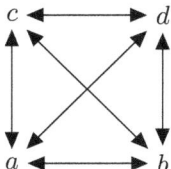

A possible calling sequence for ensuring that all secrets get shared by everyone is $ab; cd; ac; bd$. If e is also present, precede this sequence with ae, and close off with ae. Thus, in a network with four agents, all secrets can be shared in 6 calls. In general, two extra calls are sufficient for each additional agent, and we have that the number of calls for $n + 1$ agents equals $2n - 4 + 2 = 2(n + 1) - 4$. It follows that $2n - 4$ calls are always enough. It is a bit trickier to show that $2n - 4$ calls are needed: see the original [Tij71], or [Hur00] and the references given there.

In the case above the gossip procedure is regulated by an outside authority, but in distributed computing we look for procedures that do not need outside regulation. A possible distributed protocol for gossip spreading could be:

Search For Secrets

While not every agent knows all secrets, let an agent x who does not know all secrets randomly select an agent y, and let x call y.

The Search for Secrets protocol has the advantage of simplicity, but note that it does not exclude redundant calls. Here is a distributed protocol (proposed in [AvDGvdH14]) that tries to avoid such redundancy.

Learn New Secrets

While not every agent knows all secrets, let an agent x who does not know all secrets randomly select an agent y such that x does not know y's secret, and let x call y.

Note that the selection of the x that makes the call still involves a minimal role for the environment: selecting the caller. We will assume that this selection is random.

There is a vast literature on gossiping and broadcasting in networks [HHL88], and there are connections with the study of the behaviour of epidemics [EGKM04]. Distributed gossip protocols are studied in [AvDGvdH14, AGvdH15]. In essence, these protocols investigate how information spreads through a network.

In this paper we give the problem of designing and analyzing (distributed) gossip protocols a dynamic twist by assuming that when a call is established not only secrets are exchanged but also contact lists. Thus, we drop the assumption that the graph of connections is complete from the start. Calls in the gossip graph are constrained by the current distribution of numbers, and each call changes both the graph and the distribution of secrets.

2 Gossip Graphs

Consider a finite set A of agents, each with access to a set of other agents, and each carrying a unique secret. Then the access tables of the agents determine a graph.

Represent a graph G with secrets (henceforth: gossip graph) as a triple (A, N, S). A is the (finite) set of vertices or agents, $N \subseteq A^2$ and $S \subseteq A^2$ are relations on A, with Nxy expressing that

x has a link to y (or: x does know the contact details of y), and Sxy expressing that x does know the secret of y. Alternatively, we can think of N and S as functions in $A \to \mathcal{P}A$, so that N_x is the set of agents whose numbers are known by x. and S_x is the set of agents whose secrets are known by x.

In gossip graph $G = (A, N, S)$, an agent $x \in A$ is an expert if $S_x = A$, and if $B \subseteq A$, an agent $x \in A$ is a B-expert if $S_x \supseteq B$.

Represent a call from x to y as a tuple xy. The call xy is possible in $G = (A, N, S)$ if Nxy. A call xy merges the secret lists and the contact lists of x and y. Let G^{xy} be the result of this merge in G. That is, if $G = (A, N, S)$ and $x, y \in A$, then $G^{xy} = (A, N', S')$ where $N'_z = N_z$ for all $z \in A$ with $z \neq x, z \neq y$ and $N'_x = N'_y = N_x \cup N_y$, $S'_z = S_z$ for all $z \in A$ with $z \neq x, z \neq y$, and $S'_x = S'_y = S_x \cup S_y$. Alternatively, N' can be given as $N \cup (\{(x, y), (y, x)\} \circ N)$, and S' as $S \cup (\{(x, y), (y, x)\} \circ S)$.

A calling sequence σ is a finite list of calls. We define the set **S** of calling sequences for agent set A recursively as follows (assume x, y range over A):

$$\sigma ::= \epsilon \mid \sigma; xy$$

where ϵ is the empty sequence.

If $\sigma, \tau \in \mathbf{S}$ we use $\sigma; \tau$ for the concatenation of σ and τ. Let $G^\sigma = (A, N^\sigma, S^\sigma)$ be the graph that results after calling sequence σ. This is recursively defined as $G^\epsilon = G$, $G^{\sigma;xy} = (G^\sigma)^{xy}$. We define possible calling sequences, as follows: ϵ is possible on any G, and $\sigma; xy$ is possible on G iff σ is possible on G, and $N^\sigma xy$ holds.

We say that $G = (A, N, S)$ has accessible secrets if $I_A \subseteq S \subseteq N$, where $I_A = \{(a, a) \mid a \in A\}$. Thus, G has accessible secrets if every agent knows her own secret and moreover, if agent x knows the secret of y, x also knows the number of y. Note if $G = (A, N, S)$ has accessible secrets, then $I_A \subseteq N$. This may look strange, as no agent is ever going to call itself, but one can also think if this as expressing the requirement that agents know their own number.

PROPOSITION 2.1. *Let $G = (A, N, S)$, and let σ be a possible calling sequence for G. If G has accessible secrets then G^σ has accessible secrets.*

Proof. Induction on σ, using the fact that it follows from $S_x^\sigma \subseteq N_x^\sigma$ and $S_y^\sigma \subseteq N_y^\sigma$ that $S_x^\sigma \cup S_y^\sigma \subseteq N_x^\sigma \cup N_y^\sigma$, and therefore $S_x^{\sigma;xy} = S_y^{\sigma;xy} \subseteq N_x^{\sigma;xy} = N_y^{\sigma;xy}$. ⊣

PROPOSITION 2.2. *Let $G = (A, N, S)$, and let σ be a possible calling sequence for G. Then $N^\sigma \subseteq (N \cup N^{-1})^*$.*

Proof. Induction on σ. The base case is clear. For the inductive case, assume $N^\sigma \subseteq (N \cup N^{-1})^*$. Assume $\sigma; xy$ is a possible call for G. Then $(x, y) \in N^\sigma$. Notice that $N^{\sigma;xy} = N^\sigma \cup \{(x, y), (y, x)\} \circ N^\sigma$. We are done if we can show that $\{(x, y), (y, x)\} \circ N^\sigma \subseteq (N \cup N^{-1})^*$. From $(x, y) \in N^\sigma$, by ih, $(x, y) \in (N \cup N^{-1})^*$. Since $(N \cup N^{-1})^*$ is symmetric, also $(y, x) \in (N \cup N^{-1})^*$. Therefore $\{(x, y), (y, x)\} \subseteq (N \cup N^{-1})^*$. By induction hypothesis and relational reasoning, it follows from this that $\{(x, y), (y, x)\} \circ N^\sigma \subseteq (N \cup N^{-1})^* \circ (N \cup N^{-1})^* = (N \cup N^{-1})^*$. ⊣

A gossip graph $G = (A, N, S)$ is weakly connected if for all $x, y \in A$ there is an $N \cup N^{-1}$-path from x to y.

THEOREM 2.3. *If σ is a possible calling sequence for $G = (A, N, S)$, then G is weakly connected iff G^σ is weakly connected.*

Proof. Left to right is immediate. Right to left from Proposition 2.2. ⊣

THEOREM 2.4. *If $G = (A, N, S)$ satisfies $I_A = S \subseteq N$ and σ is a possible calling sequence for G, then $S^\sigma \circ N \subseteq N^\sigma$.*

Proof. Induction on σ. For the base case we have to show that $S \circ N \subseteq N$. We have $S \circ N = I_A \circ N = N \subseteq N$.

For the induction step, let σ be a possible calling sequence, and assume $S^\sigma \circ N \subseteq N^\sigma$. Let xy be a possible call in G^σ.

Let $(a, b) \in S^{\sigma;xy} \circ N$. If $(a, b) \in S^\sigma \circ N$, then by the induction hypothesis, $(a, b) \in N^\sigma$, and hence by $N^\sigma \subseteq N^{\sigma;xy}$ we get that $(a, b) \in N^{\sigma;xy}$, and done.

If $(a, b) \in S^{\sigma;xy} \circ N$ and $(a, b) \notin S^\sigma \circ N$, then we may assume (wlog) that $a = x$ and that there is some z with $S^{\sigma;xy}xz$, and Nzb.

From $S^{\sigma;xy}xz$ it follows that either $S^\sigma xz$ or $S^\sigma yz$ (either x or y knew the secret of z before the call xy).

In the former case, we have $(x, b) \in S^\sigma \circ N$, and therefore by the induction hypothesis, $(x, b) \in N^\sigma$. In the latter case, we have $(y, b) \in S^\sigma \circ N$, and therefore by the induction hypothesis, $(y, b) \in N^\sigma$.

From $(x, b) \in N^\sigma$ or $(y, b) \in N^\sigma$ it follows by the definition of $N^{\sigma;xy}$ that $(x, b) \in N^{\sigma;xy}$, and done. ⊣

A gossip graph $G = (A, N, S)$ is *complete* if it holds for all $x \in A$ that $S_x = A$. That is, a gossip graph is complete if all agents know all secrets.

A terminal point in $G = (A, N, S)$ is a point x for which $N_x \subseteq \{x\}$. That is, a terminal point is an agent that knows at most her own number. The skin of a graph $G = (A, N, S)$ is the set $\{x \in A \mid N_x \subseteq \{x\}\}$ (the set of terminal points). Let $s(G)$ be the result of skinning graph G, i.e. removing all terminal points from G. That is, $s(G) = (B, N', S')$ where $B = \{x \in A \mid N_x - \{x\} \neq \emptyset\}$, $N' = N \cap B^2$, $S' = S \cap B^2$. Note that skinning a graph is not a closure operation: there are graphs with $s(s(G)) \neq s(G)$.

N is strongly connected on $G = (A, N, S)$ if for any $x, y \in A$ there is an N-path from x to y. Call $G = (A, N, S)$ strongly connected if N is strongly connected on G.

The **Search For Secrets** protocol now takes the following shape. Note that the only change is the requirement that the caller has to know the number of the agent that gets called.

Search For Secrets

While not every agent knows all secrets, randomly select a pair xy such that Nxy and let x call y.

In some cases, the dynamics can speed up the calling. A circle with five agents $a \longrightarrow b \longrightarrow c \longrightarrow d \longrightarrow e \longrightarrow a$ needs $2n - 3 = 7$ calls before everyone knows all secrets [HHL88], but in our dynamic approach 6 calls are sufficient: $ab; cd; ea; de; ac; bc$. This shows that old questions about minimum lengths of calling sequences can receive new answers in this dynamic setting.

3 Learn New Secrets

The following protocol is studied in [AvDGvdH14, AGvdH15] in the context of totally connected graphs.

Learn New Secret Protocol

While not every agent is an expert, let an agent x that is not an expert randomly choose an agent y from the list of agents for which Nxy but not Sxy, and perform the call xy.

This is like **Learn New Secrets** from the introductory section, but with the extra requirement that the caller has to know the number of the agent that gets called.

We define LNS-permitted calling sequences, as follows: ϵ is LNS-permitted on any G, and $\sigma;xy$ is LNS-permitted on G iff σ is LNS-permitted on G and xy is LNS-permitted on G^σ. A calling sequence σ is LNS-stuck on G if σ is LNS-permitted on G, G^σ is not complete, and no call is LNS-permitted on G^σ.

Consider the spider-in-the-web example again. Trying out all the possible calling sequences reveal that they all get stuck, because of the fact that in no call xy the caller learns a useful new number. That is, all calls xy are such that if x learns the number of z, then x also learns the secret of z. In the example picture, the LNS permitted sequences are all the permutations of $ad;bd;cd$, and they all get stuck. So it makes sense to ask ourselves which graphs can be completed by some particular protocol.

It is straightforward to define and implement search algorithms for LNS-permitted calling sequences and LNS-stuck calling sequences [EG15].

The LNS protocol is successful on G if either G is complete, or there is an LNS-permitted call xy, and after any LNS-permitted call xy the LNS protocol is successful on G^{xy}. It follows that LNS is successful on G iff every sequence of LNS-permitted calls σ results in a graph G^σ that is complete, or is such that there is an LNS-permitted call, and after any LNS-permitted call xy, LNS is successful on $G^{\sigma;xy}$.

It follows from this definition that the LNS protocol is not successful on G iff there is a calling sequence σ that is LNS-stuck on G. This gives a straightforward algorithm for recognizing the gossip graphs where LNS is successful:

LNS gossip graph algorithm

Search for an LNS-stuck calling sequence in depth-first fashion, and declare success if no such calling sequence can be found [EG15].

A calling sequence σ for G is LNS-maximal if σ is LNS-permitted for G, and no calls are LNS-permitted in G^σ. A calling sequence σ for $G = (A, N, S)$ is LNS-maximal within $B \subseteq A$ if all calls in σ are within B, σ is LNS-permitted for G, and no calls within B are LNS-permitted in G^σ.

PROPOSITION 3.1. *If σ is an LNS-maximal calling sequence for G, and G has accessible secrets, then $S^\sigma = N^\sigma$.*

Proof. Let G be a gossip graph with accessible secrets, and let there be x, y with $N^\sigma xy$ and not $S^\sigma xy$. Then the call xy is LNS-permitted in G^σ, and contradiction with the LNS-maximality of σ. This shows $N^\sigma \subseteq S^\sigma$. The property $S^\sigma \subseteq N^\sigma$ follows from the fact that G has accessible secrets, and Proposition 2.1. Together, this gives $S^\sigma = N^\sigma$. ⊣

PROPOSITION 3.2. *If σ is an LNS-maximal calling sequence for G, and G satisfies $I_A = S \subseteq N$, then $S^\sigma \circ N^* = S^\sigma$.*

Proof. $S^\sigma \subseteq S^\sigma \circ N^*$ by definition of N^*.
$S^\sigma \circ N^* \subseteq S^\sigma$: let $(x,y) \in S^\sigma \circ N^*$. Then for some $k \in \mathbb{N}$, $(x,y) \in S^\sigma \circ N^k$. We get from Theorem 2.4 plus Proposition 3.1 that $S^\sigma \circ N \subseteq S^\sigma$. Applying this fact k times yields $(x,y) \in S^\sigma$. ⊣

COROLLARY 3.3. *If B is a strongly connected component of $G = (A, N, S)$ then any LNS-maximal calling sequence σ within B makes all elements of B become experts for B.*

Proof. If B is strongly connected, then $B \subseteq N^*$. If σ is LNS-maximal, then this and Proposition 3.2 implies that $S^\sigma \circ B \subseteq S^\sigma$, which means that each member of B has learnt the secret of all members of B. ⊣

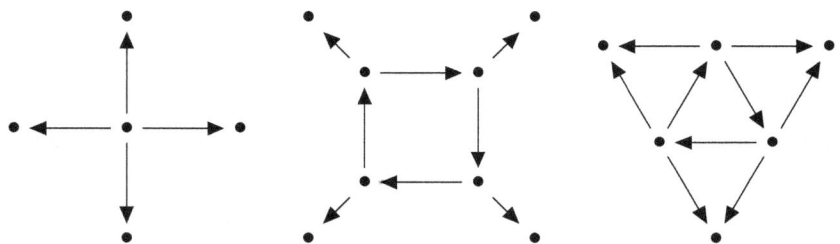

Call a graph $G = (A, N, S)$ a *sun* if $S = I_A \subseteq N$, N is weakly connected on G, and N is strongly connected on $s(G)$. The picture above gives three examples. We will show that G is a sun if and only if LNS is successful on G.

THEOREM 3.4. *The LNS protocol is successful for any sun G.*

Proof. Let $G = (A, N, S)$ be a sun. Let σ be any LNS-maximal calling sequence for G. Let $x, y \in A$. We have to show that $S^\sigma xy$.

If x is in the strongly connected core of G, then $N^* xy$. Because Sxx, also $S^\sigma xx$, and therefore $(x,y) \in S^\sigma \circ N^*$. By Proposition 3.2 it follows that $S^\sigma xy$.

If x is a terminal node, then by maximality of σ, there is some u with $(u,x) \in \sigma$. This means $N^\sigma uz$ for some z with Nzx, for u must have learnt x's number from some such z. Thus, after the call ux, x has the number of some z with Nzx, that is, $N^\sigma xz$. By LNS-maximality of σ it follows that $S^\sigma xz$. Since z is in the strongly connected core of G, it follows that $(x,y) \in S^\sigma \circ N^*$. By Theorem 3.2, $S^\sigma xy$, and we are done. ⊣

Let $\sim$ be the relation on $G = (A, N, S)$ given by $x \sim y$ iff there is an N-path from x to y and there is an N-path from y to x. Then $\sim$ is an equivalence relation, and a cell in the partition induced by $\sim$ is called a strongly connected component of G. Use $[x]_\sim$ for the strongly connected component of G that contains x. A strongly connected component B is initial in G if for all $x \in A$, $b \in B$: if Nxb then $x \in B$. Notice that a gossip graph G is strongly connected iff $\sim$ is universal on G.

THEOREM 3.5. *Let G be a connected graph with $I_A = S \subseteq N$. If G is not a sun graph then the LNS protocol is not successful on G.*

Proof. Let $G = (A, N, S)$ be gossip graph that is weakly connected but not a sun graph. Let H be an initial strongly connected component of G, and let B be its carrier set.

There are $x, y, z \in A$ with $x \in B$, $y \notin B$, $z \notin B$, Nxy, and either Nyz or Nzy. For if not, then G is a sun. Notice that $B = [x]_\sim$.

Since B is an initial strongly connected component, for all $u \in A - B$ and $v \in B$ we have $\neg Nuv$. In particular, $\neg Nzx$.

Let σ' be an LNS-maximal calling sequence for $A - B$ in G. Then not $N^{\sigma'} zx$, for otherwise $\exists v \in A - B \ \exists w \in B$ with Nvw, and contradiction with the initiality of B. Let σ'' be an LNS-maximal calling sequence for B in $G^{\sigma'}$. Then not $N^{\sigma';\sigma''} zx$, for calls in σ'' do not involve z. Let By be a calling sequence where each member of B calls y. Then not $N^{\sigma';\sigma'';By} zx$. Let σ''' be an LNS-maximal calling sequence for $A - \{z\}$ in $G^{\sigma';\sigma'';By}$. Then not $N^{\sigma';\sigma'';By;\sigma'''} zx$.

Observe that $\sigma'; \sigma''; By; \sigma'''$ is an LNS maximal sequence of calls in G. For suppose z could still make a call in $G^{\sigma';\sigma'';By;\sigma'''}$. Then the call cannot be within $A - B$, for otherwise contradiction with the fact that σ' is maximal for $A - B$ in G. The call cannot be to an agent in B, for all calls made from or to z are in σ', and z cannot have learnt a number in B from these. Thus, $\sigma'; \sigma''; By; \sigma'''$ is LNS maximal. But $G^{\sigma';\sigma'';By;\sigma'''}$ is not complete, since not $N^{\sigma';\sigma'';By;\sigma'''} zx$. ⊣

THEOREM 3.6. *For any connected graph $G = (A, N, S)$ with $I_A = S \subseteq N$ the following holds: $s(G)$ is strongly connected iff the LNS protocol is successful for G.*

Proof. Immediate from Theorems 3.4 and 3.5. ⊣

4 Further Questions

There are two ways in which the LNS protocol can be unsuccessful. On some graphs G you can find successful LNS-permitted maximal call sequences if you are lucky, but you might get stuck if you change the order of the calls. For example, in the graph $a \longrightarrow b \longrightarrow c$ the calling sequences $ab; ac; bc$ and $ab; bc; ca$ are successful, but the calling sequence $bc; ab$ gets stuck. On the other hand, in the graph $a \longrightarrow b \longleftarrow c$, any calling sequence gets stuck: we have $ab; cb$ and stuck, and $cb; ab$ and stuck, and there are no success sequences. It would be of interest to characterize the graphs where the LNS protocol always gets stuck. This question is addressed in [DvEPetal16].

Also, it would be nice to characterize expected length of the calling sequences in this dynamic setting, for given protocols. For each protocol it makes sense to ask if it still holds that $2n - 4$ calls are enough. The new dynamic setting allows us to get new answers to old questions. What is the minimum, average, maximum number of calling sequences in a gossip graph with n nodes, give property X of the edge connection, and given that distributed protocol P was used?

Acknowledgement

We thank an anonymous referee for remarks and questions that have helped us improve the paper. We also acknowledge support from ERC project EPS 313360 and from CWI, Amsterdam.

BIBLIOGRAPHY

[AGvdH15] Krzyzstof R. Apt, Davide Grossi, and Wiebe van der Hoek. Epistemic protocols for distributed gossiping. In *Proceedings of TARK 2015*, 2015.

[AvDGvdH14] M. Attamah, H. van Ditmarsch, D. Grossi, and W. van der Hoek. Knowledge and gossip. In *Proc. of 21st ECAI*, pages 21–26. IOS Press, 2014.

[DvEPetal16] Hans Ditmarsch, Jan van Eijck, Pere Pardo, Rahim Ramezanian, and François Schwarzentruber. Dynamic gossip. Technical report, arxiv, 2016. http://arxiv.org/abs/1511.00867.

[EG15] Jan van Eijck and Malvin Gattinger. Gossip. Technical report, CWI, Amsterdam, available from www.cwi.nl/~jve/papers/15/pdfs/Gossip.pdf, 2015.

[EGKM04] Patrick Th. Eugster, Rachid Guerraoui, Anne-Marie Kermarrec, and Laurent Massoulié. Epidemic information dissemination in distributed systems. *IEEE Computer*, 37(5):60–67, 2004.

[HHL88] Sandra Mitchell Hedetniemi, Stephen T. Hedetniemi, and Arthur L. Liestman. A survey of gossiping and broadcasting in communication networks. *Networks*, 18(4):319–349, 1988.

[Hur00] C. A. J. Hurkens. Spreading gossip efficiently. *NAW*, 5(1):208–210, 2000.

[Tij71] R. Tijdeman. On a telephone problem. *Nieuw Archief voor Wiskunde*, 3(19):188–192, 1971.

Variations on a Visserian Theme

ALI ENAYAT

Abstract

A first order theory T is *tight* iff for any two deductively closed extensions U and V of T (both of which are formulated in the language of T), U and V are bi-interpretable iff $U = V$. By a theorem of Visser, PA (Peano Arithmetic) is tight. Here we show that Z_2 (second order arithmetic), ZF (Zermelo-Fraenkel set theory), and KM (Kelley-Morse theory of classes) are also tight theories.

1 Introduction

The source of inspiration for this paper is located in a key result of Albert Visser [V, Corollaries 9.4 & 9.6] concerning a curious interpretability-theoretic feature of PA (Peano arithmetic), namely:

1.1. THEOREM. (Visser) *Suppose U and V are deductively closed extensions of* PA (*both of which are formulated in the language of* PA). *Then U is a retract of V if $V \subseteq U$.* **In particular, U and V are bi-interpretable iff** $U = V$.

A natural reaction to Theorem 1.1 is to ask whether the exhibited interpretability-theoretic feature of PA is shared by other theories. As shown here, the answer to this question is positive[1], in particular:

1.2. THEOREM[2]. *Theorem 1.1 remains valid if* PA *is replaced throughout by* Z_2 (*second order arithmetic*); *or by* ZF (*Zermelo-Fraenkel set theory*); *or by* KM (*Kelley-Morse theory of classes*).

In the remainder of this section we review some basic notions and results of interpretability theory in order to clarify and contextualize Theorems 1.1 & 1.2.

1.3. DEFINITIONS. Suppose U and V are first order theories, and for the sake of notational simplicity, let us assume that U and V are theories that *support a definable pairing function* and *are formulated in relational languages*. We use $\mathcal{L}_U$ and $\mathcal{L}_V$ to respectively designate the languages of U and V.

(a) An interpretation $\mathcal{I}$ of U in V, written:

$$\mathcal{I} : U \to V$$

is given by a translation τ of each $\mathcal{L}_U$-formula φ into an $\mathcal{L}_V$-formula φ^τ with the requirement that $V \vdash \varphi^\tau$ for each $\varphi \in U$, where τ is determined by an $\mathcal{L}_V$-formula $\delta(x)$ (referred to as a *domain*

---

[1]The ZF-case of Theorem 1.2 was established independently in unpublished work of Albert Visser and Harvey Friedman. I am thankful to Albert for bringing this to my attention.

[2]See Remark 2.9 for a more complete version of this theorem.

formula), and a mapping $P \mapsto_\tau A_P$ that translates each n-ary $\mathcal{L}_U$-predicate P into some n-ary $\mathcal{L}_V$-formula A_P. The translation is then lifted to the full first order language in the obvious way by making it commute with propositional connectives, and subject to:

$$(\forall x \varphi)^\tau = \forall x(\delta(x) \to \varphi^\tau) \text{ and } (\exists x \varphi)^\tau = \exists x(\delta(x) \wedge \varphi^\tau).$$

Note that each interpretation $\mathcal{I} : U \to V$ gives rise to an *inner model construction that **uniformly** builds a model* $\mathcal{M}^\mathcal{I} \models U$ *for any* $\mathcal{M} \models V$.

(b) U is *interpretable* in V, written $U \trianglelefteq V$, iff there is an interpretation $\mathcal{I} : U \to V$. U and V are *mutually interpretable* when $U \trianglelefteq V$ and $V \trianglelefteq U$.

(c) We indicate the universe of each structure with the corresponding Roman letter, e.g., the universes of structures $\mathcal{M}, \mathcal{N}$, and $\mathcal{M}^*$ are respectively M, N, and M^*. Given an $\mathcal{L}$-structure $\mathcal{M}$ and $X \subseteq M^n$ (where n is a positive integer), we say that X is $\mathcal{M}$-*definable* iff X is parametrically definable in $\mathcal{M}$, i.e., iff there is an n-ary formula $\varphi(x_1, \ldots, x_n)$ in the language $\mathcal{L}_M$ obtained by augmenting $\mathcal{L}$ with constant symbols $\overline{m}$ for each $m \in M$ such that $X = \varphi^\mathcal{M}$, where $\varphi^\mathcal{M} = \{(a_1, \cdots, a_n) \in M^n : (\mathcal{M}, m)_{m \in M} \models \varphi(\overline{a}_1, \cdots, \overline{a}_n)\}$.

(d) Suppose $\mathcal{N}$ is an $\mathcal{L}_U$-structure and $\mathcal{M}$ is an $\mathcal{L}_V$-structure. We say that $\mathcal{N}$ is *parametrically interpretable* in $\mathcal{M}$, written $\mathcal{N} \trianglelefteq_\mathrm{par} \mathcal{M}$ (equivalently: $\mathcal{M} \trianglerighteq_\mathrm{par} \mathcal{N}$) iff the universe of discourse of $\mathcal{N}$, as well as all the $\mathcal{N}$-interpretations of $\mathcal{L}_U$-predicates are $\mathcal{M}$-definable. Note that $\trianglelefteq_\mathrm{par}$ is a transitive relation.

(e) U is a *retract* of V iff there are interpretations $\mathcal{I}$ and $\mathcal{J}$ with $\mathcal{I} : U \to V$ and $\mathcal{J} : V \to U$, and a binary U-formula F such that F is, U-verifiably, an isomorphism between id_U (the identity interpretation on U) and $\mathcal{J} \circ \mathcal{I}$. In model-theoretic terms, this translates to the requirement that the following holds for every $\mathcal{M} \models U$:

$$F^\mathcal{M} : \mathcal{M} \xrightarrow{\cong} \mathcal{M}^* := \left(\mathcal{M}^\mathcal{J}\right)^\mathcal{I}.$$

(f) U and V are *bi-interpretable*[3] iff there are interpretations $\mathcal{I}$ and $\mathcal{J}$ as above that witness that U is a retract of V, and additionally, there is a V-formula G, such that G is, V-verifiably, an isomorphism between id_V and $\mathcal{I} \circ \mathcal{J}$. In particular, if U and V are bi-interpretable, then given $\mathcal{M} \models U$ and $\mathcal{N} \models V$, we have

$$F^\mathcal{M} : \mathcal{M} \xrightarrow{\cong} \mathcal{M}^* := \left(\mathcal{M}^\mathcal{J}\right)^\mathcal{I} \text{ and } G^\mathcal{N} : \mathcal{N} \xrightarrow{\cong} \mathcal{N}^* := \left(\mathcal{N}^\mathcal{I}\right)^\mathcal{J}.$$

We conclude this section with salient examples. In what follows ACA_0 (arithmetical comprehension with limited induction) and GB (Gödel-Bernays theory of classes) are the well-known subsystems of Z_2 and KM (respectively) satisfying: ACA_0 is a conservative extension of PA, and GB is a conservative extension of ZF.

---

[3]The notion of bi-interpretability has been informally around for a long time, but according to Hodges [H] it was first studied in a general setting by Ahlbrandt and Ziegler [AZ]. A closely related concept (dubbed sometimes as *synonymy*, and other times as *definitional equivalence*) was introduced by de Bouvère [D]. Synonymy is a stronger form of bi-interpretation; however, by a result of Friedman and Visser [FV], in many cases synonymy is implied by bi-interpretability, namely, when the two theories involved are sequential, and the bi-interpretability between them is witnessed by a pair of one-dimensional, identity preserving interpretations.

1.4. THEOREM. (Folklore) PA $\trianglelefteq$ ACA$_0$ and ZF $\trianglelefteq$ GB; but ACA$_0$ $\ntrianglelefteq$ PA and GB $\ntrianglelefteq$ ZF.

Proof Outline. The first two statements have routine proofs; the last two follow by combining (a) the finite axiomatizability of ACA$_0$ and GB, (b) the reflexivity of PA and ZF (i.e., they prove the consistency of each finite fragment of themselves), and (c) Gödel's second incompleteness theorem. ⊣

By classical results of Ackermann and Mycielski, the structures $(V_\omega, \in)$ and $(\mathbb{N}, +, \cdot)$ are bi-interpretable, where V_ω is the set of hereditarily finite sets. The two interpretations at work can be used to show Theorem 1.5 below. In what follows, ZF$_{\text{fin}}$ is the theory obtained by replacing the axiom of infinity by its negation in the usual axiomatization of ZF and TC is the sentence asserting "every set has a transitive closure".[4]

1.5. THEOREM. (Ackermann [A], Mycielski [My], Kaye-Wong [KW]) PA *and* ZF$_{\text{fin}}$ + TC *are bi-interpretable.*

1.6. THEOREM. (E-Schmerl-Visser [ESV, Theorem 5.1]) ZF$_{\text{fin}}$ *and* PA *are not bi-interpretable; indeed* ZF$_{\text{fin}}$ *is not even a sentential retract*[5] *of* ZF$_{\text{fin}}$ + TC.

2 Solid Theories

The notions of solidity, neatness, and tightness encapsulated in Definition 2.1 below are only implicitly introduced in Visser's paper [V]. It is not hard to see that a solid theory is neat, and a neat theory is tight. Hence to establish Theorems 1.1 and 1.2 it suffices to verify the solidity of PA, Z_2, ZF, and KM. This is precisely what we will accomplish in this section. The proof of Theorem 1.1 is presented partly as an exposition of Visser's original proof which is rather indirect since it is couched in terms of series of technical general lemmata, and partly because it provides a warm-up for the proof of Theorem 2.5 which establishes the solidity of Z_2. The proof of Theorem 2.6 establishing the solidity of ZF, on the other hand, requires a brand new line of argument. The proof of Theorem 2.7, which establishes the solidity of KM is the most complex among the proofs presented here; it can be roughly described as using a blend of ideas from the proofs of Theorems 2.5 and 2.6.

2.1. DEFINITION. Suppose T is a first order theory.

(a) T is *solid* iff the following property $(*)$ holds for all models $\mathcal{M}$, $\mathcal{M}^*$, and $\mathcal{N}$ of T:

$(*)$ If $\mathcal{M} \trianglerighteq_{\text{par}} \mathcal{N} \trianglerighteq_{\text{par}} \mathcal{M}^*$ and there is an $\mathcal{M}$-definable isomorphism $i_0 : \mathcal{M} \to \mathcal{M}^*$, then there is an $\mathcal{M}$-definable isomorphism $i : \mathcal{M} \to \mathcal{N}$.

(b) T is *neat* iff for any two deductively closed extensions U and V of T (both of which are formulated in the language of T), U is a retract of V if $V \subseteq U$.

(c) T is *tight* iff for any two deductively closed extensions U and V of T (both of which are formulated in the language of T), U and V are bi-interpretable iff $U = V$.

---

[4]More explicitly; the axioms of ZF$_{\text{fin}}$ consists of the axioms of Extensionality, Empty Set, Pairs, Union, Power set, Foundation, and ¬Infinity, plus the scheme of Replacement. Note that ZF$_{\text{fin}}$ has also been used in the literature (e.g., by the Prague school) to denote the stronger theory in which the Foundation axiom is strengthened to the Foundation *scheme*; the latter theory is deductively identical to ZF$_{\text{fin}}$ + TC in our notation.

[5]The notion of a sentential retract is the natural weakening of the notion of a retract in which the requirement of the existence of a definable isomorphism between $\mathcal{M}$ and $\mathcal{M}^*$ is weakened to the requirement that $\mathcal{M}$ and $\mathcal{M}^*$ be elementarily equivalent.

2.1.1. REMARK. A routine argument shows that if T and T' are bi-interpretable, and T is solid, then T' is also solid.

2.2. THEOREM. (Visser [V]) PA *is solid*.

Proof. Suppose $\mathcal{M}$, $\mathcal{M}^*$, and $\mathcal{N}$ are models of PA such that:

$$\mathcal{M} \trianglerighteq_{\text{par}} \mathcal{N} \trianglerighteq_{\text{par}} \mathcal{M}^*, \text{ and}$$

there is an $\mathcal{M}$-definable isomorphism $i_0 : \mathcal{M} \to \mathcal{M}^*$. A key property[6] of PA is that if $\mathcal{M}$ is a model of PA and $\mathcal{N}$ is a model of the fragment of PA known as (Robinson's) Q, then as soon as $\mathcal{N} \trianglelefteq_{\text{par}} \mathcal{M}$ there is an $\mathcal{M}$-definable initial embedding $j : \mathcal{M} \to \mathcal{N}$, i.e., an embedding j such that the image $j(\mathcal{M})$ of $\mathcal{M}$ is (1) a submodel of $\mathcal{N}$, and (2) an initial segment of $\mathcal{N}$. Hence there is an $\mathcal{M}$-definable initial embedding $j_0 : \mathcal{M} \to \mathcal{N}$ and an $\mathcal{N}$-definable initial embedding $j_1 : \mathcal{N} \to \mathcal{M}^*$.

We claim that both j_0 and j_1 are surjective. To see this, suppose not. Then $j(\mathcal{M})$ is a proper initial segment of $\mathcal{M}^*$, where j is the $\mathcal{M}$-definable embedding $j : \mathcal{M} \to \mathcal{M}^*$ given by $j := j_1 \circ j_0$. But then $i_0^{-1}(j(\mathcal{M}))$ is a proper $\mathcal{M}$-definable initial segment of $\mathcal{M}$ with no last element. This is a contradiction since $\mathcal{M}$ is a model of PA, and therefore no proper initial segment of $\mathcal{M}$ is $\mathcal{M}$-definable. Hence j_0 and j_1 are both surjective; in particular j_0 serves as the desired $\mathcal{M}$-definable isomorphism between $\mathcal{M}$ and $\mathcal{N}$. ⊣

2.2.1. COROLLARY. $\mathsf{ZF}_{\text{fin}} + \mathsf{TC}$ *is solid*.

Proof. In light of Remark 2.1.1, this is an immediate consequence of coupling Theorem 2.2 and Theorem 1.5. ⊣

Before presenting the proof of solidity of Z_2 we need to state two propositions concerning theories that prove the full scheme of induction over some specified choice of 'numbers'. The proofs of Propositions is a straightforward adaptation of the well-known proof for the special case of PA, so it is only presented in outline form.

2.3. PROPOSITION. *Let T be a theory formulated in a language $\mathcal{L}$ such that T interprets* Q *via an interpretation whose domain formula for 'numbers' is* $\mathbb{N}(x)$. *Furthermore assume the following two hypotheses*:

(a) $T \vdash \mathsf{Ind}_\mathbb{N}(\mathcal{L})$, *where* $\mathsf{Ind}_\mathbb{N}(\mathcal{L})$ *is the scheme of induction over* $\mathbb{N}$ *whose instances are universal closures of $\mathcal{L}$-formulae of the form below*:

$$(\theta(0) \wedge \forall x \, (\mathbb{N}(x) \wedge \theta(x) \to \theta(x^+))) \to \forall x \, (\mathbb{N}(x) \to \theta(x)),$$

where x^+ is shorthand for the successor of x, and θ is allowed to have suppressed parameters; these parameters are not required to lie in $\mathbb{N}$.

(b) $\mathcal{K} \models T$ *and* $\mathcal{K} \trianglerighteq_{\text{par}} \mathcal{N} \models$ Q.

Then *there is a $\mathcal{K}$-definable initial embedding* $j : \mathbb{N}^\mathcal{K} \to \mathcal{N}$.

Proof outline. Since $\mathbb{N}^\mathcal{K} \models \mathsf{Q} + \mathsf{Ind}_\mathbb{N}(\mathcal{L})$, the following definition by recursion produces the desired j.

---

[6]This important property seems to have been first noted by Feferman [F], who used it in his proof of Π_1-conservativity of $\neg\mathrm{Con}(\mathrm{PA})$ over PA.

$$j(0^{\mathbb{N}^{\mathcal{K}}}) = 0^{\mathcal{N}} \text{ and } j\left((x^+)^{\mathbb{N}^{\mathcal{K}}}\right) = (j(x)^+)^{\mathcal{N}}.$$

⊣

2.4. PROPOSITION. *Suppose $\mathcal{K}$ is an $\mathcal{L}$-structure that interprets a model of Q via an interpretation whose domain formula for 'numbers' is $\mathbb{N}(x)$ and $\mathsf{Ind}_{\mathbb{N}}(\mathcal{L})$ holds in $\mathcal{K}$. Then every $\mathcal{K}$-definable proper initial segment of $\mathbb{N}^{\mathcal{K}}$ has a last element.*

Proof. Easy: the veracity of $\mathsf{Ind}_{\mathbb{N}}(\mathcal{L})$ in $\mathcal{K}$ immediately implies that any $\mathcal{K}$-definable initial segment of $\mathbb{N}^{\mathcal{K}}$ with no last element coincides with $\mathbb{N}^{\mathcal{K}}$. ⊣

2.5. THEOREM. *Z_2 is solid.*

Proof. Following standard practice (as in [S]) models of Z_2 are represented as two-sorted structures of the form $(\mathcal{M}, \mathcal{A})$, where $\mathcal{M} \models \mathsf{PA}$, $\mathcal{A}$ is a collection of subsets of M, and $(\mathcal{M}, \mathcal{A})$ satisfies the full comprehension scheme. Suppose $(\mathcal{M}, \mathcal{A})$, $(\mathcal{M}^*, \mathcal{A}^*)$, and $(\mathcal{N}, \mathcal{B})$ are models of Z_2 such that:

$$(\mathcal{M}, \mathcal{A}) \trianglerighteq_{\mathrm{par}} (\mathcal{N}, \mathcal{B}) \trianglerighteq_{\mathrm{par}} (\mathcal{M}^*, \mathcal{A}^*),$$

and there is an $(\mathcal{M}, \mathcal{A})$-definable isomorphism

$$\widehat{i_0} : (\mathcal{M}, \mathcal{A}) \to (\mathcal{M}^*, \mathcal{A}^*).$$

Note that $\widehat{i_0}$ is naturally induced by i_0, where:

$$i_0 := \widehat{i_0} \upharpoonright M : \mathcal{M} \to \mathcal{M}^*,$$

since $\widehat{i_0}(A) = \{i_0(m) : m \in A\}$ for $A \in \mathcal{A}^*$.

It is clear that $\mathsf{Z}_2 \vdash \mathsf{Q}^{\mathbb{N}} + \mathsf{Ind}_{\mathbb{N}}(\mathcal{L})$ for $\mathcal{L} := \mathcal{L}_{\mathsf{Z}_2}$, so by Proposition 2.3, we may conclude:

(1) There is an $(\mathcal{M}, \mathcal{A})$-definable *initial embedding* $j_0 : \mathcal{M} \to \mathcal{N}$, and

(2) There is an $(\mathcal{N}, \mathcal{B})$-definable *initial embedding* $j_1 : \mathcal{N} \to \mathcal{M}^*$.

Similar to the proof of Theorem 2.2 we now argue that both j_0 and j_1 are surjective since otherwise the $(\mathcal{M}, \mathcal{A})$-definable embedding $j : \mathcal{M} \to \mathcal{M}^*$ given by $j := j_1 \circ j_0$ will have the property that $j(M)$ is a proper initial segment of $\mathcal{M}^*$, which in turn implies that $i_0^{-1}(j(M))$ is an $(\mathcal{M}, \mathcal{A})$-definable proper initial segment of $\mathcal{M}$ with no last element, which contradicts Proposition 2.4. Hence (1) and (2) can be strengthened to:

(1^+) There is an $(\mathcal{M}, \mathcal{A})$-definable *isomorphism* $k_0 : \mathcal{N} \to \mathcal{M}$, and

(2^+) There is an $(\mathcal{N}, \mathcal{B})$-definable *isomorphism* $k_1 : \mathcal{M}^* \to \mathcal{N}$.

Let $\widehat{k_0} : (\mathcal{N}, \mathcal{B}) \to (\mathcal{M}, \mathcal{A})$ be the natural extension of k_0, i.e., $\widehat{k_0}(n) := k_0(n)$ for $n \in N$, and $\widehat{k_0}(B) = \{k_0(n) : n \in B\}$ for $B \in \mathcal{B}$. Note that the $(\mathcal{M}, \mathcal{A})$-definability of k_0, along with the veracity of the comprehension scheme in $(\mathcal{M}, \mathcal{A})$ assures us that $\widehat{k_0}(B) \in \mathcal{A}$ for each $B \in \mathcal{B}$. Therefore $\widehat{k_0}$ is an *embedding*. Using an identical reasoning, since $(\mathcal{M}^*, \mathcal{A}^*)$ is $(\mathcal{N}, \mathcal{B})$-definable by assumption, we can extend k_1 to an embedding $\widehat{k_1} : (\mathcal{M}^*, \mathcal{A}^*) \to (\mathcal{N}, \mathcal{B})$. Let $\widehat{k} := \widehat{k_0} \circ \widehat{k_1} \circ \widehat{i_0}$. Then:

(3) $\widehat{k} : (\mathcal{M}, \mathcal{A}) \to (\mathcal{M}, \mathcal{A})$ and $\widehat{k}$ is an $(\mathcal{M}, \mathcal{A})$-definable embedding.

The proof of Theorem 2.5 will be complete once we verify that $\widehat{k_0}$ is surjective. Since we already know that k_0 is surjective, it suffices to check that $\mathcal{A} = \widehat{k_0}(\mathcal{B}) := \{\widehat{k_0}(B) : B \in \mathcal{B}\}$. Observe that the restriction $k : \mathcal{M} \to \mathcal{M}$ of $\widehat{i}$ to 'numbers' is an automorphism of $\mathcal{M}$, thanks to (1^+), (2^+), and the assumption that $\widehat{i_0}$ is an isomorphism. But since $\widehat{k}$ is $(\mathcal{M}, \mathcal{A})$-definable $i(m) = m$ for all $m \in M$, thanks to the veracity of $\mathsf{Ind}_\mathbb{N}(\mathcal{L})$ in $(\mathcal{M}, \mathcal{A})$, for $\mathcal{L} = \mathcal{L}_{\mathsf{Z}_2}$, which in turn implies that $\widehat{k}$ is just the identity automorphism on $(\mathcal{M}, \mathcal{A})$. Hence $\widehat{k_0}$ and $\widehat{k_1}$ are both surjective. ⊣

In the following corollary, $\widetilde{\mathsf{ZF}}$ is the result of substituting the Replacement scheme in the usual axiomatization of ZF (e.g., as in [K]) with the scheme of Collection, whose instances consist of universal generalizations of formulae of the form $(\forall x \in a \; \exists y \; \varphi(x, y)) \to (\exists b \; \forall x \in a \; \exists y \in b \; \varphi(x, y))$, where the parameters of φ are suppressed.

2.5.1. COROLLARY. *The following theory T is solid*:

$$T := \widetilde{\mathsf{ZF}} \backslash \{\text{Power Set}\} + \forall x \; |x| \leq \aleph_0.$$

Proof. In light of Remark 2.1.1, this is an immediate consequence of Theorem 2.5 and the well-known bi-interpretability of T with $\mathsf{Z}_2 + \Pi^1_\infty$-AC, where Π^1_∞-AC is the scheme of choice.[7] ⊣

2.6. THEOREM. ZF *is solid*.

Proof. Suppose $\mathcal{M}$, $\mathcal{M}^*$, and $\mathcal{N}$ are models of ZF such that:

$$\mathcal{M} \trianglerighteq_{\text{par}} \mathcal{N} \trianglerighteq_{\text{par}} \mathcal{M}^*,$$

and there is an $\mathcal{M}$-definable isomorphism $i_0 : \mathcal{M} \to \mathcal{M}^*$. Since $\mathcal{M}$ injects M into M^* via i_0, and $M^* \subseteq N$, we have:

(1) N is a proper class as viewed from $\mathcal{M}$.

Let $E := \in^{\mathcal{M}^*}$. E is both extensional and well-founded as viewed from $\mathcal{N}$; extensionality trivially follows from the assumption that $\mathcal{M}^* \models \mathsf{ZF}$, and well-foundedness can be easily verified using the assumptions that $\mathcal{M} \trianglerighteq_{\text{par}} \mathcal{N}$ and i_0 is an $\mathcal{M}$-definable isomorphism between $\mathcal{M}$ and $\mathcal{M}^*$. We wish to show that $E := \in^{\mathcal{M}^*}$ is *set-like*[8] as viewed from $\mathcal{N}$, i.e., for every $c \in M^*$, c_E is a set (as opposed to a proper class) of $\mathcal{N}$, where $c_E := \{x \in M^* : xEc\}$. This will take some effort to establish. We will present the argument in full detail, especially because a natural adaptation of the same argument will also work in one of the stages of the proof of Theorem 2.7 (establishing the solidity of KM), and will therefore be left to the reader. We will first show that E is set-like when restricted to $\mathbf{Ord}^{\mathcal{M}^*}$.[9] To this end, let $\delta \in \mathbf{Ord}^{\mathcal{M}^*}$, $\delta_E := \{m \in M^* : mE\delta\}$, and consider the $\mathcal{N}$-definable ordered structure Δ :

---

[7]This bi-interpretability was first explicitly noted by Mostowski in the context of the so-called β-models of $\mathsf{Z}_2 + \Pi^1_\infty$-AC (which correspond to well-founded models of T). See [S, Theorem VII.3.34] for a refined version of this bi-interpretability result.

[8]In the context of ZF, the extension of a binary formula $R(x, y)$ is set-like iff for every set s there is a set t such that $t = \{x : R(x, s)\}$.

[9]Note that if the axiom of choice holds in $\mathcal{M}^*$, then by Zermelo's well-ordering theorem, from the point of view of $\mathcal{N}$ the set-likeness of E when restricted to $\mathbf{Ord}^{\mathcal{M}^*}$ immediately implies the set-likeness of E.

$$\Delta := \left(\delta_E, E \cap \delta_E^2\right).$$

It is clear, thanks to i_0, that $\mathcal{N}$ views Δ as a *well-founded* linear order in the strong sense that every nonempty $\mathcal{N}$-definable subclass of δ_E has an E-least member. In particular, Δ is a linear order in which every element other than the last element (if it exists) has an immediate successor. Given $\gamma \in \mathbf{Ord}^{\mathcal{N}}$ let $o(\Delta) \geq \gamma$ be an abbreviation for the statement:

"there is some set f such that f is the (graph of) an order preserving function between $(\gamma, \in)$ and an initial segment of Δ",

and let $o(\Delta) \geq \mathbf{Ord}$ abbreviate "$\forall \gamma \in \mathbf{Ord}\ o(\Delta) \geq \gamma$". We wish to show that the statement $o(\Delta) \geq \mathbf{Ord}$ does not hold in $\mathcal{N}$. Suppose it does. Then arguing in $\mathcal{N}$, for each $\gamma \in \mathbf{Ord}$ there is an order-preserving map f_γ which embeds $(\gamma, \in)$ onto an initial segment of Δ. Moreover, such an f_γ is unique since it is a theorem of ZF that no ordinal has a nontrivial automorphism. Hence if $\gamma \in \gamma'$, then $f_\gamma \subseteq f_{\gamma'}$ and therefore $f := \cup \{f_\gamma : \gamma \in \mathbf{Ord}\}$ serves as an order-preserving $\mathcal{N}$-definable injection of $\mathbf{Ord}^{\mathcal{N}}$ onto an initial segment of Δ. Invoking the assumption $\mathcal{M} \trianglerighteq_{\text{par}} \mathcal{N}$ this shows that $\mathcal{M}$ must view $\mathcal{N}$ as well-founded because the map $\rho^{\mathcal{N}} : (N, \in^{\mathcal{N}}) \to (\mathbf{Ord}, \in)^{\mathcal{N}}$ (where ρ is the usual rank function) is $\in^{\mathcal{N}}$-preserving and $\mathcal{N}$-definable, and therefore $\mathcal{M}$-definable since $\mathcal{M} \trianglerighteq_{\text{par}} \mathcal{N}$. This allows us to conclude that:

(2) $\mathcal{M}$ views $(N, \in^{\mathcal{N}})$ as a well-founded extensional structure of ordinal height at most $i_0^{-1}(\delta) \in \mathbf{Ord}^{\mathcal{M}}$.

At this point we wish to invoke an appropriate form of Mostwoski's collapse theorem in order to show that (2) implies that N is a *set* from the point of view of $\mathcal{M}$. To this end, consider KP (Kripke-Platek set theory) whose axioms consist of Extensionality, Empty Set, Pairs, Union, Π_1-Foundation, and Σ_0-Collection[10]. It is well-known that KP is finitely axiomatizable, and that, provably in KP, ρ (the rank function) is an $\in$-homomorphism of the universe onto the class $\mathbf{Ord}$ of ordinals. Let KPR (Kripke-Platek set theory with ranks) be the strengthening of KP with the axiom that states that $\{x : \rho(x) < \alpha\}$ is a set for each $\alpha \in \mathbf{Ord}$. Theorem 2.6.1 below can be either seen as a scheme of theorems of ZF, or a single theorem of Gödel-Bernays theory of classes.

2.6.1. THEOREM. *If* KPR *holds in* $\mathcal{N}$, *and* $\mathbf{Ord}^{\mathcal{N}} \cong \alpha \in \mathbf{Ord}$, *then* $\mathcal{N}$ *is isomorphic to a transitive substructure of* $(V_\alpha, \in)$.

Proof outline. Let $h : \alpha \to \mathbf{Ord}^{\mathcal{N}}$ witness the isomorphism of α and $\mathbf{Ord}^{\mathcal{N}}$, and for $\gamma < \alpha$ let $\mathcal{N}_\gamma := (V_{h(\gamma)}, \in)^{\mathcal{N}}$. A routine induction on $\gamma < \alpha$ shows that there is a unique embedding $j_\gamma : \mathcal{N}_\gamma \to (V_\gamma, \in)$ whose range is transitive. This implies that if $\delta < \gamma < \alpha$, then $j_\delta \subseteq j_\gamma$. It is then easy to verify that $j_\alpha : \mathcal{N} \to (V_\alpha, \in)$ is an embedding with a transitive range, where $j_\alpha := \cup \{j_\gamma : \gamma < \alpha\}$. ⊣

By coupling (2) with Theorem 2.6.1 we can conclude that N forms a set in $\mathcal{M}$, thus contradicting (1). This concludes our verification of the failure of $o(\Delta) \geq \mathbf{Ord}$ within $\mathcal{N}$.

---

[10] It is well-known that Σ_1-Collection is provable in KP, which enables KP to carry out Σ_1-recursions. Also note that the formulation of KP in many references (including Barwise's monograph [B]) that focus on admissible set theroy includes the full scheme of Foundation since admissible sets are transtive and automatically satisfy Π_∞-Foundation. Our forumlation of KP is taken from Mathias' paper [Ma].

The failure of $o(\Delta) \geq \mathbf{Ord}$ in $\mathcal{N}$ allows us to choose $\gamma_0 \in \mathbf{Ord}^{\mathcal{N}}$ such that $\mathcal{N}$ views γ_0 to be the first ordinal γ such that $o(\Delta) \geq \gamma$ is false. We claim that γ_0 is a successor ordinal of $\mathbf{Ord}^{\mathcal{N}}$. If not, then, arguing in $\mathcal{N}$, for each $\beta \in \gamma_0$ there is a unique order-preserving map f_β which maps $(\beta, \in)$ onto an initial segment of Δ, and $f_\beta \subseteq f_{\beta'}$ whenever $\beta \in \beta' \in \gamma_0$, then $f_\beta \subseteq f_{\beta'}$. Therefore $\cup \{f_\gamma : \gamma \in \gamma_0\}$ serves as an order-preserving map between $(\gamma_0, \in)$ and an initial segment of Δ, contradicting the choice of γ_0. Hence $\gamma_0 = \beta_0 + 1$ for some $\beta_0 \in \mathbf{Ord}^{\mathcal{N}}$. This makes it clear that:

(3) f_{β_0} is a bijection between β_0 and δ_E,

since if the range of f_{β_0} is not all of δ_E, then the range $\mathrm{ran}(f_{\beta_0})$ of f_{β_0} is a proper initial segment of Δ, and f_{β_0} could be extended to an order-preserving map f_{γ_0} with domain γ_0 by setting:

$$f_{\gamma_0}(\beta_0) = \min(\delta_E \backslash \mathrm{ran}(f_{\beta_0})).$$

Thanks to (3), we now know that, as viewed by $\mathcal{N}$, E is set-like when restricted to $\mathbf{Ord}^{\mathcal{M}^*}$. To verify the set-likeness of E in $\mathcal{N}$ it is sufficient to show that s_E forms a set in $\mathcal{N}$, where $s_E := \{m \in M^* : mEs\}$ and $s := V_\delta^{\mathcal{M}^*}$ for some $\delta \in \mathbf{Ord}^{\mathcal{M}^*}$ such that KPR holds in $V_\delta^{\mathcal{M}^*}$, since such ordinals δ are cofinal in $\mathbf{Ord}^{\mathcal{M}^*}$ by the Reflection Theorem of ZF. Consider the $\mathcal{N}$-definable structure Σ:

$$\Sigma := \left(s_E, E \cap s_E^2\right).$$

Since Σ is a model of KPR whose set of ordinals is isomorphic to β_0, by Theorem 2.6.1 (applied within $\mathcal{N}$) there is an $\mathcal{N}$-definable embedding of Σ onto a (transitive) subset of $V_{\beta_0}^{\mathcal{N}}$. This makes it evident that s_E forms a set in $\mathcal{N}$. Combined with (2) this allows us to conclude:

(4) E is extensional, set-like, and well-founded within $\mathcal{N}$.

At this point we invoke the Class-form of Mostowski's Collapse Theorem:

2.6.2. THEOREM. [K, Theorem 5.14] *Suppose E is a well-founded, set-like class, and extensional on a class M^*; then there is a transitive class S and a 1-1 map G from M^* onto S such that G is an isomorphism between (M^*, E) and $(S, \in)$.*

Theorem 2.6.2 together with (4) assure us of the existence of an $\mathcal{N}$-definable $S \subseteq N$ such that S is transitive from the point of view of $\mathcal{N}$, and which has the property that there is an $\mathcal{N}$-definable isomorphism i_1, where

$$i_1 : \mathcal{M}^* \to (S, \in)^{\mathcal{N}}.$$

Finally, we verify that $S = N$. We first note that S must be a proper class in the sense of $\mathcal{N}$, since otherwise $\mathcal{N}$ would be able to define the satisfaction predicate for $(S, \in)^{\mathcal{N}}$, which coupled with the assumption that $\mathcal{N}$ is interpretable in $\mathcal{M}$, and $i := i_1 \circ i_0$ is an $\mathcal{M}$-definable isomorphism between $\mathcal{M}$ and $(S, \in)^{\mathcal{N}}$, would result in $\mathcal{M}$ being able to define a satisfaction predicate for itself, which contradicts (an appropriate version of) *Tarski's Undefinability of Truth Theorem*[11]. The transitivity of S coupled with the fact that S is a proper class in $\mathcal{N}$ together imply that S contains all of the ordinals of $\mathcal{N}$. Therefore, if $S \neq N$, then arguing in $\mathcal{N}$, let V_α^S be V_α in the sense of $(S, \in)$ and let

---

[11] For a structure $\mathcal{M}$ let:
$\mathrm{Th}^+(\mathcal{M}) = \{(\ulcorner \sigma \urcorner, a) : \mathcal{M} \models \sigma(a)\}$, and $\mathrm{Th}^-(\mathcal{M}) = \{(\ulcorner \sigma \urcorner, a) : \mathcal{M} \models \neg \sigma(a)\}$.
With the above notation in mind, the version of Tarski's theorem that is invoked here says that if $\mathcal{M}$ is a structure that interprets Q and is endowed with a pairing function, then $\mathrm{Th}^+(\mathcal{M})$ and $\mathrm{Th}^-(\mathcal{M})$ are $\mathcal{M}$-inseparable, i.e., there is no $\mathcal{M}$-definable D such that $\mathrm{Th}^+(\mathcal{M}) \subseteq D$ and $\mathrm{Th}^-(\mathcal{M}) \cap D = \varnothing$.

$\alpha_0 =$ the first ordinal α such that $V_\alpha = V_\alpha^S$, but $V_{\alpha+1} \setminus V_{\alpha+1}^S \neq \emptyset$.

This makes it clear, in light of the assumption that $\mathcal{M} \trianglerighteq_{\text{par}} \mathcal{N}$, and the fact that i is an isomorphism between $\mathcal{M}$ and $(S, \in)^\mathcal{N}$, that we have a contradiction at hand since $\mathcal{M}$ believes that $\mathcal{N}$ sees a 'new subset' of $V_{i^{-1}(\alpha_0)}$ of $\mathcal{M}$ that is missing from $\mathcal{M}$. Hence $S = N$ and we may conclude that i is an $\mathcal{M}$-definable isomorphism between $\mathcal{M}$ and $\mathcal{N}$. ⊣

2.7. THEOREM. KM *is solid.*

Proof. Models of KM can be represented as two-sorted structures of the form $(\mathcal{M}, \mathcal{A})$, where $\mathcal{M} \models$ ZF; $\mathcal{A}$ is a collection of subsets of M; and $(\mathcal{M}, \mathcal{A})$ satisfies the full comprehension scheme. Suppose $(\mathcal{M}, \mathcal{A})$, $(\mathcal{M}^*, \mathcal{A}^*)$, and $(\mathcal{N}, \mathcal{B})$ are models of KM such that:

$$(\mathcal{M}, \mathcal{A}) \trianglerighteq_{\text{par}} (\mathcal{N}, \mathcal{B}) \trianglerighteq_{\text{par}} (\mathcal{M}^*, \mathcal{A}^*),$$

and there is an $(\mathcal{M}, \mathcal{A})$-definable isomorphism

$$\widehat{i_0} : (\mathcal{M}, \mathcal{A}) \to (\mathcal{M}^*, \mathcal{A}^*).$$

As in the proof of Theorem 2.5 we note that $\widehat{i_0}$ is naturally induced by i_0, where:

$$i_0 := \widehat{i_0} \restriction M : \mathcal{M} \to \mathcal{M}^*,$$

since $\widehat{i_0}(A) = \{i_0(m) : m \in A\}$ for $A \in \mathcal{A}^*$.

N forms a proper class in $(\mathcal{M}, \mathcal{A})$ since if N forms a set, then so does $\mathcal{B}$, and $\widehat{i_0}$ is an $(\mathcal{M}, \mathcal{A})$-definable bijection between $M \cup \mathcal{A}$ and a subset of $N \cup \mathcal{B}$. Let $E := \in^{\mathcal{M}^*}$. Clearly E is extensional. Furthermore, with the help of i_0 and the assumption $(\mathcal{M}, \mathcal{A}) \trianglerighteq_{\text{par}} (\mathcal{N}, \mathcal{B})$ it is easy to see that E is well-founded from the point of view of $(\mathcal{N}, \mathcal{B})$. The reader is asked to verify that an argument very similar to the one used in the proof of Theorem 2.6 shows that E is also set-like in the sense of $(\mathcal{N}, \mathcal{B})$. Theorem 2.6.2 can then be invoked to obtain an $(\mathcal{N}, \mathcal{B})$-definable isomorphism

$$k_1 : \mathcal{M}^* \to (S, \in^\mathcal{N})$$

for some $(\mathcal{N}, \mathcal{B})$-definable transitive $S \subseteq N$. The verification that $S = N$ is identical to the corresponding part in the proof of Theorem 2.6 (and in particular uses Tarski's undefinability of truth theorem). Let $k_0 := i_0^{-1} \circ k_1^{-1}$. Clearly:

(5) $k_0 : \mathcal{N} \to \mathcal{M}$ is an $(\mathcal{M}, \mathcal{A})$-definable isomorphism, and

(6) $k_1 : \mathcal{M}^* \to \mathcal{N}$ is an $(\mathcal{N}, \mathcal{B})$-definable isomorphism.

Borrowing a notation from the proof of Theorem 2.5, let $\widehat{k_0} : (\mathcal{N}, \mathcal{B}) \to (\mathcal{M}, \mathcal{A})$ be the natural extension of k_0, and $\widehat{k_1} : (\mathcal{M}^*, \mathcal{A}^*) \to (\mathcal{N}, \mathcal{B})$ be the natural extension of k_1. Note that both $\widehat{k_0}$ and $\widehat{k_1}$ are *embeddings*. Let $\widehat{k} := \widehat{k_0} \circ \widehat{k_1} \circ \widehat{i_0}$; it is clear that:

(7) $\widehat{k} : (\mathcal{M}, \mathcal{A}) \to (\mathcal{M}, \mathcal{A})$ and $\widehat{k}$ is an $(\mathcal{M}, \mathcal{A})$-definable embedding.

Observe that (5) and (6), together with the assumption that i_0 is an isomorphism imply that the restriction $k : \mathcal{M} \to \mathcal{M}$ of $\widehat{k}$ to 'sets' is an automorphism of $\mathcal{M}$. But since $\widehat{k}$ is $(\mathcal{M}, \mathcal{A})$-definable,

$k(m) = m$ for all $m \in M$, thanks to the veracity of the scheme of $\in$-induction[12] in KM. This shows that $\widehat{k}$ is the identity map and in particular it is surjective, which in turn implies that $\widehat{k_0}$ and $\widehat{k_1}$ are both surjective. This makes it clear that there is an $(\mathcal{M}, \mathcal{A})$-definable isomorphism between $(\mathcal{M}, \mathcal{A})$ and $(\mathcal{N}, \mathcal{B})$. ⊣

Recall that $\widetilde{\mathsf{ZF}}$ was defined earlier, just before Corollary 2.5.1.

2.8.1. COROLLARY. *The following theory T is solid*:

$$T := \widetilde{\mathsf{ZF}} \setminus \{\text{Power Set}\} + \text{``}\exists \kappa \ (\kappa \text{ is strongly inaccessible, and } \forall x \ |x| \leq \kappa\text{''}).$$

Proof. In light of Remark 2.1.1, this follows from Theorem 2.8 and the well-known bi-interpretability of T with $\mathsf{KM} + \Pi^1_\infty\text{-AC}$, where $\Pi^1_\infty\text{-AC}$ is the scheme of Choice.[13] ⊣

2.9. REMARK. An examination of the proofs in this section make it clear that for each positive integer n, the theories Z_n (n-th order arithmetic) and KM_n (n-th order Kelley-Morse theory of classes) are solid theories (where $\mathsf{Z}_1 := \mathsf{PA}$, and $\mathsf{KM}_1 := \mathsf{ZF}$). This observation, in turn, implies that the theory of types Z_ω (with full comprehension) whose level-zero objects form a model of PA (equivalently $\mathsf{ZF}_{\mathsf{fin}} + \mathsf{TC}$), and the theory of types KM_ω whose level-zero objects form a model of ZF are also solid theories. Thus, the list of theories whose solidity is established in this section can be described (up to bi-interpretability) as $\{\mathsf{Z}_n : 1 \leq n \leq \omega\} \cup \{\mathsf{KM}_n : 1 \leq n \leq \omega\}$.

3 Examples and Questions

All of the theories T whose solidity was established in Section 2 are sequential[14] theories which have an interpretation $\mathbb{N}$ for 'numbers' for which the full scheme $\mathsf{Ind}_\mathbb{N}(\mathcal{L}_T)$ of induction is T-provable, so one may ask whether the T-provability of $\mathsf{Ind}_\mathbb{N}(\mathcal{L}_T)$ within a sequential theory is a sufficient condition for solidity. A simple counterexample gives a negative answer: let $\mathsf{PA}(\mathsf{G})$ be the natural extension of PA in which the induction scheme is extended to formulae in the language obtained by adding a unary predicate G to the language of arithmetic. To see that $\mathsf{PA}(\mathsf{G})$ is not solid, consider the extensions T_1 and T_2 of $\mathsf{PA}(\mathsf{G})$, where:

$$T_1 := \mathsf{PA}(\mathsf{G}) + \forall x(\mathsf{G}(x) \leftrightarrow x = 1) \text{ and } T_2 := \mathsf{PA}(\mathsf{G}) + \forall x(\mathsf{G}(x) \leftrightarrow x = 2).$$

Clearly the deductive closures of T_1 and T_2 are distinct, and yet it is easy to see that T_1 and T_2 are bi-interpretable. This shows that $\mathsf{PA}(\mathsf{G})$ is not tight, and therefore not solid.

With the help of [ESV, Theorem 4.9 & Remark 4.10] one can also show that the theory $\mathsf{ZF}_{\mathsf{fin}}$ is not tight, even though as shown in Corollary 2.2.1 its strengthening by TC is a solid theory. Another example of a theory that fails to be tight is $\mathsf{ZF} \setminus \{\text{Foundation}\}$. To see this, consider $T_1 := \mathsf{ZF}$, and

---

[12] The scheme of $\in$-induction consists of the universal closures of formulas of the form $\forall y \ (\forall x \in y \ \theta(x) \to \theta(y)) \to \forall z \ \theta(z)$, where the parameters in θ are suppressed. It is easy to see that the scheme of $\in$-induction is equivalent to the class-form of Foundation, which asserts that every nonempty definable collection of sets has an $\in$-minimal element. The class-form of Foundation follows from the set-form of Foundation and the comprehension scheme of KM: suppose a class C is nonempty, and let α_0 be the first ordinal α such that $V_\alpha \cap C \neq \varnothing$. Then an $\in$-minimal member of $V_{\alpha_0} \cap C$ is also an $\in$-minimal member of C.

[13] This bi-interpretability was first noted by Mostowski; a modern account is given in a recent paper of Antos & Friedman [AF, section 2], where $\mathsf{KM} + \Pi^1_\infty\text{-AC}$ is referred to as MK^*, and T is referred to as SetMK^*.

[14] A sequential theory is a theory that has access to a definable 'β-function' for coding finite sequences of objects in the domain of discourse.

$$T_2 := \mathsf{ZF}\setminus\{\mathsf{Foundation}\} + \exists! x(x = \{x\} + \forall t \exists \alpha(t \in V_\alpha(x)),$$

where α ranges over ordinals, and

$$V_0(x) := x, \ V_{\alpha+1}(x) := \mathcal{P}(V_\alpha(x)), \text{ and } V_\alpha(x) := \bigcup_{\beta < \alpha} V_\beta(x) \text{ for limit } \alpha.$$

Then T_1 and T_2 are extensions of $\mathsf{ZF}\setminus\{\mathsf{Foundation}\}$ with distinct deductive closures, and yet, the bi-interpretability of T_1 and T_2 can be established by well-known methods: the relevant interpretations are $\mathcal{I}$ and $\mathcal{J}$, where $\mathcal{I}$ is the classic von Neumann interpretation of ZF in $\mathsf{ZF}\setminus\{\mathsf{Foundation}\}$, and $\mathcal{J}$ is the classic Rieger-Bernays interpretation that adds a single 'Quine atom' (i.e., a set s such that $s = \{s\}$) to a model of ZF.

However, we do not know whether $T \vdash \mathsf{Q}^\mathbb{N} + \mathsf{Ind}_\mathbb{N}(\mathcal{L}_T)$ for every solid sequential theory (for an appropriate choice of numbers $\mathbb{N}$). This motivates the following question, since by a general result of Montague [Mo] the T-provability of $\mathsf{Q}^\mathbb{N} + \mathsf{Ind}_\mathbb{N}(\mathcal{L}_T)$ implies that T is not finitely axiomatizable.

3.1. QUESTION. *Is there a consistent sequential finitely axiomatized theory that is solid?*

The question below arises from reflecting on the results of Section 2 and noting that the proofs of solidity of each of the theories T established in Section 2 uses the 'full power' of T.

3.2. QUESTION. *Is there an example T of one of the theories whose solidity is established in Theorem 1.2, and some **solid** $T_0 \subseteq T$ such that the deductive closure of T_0 is a proper subset of the deductive closure of T?*

Acknowledgement

It is a pleasure and an honor to present this paper in a volume that celebrates Albert Visser's scholarship; I am grateful to Albert for bringing his Theorem 1.1 to my attention. Thanks also to Andrés Caicedo and Radek Honzík, whose interest in the ZF-case of Theorem 1.2 provided additional impetus for writing up the results here; and to the anonymous reviewer for invaluable help in weeding out infelicities of an earlier draft. Hats off to Jan, Joost and Rosalie for their dedication in bringing this volume to fruition.

BIBLIOGRAPHY

[A] W. Ackermann, *Zur widerspruchshfreiheit der zahlentheorie*, **Math. Ann.** 117 (1940), pp. 162-194.
[AZ] G. Ahlbrandt & M. Ziegler, *Quasi-finitely axiomatizable totally categorical theories*, **Ann. Pure Appl. Logic** 30 (1986), pp. 63-82.
[AF] C. Antos & S.D. Friedman, *Hyperclass Forcing in Morse-Kelley Class Theory*, manuscript available at **Math. ArXiv.** (2015).
[B] J. Barwise, **Admissible Sets and Structures**, Springer-Verlag, Berlin, 1975.
[D] K. L. de Bouvère, *Synonymous Theories*, In **Theory of Models** (edited by J.W. Addison, L. Henkin, and A. Tarski), Proceedings of the 1963 International Symposium at Berkeley, North Holland, Amsterdam, 1965, pp. 402-406, 1965.
[ESV] A. Enayat, J. Schmerl, & A. Visser, *ω-models of finite set theory*, in **Set Theory, Arithmetic, and Foundations of Mathematics: Theorems, Philosophies** (edited by J. Kennedy and R. Kossak), Cambridge University Press, 2011, pp. 43-65.
[F] S. Feferman, *Arithmetization of metamathematics in a general setting*, **Fund. Math.** 49 (1960), pp. 35-92.
[FV] H.M. Friedman & A. Visser, *When bi-interpretability implies synonymy*, available through Utrecht Preprint series (2014).
[H] W. Hodges, **Model theory**, Cambridge University Press, Cambridge, 1993.

[KW] R. Kaye & T. L. Wong, *On interpretations of arithmetic and set theory*, **Notre Dame J. Formal Logic**, 48 (2007), pp. 497-510.
[K] K. Kunen, **Set theory**, North-Holland Publishing Co., Amsterdam, 1983.
[Ma] A.R.D. Mathias, *The strength of Mac Lane set theory,* **Ann. Pure Appl. Logic** 110 (2001), pp 107–234.
[Mo] R. Montague, *Semantical closure and non-finite axiomatizability. I*, in **Infinitistic Methods** (Proc. Sympos. Foundations of Math., Warsaw, 1959). Pergamon, Oxford; Państwowe Wydawnictwo Naukowe, Warsaw, 1961, pp. 45-69.
[My] J. Mycielski, *The definition of arithmetic operations in the Ackermann model* (Russian), **Algebra i Logika Sem.** 3 no. 5-6 (1964), pp. 64–65.
[S] S. Simpson, **Subsystems of Second Order Arithmetic**, Springer, Heidelberg 1999.
[V] A. Visser, *Categories of theories and interpretations,* **Logic in Tehran**, Lecture Notes in Logic, vol. 26, Association for Symbolic Logic, La Jolla, CA, 2006, pp. 284–341.

Reduction Cycles in Lambda Calculus and Combinatory Logic

JÖRG ENDRULLIS, JAN WILLEM KLOP, ANDREW POLONSKY

We dedicate our paper to our friend and colleague Albert Visser, for his 65th birthday, hoping that it will appeal to his love for monsters (teratology) and cyclic λ-terms.

Abstract

We analyse the existence of cyclic reductions in λ-calculus and Combinatory Logic with special attention for pure cycles, that is, deterministic cycles for which every reduct contains precisely one redex.

*The ourobouros. Engraving by Lucas Jennis,
in the alchemical tract titled De Lapide Philosophico.*

1 Introduction

We are interested in cyclic reductions in λ-calculus ($\lambda\beta$ and $\lambda\beta\eta$), in Section 2, as well as Combinatory Logic (based on $\{\,\mathsf{I},\mathsf{K},\mathsf{S}\,\}$ or other bases), in Section 3.[1] A *cycle* is a reduction of the form $M_0 \to M_1 \to \cdots \to M_n \equiv M_0$ where $n \geq 1$. We call the cycle a *loop* if $n = 1$. As the main results of this paper, we consider the following two contributions:

(i) A heuristic investigation of pure two-cycles (bicycles) in $\lambda\beta$-calculus (and $\lambda\beta\eta$-calculus) followed by a complete characterisation of such bicycles.[2]

(ii) A simplified proof of the theorem that in Combinatory Logic based on $\{\,\mathsf{I},\mathsf{K},\mathsf{S}\,\}$ there are no pure cycles, and even stronger, that finite reduction graphs are acyclic.

There are the following reasons that motivate our attention for loops and cycles:

(a) Cycles, loops and pure minimal cycles provide challenging combinatorial questions, with a wide range, from nontrivial exercises to hard questions that at present are beyond our reach.

(b) Loops (one-step cycles) are relevant for infinitary rewriting [17, 21]. The Loop Lemma [16, 28, 33] shows that (for rewrite systems having left-hand sides with bounded depth) every rewrite sequence that is weakly convergent but not strongly convergent must contain a (typically infinite) looping term. The looping terms in infinitary λ-calculus have been classified in [25].

(c) Cycles are important in the study of reduction strategies, e.g. Church-Rosser strategies [8, 29]; cycles provide an obstacle, as traps from which a strategy (which in the usual definition is without memory) cannot recover [29].

(d) Minimal and pure cycles can be perceived as things of beauty, appealing to our aesthetic feelings.

(e) Cycles are connected with notions of recurrent terms, and the crucial notions of unsolvable and mute terms; see Figure 1.1.

- A term M is *recurrent* [9] if every reduct of M reduces back to M. Equivalently, M has a cycle $M \to^+ \overline{M} \to^+ M$ where $\overline{M}$ is the complete development of M. A term M is *one-step recurrent* [26, 27, 32] if every one-step reduct of M reduces back to M. The term M is *root-recurrent* if it is recurrent and reduces to a redex.

- A term M is *mute* [5, 6, 10] (also known as *root active*) if it has no root normal form, that is, a term that does not reduce to a redex. A *root normal form* is either a variable, an abstraction or an application $N_1 N_2$ where N_1 does not reduce to an abstraction. Equivalently, M is mute if it admits a reduction containing infinitely many root steps.

- A term M is *easy* [5, 6] if it can be consistently equated to any term.

---

[1] For general abstract reduction systems our current questions and notions are trivial or irrelevant. For other versions of λ-calculus and Combinatory Logic they are interesting, for instance λY (simply typed λ-calculus with recursion operators) and higher-order rewrite systems such as CRSs or HRSs.

[2] According to Katie Melua there are nine million bicycles in Beijing. In λ-calculus there are infinitely many.

- A term M is a *weak head normal form (whnf)* [1] if it is either an abstraction or of the form $xM_1\ldots M_n$ for some variable x and terms $M_1,\ldots,M_n$.
- A term M is a *head normal form (hnf)* if $M \equiv \lambda x_1\ldots x_n.yM_1\ldots M_m$ for some variables $x_1,\ldots,x_n,y$ and terms $M_1,\ldots,M_m$.
- A term M is a *unsolvable* [2] if it has no head normal form.
- A term is *productive* [14, 15, 18], if it can be fully evaluated to a $\bot$-free Böhm Tree. Such terms correspond to the strongly normalising terms in the infinitary λ-calculus.

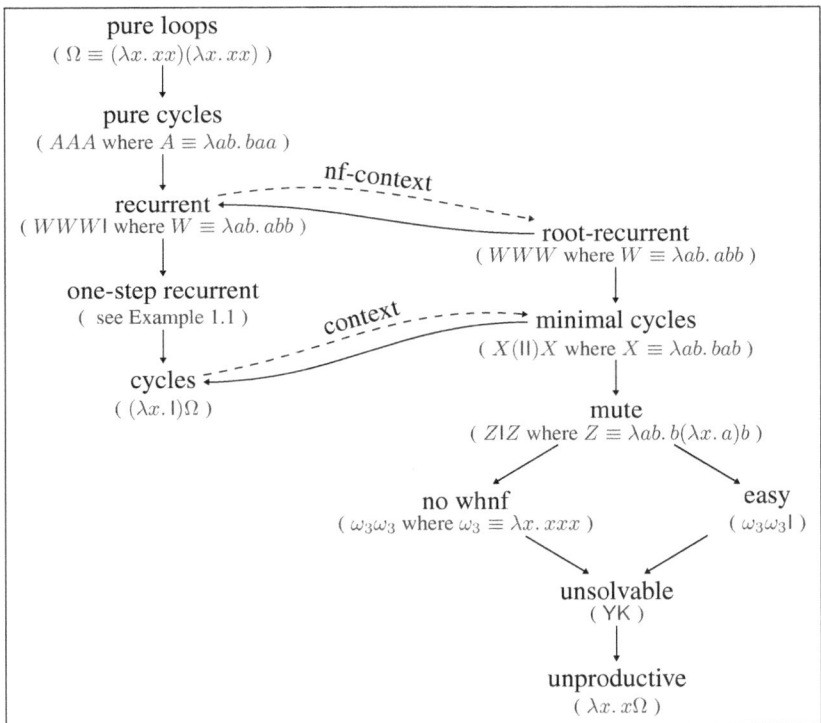

Figure 1.1. *Spectrum of relations between notions of cyclicity and undefinedness; next to the strict implications downward and to the left (solid arrows), there are relations as in Proposition 1.2 asserting a contextual decomposition (dashed arrows). Some of the example terms in this overview are reminiscent of, but not identical to, the terms in [2, Exercise 3.5.5].*

EXAMPLE 1.1. *The following is an example of a one-step recurrent term which is not recurrent from [32]:*

$$M \equiv XXYZ \qquad X \equiv \lambda xyz.\, xxy(yz) \qquad Y \equiv \lambda xz.\, x(xz) \qquad Z \equiv \lambda z.\, \mathsf{I}(\mathsf{I}z)$$

In the above figure, we can also go from the left side to the right, using the following simple fact.

PROPOSITION 1.2.

(i) *Let M be recurrent. Then there exists a context $C[x_1,\ldots,x_k]$ which is a normal form, and root-recurrent terms $M_1,\ldots,M_k$ such that $M \equiv C[M_1,\ldots,M_k]$.*

(ii) *Let N be a cycle. Then there is a context $C[x_1,\ldots,x_k]$ with $k \geq 1$ and root-cycles $N_1,\ldots,N_k$ such that $N \equiv C[N_1,\ldots,N_k]$.*

The proof of the above proposition is a straightforward induction on the term structure of M.

We note that it is decidable whether a finite λ-term admits a loop, or a cycle of length n (for fixed $n \in \mathbb{N}$). However, it is not decidable whether a term admits a cyclic reduction; more precisely, this property is Σ_1^0-complete. This is to be contrasted to the more standard notions of meaninglessness that appear in Figure 1.1, which are Π_2^0-complete. (See [12, 13] for degrees of undecidability of various properties of rewrite systems.)

Thus, cyclic terms may be seen as a special class of meaningless terms, the membership of which can be known in finite time — being such a term is then a *positive* property. This sits well with experience, for playing with cyclic terms is a most entertaining pastime!

2 Pure Cycles in $\lambda\beta$-calculus

Pure cycles are reduction graphs that just consist out of one cycle. They are interesting and beautiful objects, just as one might appreciate magic squares. They are also interesting from a combinatorial point of view, presenting challenging puzzles.

We are considering cyclic reductions that are *deterministic*, that is, in every step there is only one redex present, so only one step possible:

$$M_0 \to M_1 \to M_2 \to \cdots \to M_n \equiv M_0 \qquad (1)$$

If $n = 1$, we sometimes call the cycle a reduction *loop*. For $n = 2$ and $n = 3$, we call the cycle a *bicycle* or *tricycle*, respectively. A cycle of length n will also be called an n-cycle, so a loop is a 1-cycle. Instead of 'deterministic' we will also call such a cycle *pure*.

Note that of course, we can *rotate* cycles, e.g. the cycle

$$M_3 \to M_4 \to M_5 \to \cdots \to M_n \to M_1 \to M_2 \to M_3$$

is a rotation of the cycle (1) above. We will identify cycles differing only by a rotation.

We will in this paper restrict to 'minimal' cycles defined as follows.

DEFINITION 2.1. *A cycle $M_0 \to M_1 \to M_2 \to \cdots \to M_n \equiv M_0$ is minimal if no proper subterm of M_0 admits a cycle, and it is not an unwinding of a shorter cycle, that is, $M_i \not\equiv M_j$ for all $0 \leq i < j < n$.*

Other notions of minimality of cycles are possible. Since we are only interested in minimal pure cycles, this definition is sufficient for us.

Note that all minimal cycles have the same Böhm Tree $\bot$. Hence, like fixed point combinators, minimal cycles cannot be distinguished by comparing their Böhm Trees. They can however, frequently be discriminated using (atomic) clocked Böhm Trees [19, 20, 22, 23].

LEMMA 2.2. *If a term M has only one redex R, then either R is the head redex, or M is a head normal form.*

Proof. Suppose M has only one redex R. Suppose that M is not a head normal form. Then M has the form $M \equiv \lambda \vec{x}. R'\vec{N}$ for some redex R'. Because M has only one redex, R' must be R. ⊣

PROPOSITION 2.3. *Every minimal pure cycle is a head reduction.*

Proof. Let $\mathcal{C} \equiv M_0 \to M_1 \to M_2 \to \cdots \to M_n \equiv M_0$ be a minimal pure cycle. By minimality, $\mathcal{C}$ must contain a root-step, and consequently none of the terms $M_0, M_1, \ldots, M_n$ is a head normal form. Thus by purity and Lemma 2.2 it follows that each step in $\mathcal{C}$ is a head step. ⊣

Loops in λ-calculus

It is not too hard to characterize the 1-cycles (loops) in λ-calculus, but maybe somewhat tedious. Part of the exercise's solution can be found in the literature.

THEOREM 2.4.

(i) *The only redex $(\lambda x. A)B$ in λ-calculus reducing in one step to itself:*

$$(\lambda x. A)B \to A[x := B] \equiv (\lambda x. A)B$$

is $(\lambda x. xx)(\lambda x. xx) \equiv \Omega$.

(ii) *The only λ-terms in λ-calculus reducing in one step to themselves are the terms $C[\Omega]$, for some context $C[\Box]$. In other words, Ω is the only pure minimal loop in λ-calculus.*

Proof. (i) The proof of this exercise was given by Bruce Lercher in [31].

(ii) The proof of this part of the exercise was independently given by Albert Visser [34] The exercise occurs as [2, Exercise 3.5.6] without proof.

⊣

REMARK 2.5. *Note that also in the $\lambda\beta\eta$-calculus Theorem 2.4 holds since an η-step reduces the size of the term.*

Pure bicycles in λ-calculus

We will now move from considering 1-cycles (loops) to 2-cycles (bicycles). The characterisation of the minimal pure bicycles in $\lambda\beta$-calculus is non-trivial. In fact, the characterisation by hand readily succeeds (although case (iv) below was first found with computer assistance and a search program). The hard part however, is the proof that this characterisation is indeed complete The lengthy proof requiring a sizeable tree of case distinctions, is included in the technical report [24].

THEOREM 2.6. *The following is a classification of all minimal pure bicycles in $\lambda\beta$-calculus:*

(i) *Let $A \equiv \lambda ab. baa$. Then*

$$AAA \to (\lambda b. bAA)A \to AAA. \tag{2}$$

(ii) Let $A \equiv \lambda ab.\, bba$. Then

$$AAA \to (\lambda b.\, bbA)A \to AAA\,. \tag{3}$$

(iii) Let $A \equiv \lambda ab.\, bab$. Then

$$A \bigcirc A \to (\lambda b.\, b \bigcirc b)A \to A \bigcirc A \tag{4}$$

is a minimal pure cycle for any λ-term $\bigcirc$ in normal form. We call $\bigcirc$ a 'white hole' as it has no influence on the reduction.

(iv) Let $\bigcirc$ be a λ-term in normal form, without free occurrence of the variable a and such that the variable b does not occur as left argument of an application.

Define $A \equiv \lambda ab.\, b \bigcirc b$, then

$$A\,(\bigcirc[b{:=}A])\,A \to (\lambda b.\, b \bigcirc b)A \to A\,(\bigcirc[b{:=}A])\,A \tag{5}$$

is a minimal pure cycle.

REMARK 2.7. *We note the following.*

(a) *Actually, our classification has yielded infinitely many solutions due to clauses (iii) and (iv) where we see the presence of a white hole.*

We give two examples of clause (iv):

- *For $\bigcirc \equiv b$ we get $A \equiv \lambda ab.\, bbb$ and*

$$AAA \to (\lambda b.\, bbb)A \to AAA\,. \tag{6}$$

- *For $\bigcirc \equiv z(\lambda x.\, x)b$ we get $A \equiv \lambda ab.\, b(z(\lambda x.\, x)b)b$ and*

$$A(z(\lambda x.\, x)A)A \to (\lambda b.\, b(z(\lambda x.\, x)b)b)A \to A(z(\lambda x.\, x)A)A \tag{7}$$

is a minimal pure cycle.

(b) *The term $\Omega\Omega$ does not fall into the classification as it is not minimal.*

(c) *The following term seems to be another minimal pure bicycle:*

$$A \equiv \lambda x.\, \omega_3 \qquad \text{where} \qquad \omega_3 \equiv \lambda x.\, xxx$$

Then $\omega_3 A \to AAA \to \omega_3 A$. However, this is in fact the bicycle (6).

We see here for the first time the emergence of a notion that might be interesting in itself, namely the concept of 'white holes'. A white hole can be seen as a dummy subterm that 'does nothing' and can be exchanged for anything. Here we have seen a cycle with a single white hole, later we will see cycles with multiple white holes.

One could wonder if more pure bicycles arise when we adopt the η-rule. However, this is not the case.

PROPOSITION 2.8. *The only minimal pure bicycles in $\lambda\beta\eta$-calculus are the ones of the $\lambda\beta$-calculus.*

Proof. Assume that t is a λ-term admitting a minimal pure 2-cycle $t \to s \to t$ in $\lambda\beta\eta$-calculus such that (without loss of generality) $t \to s$ is an η-contraction. Then the step $s \to t$ must be a β-step since η-steps reduce the size of the term. By purity, the contraction of the η-redex in $t \to s$ must create the β-redex contracted in $s \to t$. Then it follows that the η-step must be of the form $t \equiv C[(\lambda x. t'x)t''] \to C[t't''] \equiv s$ where t' is an abstraction without free occurrences of x. However, then there is also a β-redex $(\lambda x. t'x)t''$ contradicting that the cycle is pure. ⊣

If we drop the purity requirement, there are new bicycles arising in $\lambda\beta\eta$-calculus. For example, let $A \equiv \lambda x. \lambda y. xxy$, then

$$AA \to_\beta \lambda y. AAy \to_\eta AA$$

is a bicycle in $\lambda\beta\eta$-calculus. However, the cycle is not pure as it is not deterministic (the reduction graph of AA is not a line).

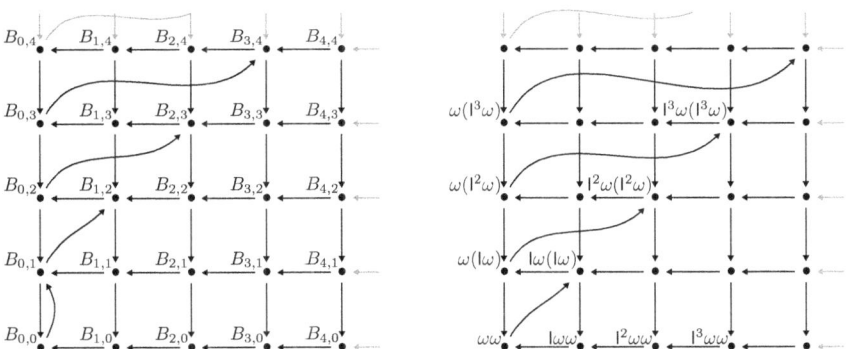

Figure 2.1. *The β-reduction graph of $B_{0,0} \equiv (\lambda x.x(\lambda y.xy))(\lambda x.x(\lambda y.xy))$ (on the left). Note the remarkable resemblance with the Combinatory Logic reduction graph of $\omega\omega$ where $\omega \equiv \mathsf{SII}$ (on the right). However, the former contains a bicycle.*

In this figure the following notation is used: $B_0 \equiv B$, $B_{n+1} \equiv \lambda y. B_n y$, $B_{n,m} \equiv B_n B_m$. Note that only the β-reduction graph is displayed; the $\beta\eta$-reduction graph would have 6 extra points in a cluster attached to the origin $B_{0,0}$.

The term $\mathsf{SII}(\mathsf{SII})$ is the shortest term in Combinatory Logic that admits a cyclic reduction. Except for $\Omega \equiv (\lambda x. xx)(\lambda x. xx)$, the term $B_{0,0}$ is the shortest such term in $\lambda\beta$-calculus.

An interesting (non-pure) 2-cyclic term has been found in [10] as an example of a non-regular mute term:

$$BB \to B(\lambda y. By) \to BB \qquad \text{where } B \equiv \lambda x. x(\lambda y. xy) \qquad (8)$$

The reduction graph of this term is displayed in Figure 2.1.

Remarkably, all the bicycle terms are of the shape ABA, a repeated application with identical first and last element. Let us call such terms Möbius-terms as the first and the last A differ by a 'twist', a left argument of an application versus a right argument.

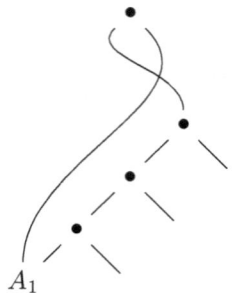

Figure 2.2. *Möbius-terms.*

DEFINITION 2.9. *A term $A_1 A_2 \cdots A_n$ with $A_1 \equiv A_n$ is called a* **Möbius-term**.

Pure tricycles in λ-calculus

Let us now turn to the case of $n = 3$ or tricycles.
An example of a pure tricycle is the term:

$$|AA| \qquad \text{where} \qquad A \equiv \lambda ab.\, baab$$

It is not hard to write down several dozens of tricycles, namely

$$AAAA \to^3 AAAA \qquad \text{where} \qquad A \equiv \lambda xyz.\, z\, w$$

with w a word over $\{x, y, z\}$ of length 3. E.g. $A \equiv \lambda xyz.\, zzzz$ or $A \equiv \lambda xyz.\, zxyz$ and so on. There are 27 of such terms, they all yield pure tricycles.

For several of these, we can obtain more tricycles by allowing white holes in the starting term. For example, let $A \equiv \lambda xyz.zxzy$, then

$$A \bigcirc AA \;\to\; (\lambda yz.z \bigcirc zy)AA \;\to\; (\lambda z.z \bigcirc zA)A \;\to\; A \bigcirc AA$$

Here $\bigcirc$ can be an arbitrary term in normal form. We can find even more tricycles by admitting white holes in the $A \equiv \lambda xyz.\, z\, w$ itself. They are all Möbius-terms. There will be expectedly, some that have the possibility of refined white holes in analogy of the bicycle case Theorem 2.6 (iv) above. At this stage one could venture that all tricycles contain Möbius-terms. But no!

EXAMPLE 2.10. *Let $A \equiv \lambda ab.\, bbaa$ and $B \equiv \lambda abc.\, cba$. Then*

$$AAB \;\to\; (\lambda b.\, bbAA)B \;\to\; BBAA \;\to\; (\lambda bc.\, cbB)AA \;\to\; (\lambda c.\, cAB)A \;\to\; AAB$$

Here none of the terms in the cycle is a Möbius-term.

We do not attempt to state and prove a complete characterisation of pure tricycles in $\lambda\beta$-calculus; it could be a challenging task for an automated proof assistant.

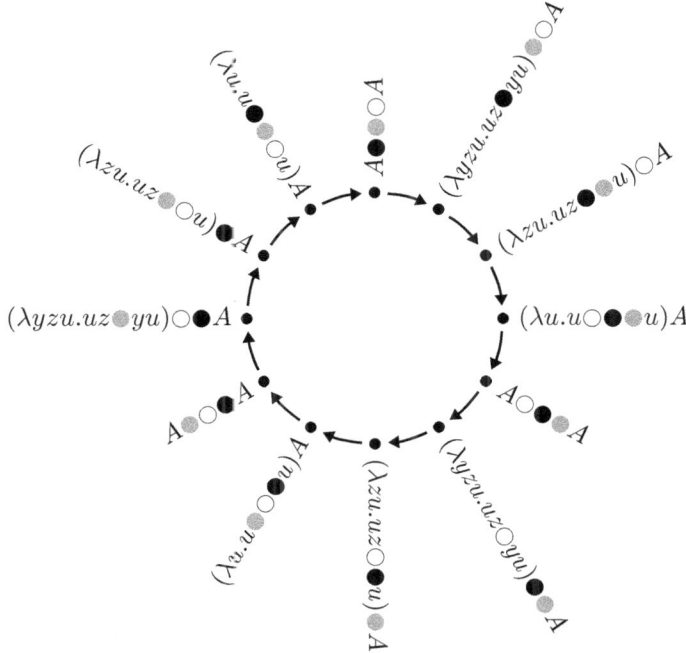

Figure 2.3. *The clock for Albert from a different perspective. Here $A \equiv \lambda xyzu.\,uzxyu$.*

Pure n-cycles

Let us now turn to pure n-cycles in $\lambda\beta$-calculus for $n > 3$. Analogous to the case for $n = 2, 3$, we can easily find pure n-cycles of the form $A \ldots A \to^n A \ldots A$ (with $n + 1$ A's). But there are more 'construction principles'. A complete classification for $n > 3$ is not feasible at the present state of the theory. We will only make a few remarks hinting at the rich complexity underlying the structure of general pure n-cycles. We found at least three 'phenomena' that cause the complexity:

(i) The first feature that we already encountered is the emergence of 'white holes' turning some pure cycles into *schemes* for pure cycles. As a complication, we found that there may be "refined white holes", involving a substitution, as in Theorem 2.6 (iv) of our bicycle characterisation.

(ii) The second feature is that pure n-cycles may incorporate a "permutative" effect. We illustrate this with a particular pure 12-cycle, a clock that we dedicate to Albert (a coloured version is included in the technical report [24]):

$$
\begin{aligned}
A\bullet\bullet\circ A &\to \\
(\lambda yzu.uz\bullet yu)\bullet\circ A &\to \\
(\lambda zu.uz\bullet\bullet u)\circ A &\to \\
(\lambda u.u\circ\bullet\bullet u)A &\to_r
\end{aligned}
$$

$$A\bigcirc\bullet\bullet A \to$$
$$(\lambda yzu.uz\bigcirc yu)\bullet\bullet A \to$$
$$(\lambda zu.uz\bigcirc\bullet u)\bullet A \to$$
$$(\lambda u.u\bullet\bigcirc\bullet u)A \to_r$$
$$A\bullet\bigcirc\bullet A \to$$
$$(\lambda yzu.uz\bullet yu)\bigcirc\bullet A \to$$
$$(\lambda zu.uz\bullet\bigcirc u)\bullet A \to$$
$$(\lambda u.u\bullet\bullet\bigcirc u)A \to_r$$
$$A\bullet\bullet\bigcirc A$$

Here $A \equiv \lambda xyzu.\,uzxyu$. Note that it contains Möbius-terms. Also note that there are three root steps. The three white holes (coloured for better traceability) are cyclically permuted in the first 4 steps, and thus restored to their original position after 12 steps.

(iii) The third feature causing some complexity to the structure of pure n-cycles, is the "estafette" effect. For several of the examples above for bicycles and tricycles, the reduction seems to pass over 'control' from one constituent subterm in the cyclic term to the next. However, there is a more complicated mechanism possible where control jumps to a constituent subterm further on. For example, for $A \equiv \lambda abc.\,cab$ and $B \equiv \lambda abc.\,cabbc$, we have:

$$\widehat{A \bigcirc BBA} \to^3 B \bigcirc BA \to^3 A \bigcirc BBA$$

Here A passes the control to B and B passes control back to A.

With these general explorations we end our search for pure cycles in $\lambda\beta$-calculus.

3 Pure Cycles in Combinatory Logic

After our consideration of pure cycles in the $\lambda\beta$-calculus, and briefly in the $\lambda\beta\eta$-calculus, the obvious question is what pure cycles there are in Combinatory Logic, CL. Here we first consider CL with the well-known basis $\{S, K\}$ or $\{S, K, I\}$. Surprisingly, there are no pure cycles here. It should be mentioned that this is basis-dependent, other bases may admit pure cycles, and we will give two examples later on.

The set of ground terms $Ter^0(\text{CL})$ consists of S, K and I and terms that can be built by application: $M \cdot N \in Ter^0(\text{CL})$ whenever $M, N \in Ter^0(\text{CL})$. We adopt the usual bracketing convention (association to the left), writing MN as shorthand for $M \cdot N$.

In Combinatory Logic, we have the following rewrite rules

$$S\,abc \to ac(bc) \qquad\qquad K\,ab \to a \qquad\qquad I\,a \to a$$

Let M be a term. The set of *positions* $Pos(M) \subseteq \{1,2\}^*$ of M is defined by:

$$Pos(\mathsf{S}) = Pos(\mathsf{K}) = Pos(\mathsf{S}) = \{\varepsilon\}$$
$$Pos(MN) = \{\,1p \mid p \in Pos(M)\,\} \cup \{\,2p \mid p \in Pos(N)\,\}$$

The *subterm* $M|_p$ of M at position $p \in Pos(M)$ is defined by:

$$M|_\varepsilon = M \qquad (MN)|_1 = M \qquad (MN)|_2 = N$$

The set of *spine positions of M* consists of all positions from $Pos(M)$ of the form $11\cdots 1$. A *spine redex* is a redex on a spine position.

A term M is in *head normal form* if it contains no spine redex.

LEMMA 3.1. *A term M is in head normal form if and only if M is of one of the forms*

$$M \equiv \mathsf{S}AB \qquad M \equiv \mathsf{S}A \qquad M \equiv \mathsf{S} \qquad M \equiv \mathsf{K}A \qquad M \equiv \mathsf{K} \qquad M \equiv \mathsf{I}$$

for some terms A, B. As a consequence, any reduct of a term in head normal form is in head normal form.

For positions $p, q \in \{1, 2\}^*$ we say p *is left of* q, denoted $p <_\ell q$, if either

(i) $pp' = q$ for some $p' \in \{1, 2\}^+$ (that is, p is a strict prefix of q), or

(ii) $p = c1p'$ and $q = c2q'$ for some $c, p', q' \in \{1, 2\}^*$.

Let R be a redex at position p in a term M. Then R is called *leftmost* if M contains no redex at a position left of p.

LEMMA 3.2. *If the leftmost redex in a term M is not a spine redex, then M is in head normal form.*

Proof. Let the leftmost redex R be at position p in M. If p is not a spine position, it must contain an occurrence of 2. That is $p = p'2p''$ for some $p' \in \{1\}^*$ and $p'' \in \{1, 2\}^*$. Then for every spine position q, we have $q <_\ell p$ since:

- If $|q| \leq |p'|$, then q is a strict prefix of p.
- If $|q| > |p'|$, $q = p'1q''$ and $p = p'2p''$ for some $q'' \in \{1, 2\}^*$.

Thus M cannot contain a spine redex, otherwise R is not leftmost. ⊣

We define $r(M)$ as the length of the rightmost branch of M, that is:

$$r(\mathsf{S}) = 0 \qquad r(\mathsf{K}) = 0 \qquad r(\mathsf{I}) = 0 \qquad r(MN) = 1 + r(N)$$

PROPOSITION 3.3. *Let $M \to N$ via the contraction of the spine redex R.*

(i) *If R is not a root-step, then $r(M) = r(N)$.*

(ii) *If R is a root-step, then we have one of the following cases:*

    (a) $M \equiv \mathsf{S}ABC$ *for some A, B, C. Then $r(M) < r(N)$.*

    (b) $M \equiv \mathsf{K}AB$ *for some A, B.*

    (c) $M \equiv \mathsf{I}A$ *for some A.*

Note that if $r(M) > r(N)$, then we must be in case (b) or (c).

Proof. Case (i) is immediate. For case (ii), it is clear that a root step must be of one of the forms (a), (b) or (c). For (a) we have $M \equiv \mathsf{S}ABC$ and it follows that $N = AC(BC)$ and $r(M) = 1 + r(C) < 2 + r(C) = r(N)$. Thus, indeed, (b) and (c) are the only possibility for $r(M) > r(N)$. ⊣

We use $\to_{lm}$ to denote a leftmost reduction step, that is, a step arising from the contraction of the leftmost redex. We write $\mathcal{R}_{lm}(M)$ for the leftmost reduction starting from M, that is, contracting the leftmost redex in each step.

THEOREM 3.4. *If $\mathcal{R}_{lm}(M)$ contains a cycle, then $\mathcal{G}(M)$ is infinite.*

Proof. Assume that $\mathcal{R}_{lm}(M)$ contains a cycle, that is, $M \twoheadrightarrow_{lm} N \to^+_{lm} N$ for some N. Let O be a the smallest subterm of N that admits a cyclic leftmost reduction, say $O \equiv O_0 \to_{lm} O_1 \to_{lm} \cdots \to_{lm} O_m \equiv O$. By the minimality of O, this reduction must contain a root step $O_i \to_{lm} O_{i+1}$. We distinguish cases:

(i) For $O_i \equiv \mathsf{S}ABC$. Then $r(O_i) < r(O_{i+1})$ by Proposition 3.3.

(ii) For $O_i \equiv \mathsf{K}AB$ and $O_i \equiv \mathsf{I}A$, we have $Q_{i+1} \equiv A$. Due to the cycle, we have $O_{i+1} \to^+ O_i$ and thus $A \to^+ \mathsf{K}AB$ or $A \to^+ \mathsf{I}A$, respectively. Then the reduction graph $\mathcal{G}(M)$ is infinite.

Since the cycle contains a root step, it follows that none of the terms $O_0, \ldots, O_m$ on the cycle is in head normal form. By Lemma 3.2 it follows that every redex contracted in the cycle is a spine redex.

We know that the cycle contains a root step. Note that the cycle must contain root steps of form (ii). For otherwise, by Proposition 3.3, no step on the cycle increases $r(\cdot)$ and there is at least one step of form (i) that decreases $r(\cdot)$. ⊣

Actually, we can strengthen this theorem as follows, as done in [30]. There standardisation for CL-reductions is used, together with a pumping-like argument. The new argument presented below is drastically shorter, but employs a theorem (not available at the time of writing [30]) that requires some effort.

THEOREM 3.5. *Finite reduction graphs $\mathcal{G}(M)$ are acyclic.*

Proof. Assume that $\mathcal{G}(M)$ is finite and contains a cycle $N \to^+ N$. Then we can pick a subterm O of N such that O has a minimal cycle. This minimal cycle must contain a root step and consequently O admits an infinite head reduction.[3] Since $\mathcal{G}(M)$ is finite, the infinite head reduction of O must ends in a cycle. However, then Theorem 3.4 contradicts finiteness of $\mathcal{G}(M)$. ⊣

COROLLARY 3.6. *There are no pure cycles in Combinatory Logic based on $\{\mathsf{S},\mathsf{K},\mathsf{I}\}$.*

Proof. If there would be such a cycle, it is a leftmost reduction, because each term M on the cycle contains only one redex. According to Theorem 3.4 $\mathcal{G}(M)$ is infinite, in contradiction with the purity of the cycle. ⊣

---

[3] Here we use the theorem that having an infinite reduction with infinitely many head steps, is equivalent with having an infinite head reduction; so head reduction steps can be made consecutive. A proof is in [7, Theorem 2.1]. The theorem is also mentioned in [2, Exercise 13.6.13, page 351], but without proof.

Let us mention that in CL(S) there are no cycles whatsoever [3]. Moreover, there are no ground spirals, that is, there exists no ground term t such that $t \to^+ C[t]$, see [3, 35]. It is *open* whether there are open terms that admit spiralling reductions $t \to^+ C[t\sigma]$. The absence of ground spirals also holds for CL(δ) where $\delta ab \to b(ab)$, see [4].

EXAMPLE 3.7. *The basis* $\{B, C, W, K\}$ *for CL. This basis is mentioned e.g. in [11]. The reduction rules are*

$$Babc \to a(bc) \qquad Cabc \to acb \qquad Wab \to abb \qquad Kab \to a$$

We have the pure loop $WWW \to WWW$.

EXAMPLE 3.8. *The basis* $\{I, S, C, B\}$ *for* CL_I, *the λI-version of CL. Let* $D \equiv S(CI)(CI)$, *then we have the pure cycle*

$$DD \to CID(CID) \to I(CID)D \to CIDD \to IDD \to DD$$

CONJECTURE 3.9. *In Combinatory Logic with basis* $\{S, K, I\}$, *head reduction is acyclic.*

Here we note that in CL(S, K, I) there is no term D such that $Dx \twoheadrightarrow_{\text{head}} xx$, as proved by Waldmann in [35]; if there was, the conjecture was instantly refuted.

QUESTION 1. *What combinators can be realised in CL(S, K, I) with only head reduction? As stated, the duplicator D cannot be realised.*

QUESTION 2. *Another basis for CL_I is* $\{I, J\}$ *with* $Jabcd \to ab(adc)$. *Are there pure cycles in this basis?*

BIBLIOGRAPHY

[1] S. Abramsky and C.-H. L. Ong. Full Abstraction in the Lazy Lambda Calculus. *Information and Computation*, 105(2):159–267, 1993.
[2] H. P. Barendregt. *The Lambda Calculus, its Syntax and Semantics*, volume 103 of *Studies in Logic and The Foundations of Mathematics*. North-Holland, revised edition, 1984.
[3] H.P. Barendregt, J. Endrullis, J.W. Klop, and J. Waldmann. Dance of the Starlings. 2015. To be published on arXiv.org.
[4] H.P. Barendregt, J. Endrullis, J.W. Klop, and J. Waldmann. Songs of the Starling and the Owl. 2015. To be published on arXiv.org.
[5] A. Berarducci. Infinite λ-Calculus and Non-Sensible Models. In *Logic and Algebra (Pontignano, 1994)*, pages 339–377. Dekker, New York, 1996.
[6] A. Berarducci and B. Intrigila. Some New Results on Easy lambda-Terms. *Theoretical Computer Science*, 121(1&2):71–88, 1993.
[7] A. Berarducci and B. Intrigila. Church–Rosser λ-theories, Infinite λ-calculus and Consistency Problems. *Logic: From Foundations to Applications*, pages 33–58, 1996.
[8] J. A. Bergstra and J. W. Klop. Church-Rosser Strategies in the Lambda Calculus. *Theoretical Computer Science*, 9:27–38, 1979.
[9] C. Böhm and S. Micali. Minimal Forms in lambda-Calculus Computations. *Journal of Symbolic Logic*, 45(1):165–171, 1980.
[10] A. Bucciarelli, A. Carraro, G. Favro, and A. Salibra. A graph-easy class of mute lambda-terms. In *Proc. of the 15th Italian Conference on Theoretical Computer Science*, volume 1231, pages 59–71. CEUR-WS.org, 2014.
[11] H .B. Curry and R. Feys. *Combinatory Logic*, volume I. North-Holland, 1958.
[12] J. Endrullis, H. Geuvers, J. G. Simonsen, and H. Zantema. Levels of Undecidability in Rewriting. *Information and Computation*, 209(2):227–245, 2011.
[13] J. Endrullis, H. Geuvers, and H. Zantema. Degrees of Undecidability in Term Rewriting. In *Proc. Conf. on Computer Science Logic (CSL 2009)*, volume 5771 of *LNCS*, pages 255–270. Springer, 2009.

[14] J. Endrullis, C. Grabmayer, and D. Hendriks. Data-Oblivious Stream Productivity. In *Proc. Conf. on Logic for Programming Artificial Intelligence and Reasoning (LPAR 2008)*, number 5330 in LNCS, pages 79–96. Springer, 2008.
[15] J. Endrullis, C. Grabmayer, D. Hendriks, J. W. Klop, and R. C. de Vrijer. Proving Infinitary Normalization. In *Postproc. Conf. on Types for Proofs and Programs (TYPES 2008)*, volume 5497 of *Lecture Notes in Computer Science*, pages 64–82. Springer, 2009.
[16] J. Endrullis, C. Grabmayer, and R. C. de Vrijer. Equivalence of Two Versions of SN^∞ (w-SN^∞ and s-SN^∞). 2008.
[17] J. Endrullis, H. Hvid Hansen, D. Hendriks, A. Polonsky, and A. Silva. A Coinductive Framework for Infinitary Rewriting and Equational Reasoning. In *Proc. Conf. on Rewriting Techniques and Applications (RTA 2015)*, Leibniz International Proceedings in Informatics. Schloss Dagstuhl, 2015.
[18] J. Endrullis and D. Hendriks. Lazy Productivity via Termination. *Theoretical Computer Science*, 412(28):3203–3225, 2011.
[19] J. Endrullis, D. Hendriks, and J. W. Klop. Modular Construction of Fixed Point Combinators and Clocked Böhm Trees. In *Proc. Symp. on Logic in Computer Science (LICS 2010)*, pages 111–119. IEEE Computer Society, 2010.
[20] J. Endrullis, D. Hendriks, J. W. Klop, and A. Polonsky. Clocks for functional programs. In *The Beauty of Functional Code*, pages 97–126. Springer, 2013.
[21] J. Endrullis, D. Hendriks, and J.W. Klop. Highlights in Infinitary Rewriting and Lambda Calculus. *Theoretical Computer Science*, 464:48–71, 2012.
[22] J. Endrullis, D. Hendriks, J.W. Klop, and A. Polonsky. Discriminating Lambda-Terms Using Clocked Böhm Trees. *Logical Methods in Computer Science*, 10(2:4):1–29, 2014.
[23] J. Endrullis, D. Hendriks, J.W. Klop, and A. Polonsky. Clocked Lambda Calculus. *Mathematical Structures in Computer Science*, 2015. In print.
[24] J. Endrullis, J.W. Klop, and A. Polonsky. Reduction Cycles in Lambda Calculus and Combinatory Logic (Coloured and Extened Version). *ArXiv*, 2016.
[25] J. Endrullis and A. Polonsky. Infinitary Rewriting Coinductively. In *Proc. Types for Proofs and Programs (TYPES 2012)*, volume 19 of *Leibniz International Proceedings in Informatics*, pages 16–27. Schloss Dagstuhl, 2013.
[26] S. Hirokawa. One-step recurrent terms in λ-calculus. Technical report.
[27] S. Hirokawa. Some properties of one-step recurrent terms in lambda calculus. Technical report.
[28] J. R. Kennaway, J. W. Klop, M. R. Sleep, and F.-J. de Vries. An Infinitary Church–Rosser Property for Non-Collapsing Orthogonal Term Rewriting Systems. Technical Report CS-R9043, CWI, 1990.
[29] J. W. Klop, V. van Oostrom, and F. van Raamsdonk. Reduction Strategies and Acyclicity. In *Rewriting, Computation and Proof*, volume 4600 of *Lecture Notes in Computer Science*, pages 89–112. Springer, 2007.
[30] J.W. Klop. Reduction cycles in combinatory logic. In R. Hindley and J.P. Seldin, editors, *Essays on Combinatory Logic, Algebra, Lambda Calculus and Formalism*. Academic Press, 1980.
[31] B. Lercher. Lambda-Calculus Terms That Reduce To Themselves. *Notre Dame Journal of Formal Logic*, 17(2):291–292, 1976.
[32] S. Sekimoto and S. Hirokawa. One-step recurrent terms in lambda-beta-calculus. *Theor. Comput. Sci.*, 56:223–231, 1988.
[33] J. G. Simonsen. Weak Convergence and Uniform Normalization in Infinitary Rewriting. In *Proc. Conf. on Rewriting Techniques and Applications (RTA 2009)*, volume 6 of *Leibniz International Proceedings in Informatics*, pages 311–324. Schloss Dagstuhl, 2010.
[34] Albert Visser, 1976. Personal communication.
[35] J. Waldmann. The Combinator S. *Information and Computation*, 159(1–2):2–21, 2000.

Linear Depth Increase of Lambda Terms in Leftmost-Outermost Rewrite Sequences

CLEMENS GRABMAYER

Abstract

Accattoli and Dal Lago have recently proved that the number of steps in a leftmost-outermost β-reduction rewrite sequence to normal form provides an invariant cost model for the Lambda Calculus. They sketch how to implement leftmost-outermost rewrite sequences on a reasonable machine, with a polynomial overhead, by using simulating rewrite sequences in the linear explicit substitution calculus.

I am interested in an implementation that demonstrates this result, but uses graph reduction techniques similar to those that are employed by runtime evaluators of functional programs. As a crucial stepping stone I prove here the following property of leftmost-outermost β-reduction rewrite sequences in the Lambda Calculus: For every λ-term M with depth d it holds that in every step of a leftmost-outermost β-reduction rewrite sequence starting on M the term depth increases by at most d, and hence that the depth of the n-th reduct of M in such a rewrite sequence is bounded by $d \cdot (n+1)$.

Dedicated to Albert Visser on the occasion of his retirement,
with much gratitude for my time in his group in Utrecht,
and with my very best wishes for the future!

1 Introduction

Recently Accattoli and Dal Lago [1, 2] have proved that the number of steps in a leftmost-outermost rewrite sequence to normal form provides an invariant cost model for the Lambda Calculus. That is, there is an implementation I on a reasonable machine (e.g., a Turing machine, or a random access machine) of the partial function that maps a λ-term to its normal form whenever that exists such that I has the following property: there is a bivariate integer polynomial $p(x, y)$ such that if a λ-term M of size m has a leftmost-outermost rewrite sequence of length n to a normal form N, then I obtains a compact representation of N from M in time bounded by $p(n, m)$. Accattoli and Dal Lago first simulate leftmost-outermost rewrite sequences by 'leftmost-outermost useful' rewrite sequences in the linear explicit substitution calculus, using the restriction that only those steps are performed that facilitate the simulation of leftmost-outermost β-reduction steps. Subsequently they show that such rewrite sequences can be implemented on a reasonable machine.

I am interested in obtaining a graph rewriting implementation for leftmost-outermost β-reduction in the Lambda Calculus that demonstrates this result, but that is close in spirit to graph reduction as it is widely used for the implementation of functional programming languages. In particular, my aim is to describe a port graph grammar [7] implementation that is based on TRS (term rewrite system) representations of λ-terms. Such λ-term representations correspond closely to supercombinator systems that are obtained by lambda-lifting, or fully-lazy lambda-lifting, as first described by to Hughes [6].

That such an implementation is feasible by employing subterm sharing is suggested by the following property of (plain, unshared) leftmost-outermost β-reduction rewrite sequences in the Lambda Calculus, which will be shown here. The depth increase in the steps of an arbitrarily long leftmost-outermost rewrite sequence from a λ-term M is uniformly bounded by $|M|$, the depth of M. As a consequence, for the depth of the n-th reduct L_n of a λ-term M in a leftmost-outermost rewrite sequence it holds: $|L_n| \leq |M| \cdot (n + 1)$.

In the terminology of [1, 2] this property shows that leftmost-outermost rewrite sequences do not cause 'depth explosion' in λ-terms. General $\to_\beta$ rewrite sequences do not enjoy this property, as along them the depth can increase exponentially, which is shown by the example below.

EXAMPLE 1.1 ('depth-exploding' family, from Asperti and Lévy [3]). Consider the following λ-terms:

$$M_0 := xx \qquad M_{i+1} := \textit{two}(\lambda x.M_i)x \qquad N_0 := M_0 = xx \qquad N_{i+1} := N_i[x := N_i]$$

where $\textit{two} := \lambda x.\lambda y.x(xy)$ is the Church numeral for 2. By induction on i it can be verified that:

$$|M_i| = \left\{ \begin{array}{r} 1 \text{ if } i = 0 \\ 3(i+1) \text{ if } i \geq 1 \end{array} \right\} \in O(i) \qquad M_i \to_\beta^{4i} N_i \qquad |N_i| = 2^i$$

and that the syntax tree of N_i is the complete binary application tree of depth $|N_i| = 2^i$ with at depth 2^i occurrences of x. The induction step for the statement on the rewrite sequence can be performed as follows:

$$M_{i+1} = \textit{two}(\lambda x.M_i)x \to_\beta^{4i} \textit{two}(\lambda x.N_i)x = (\lambda x.\lambda y.x(xy))(\lambda x.N_i)x \qquad \text{(by the ind. hyp.)}$$
$$\to_\beta (\lambda y.(\lambda x.N_i)((\lambda x.N_i)y))x$$
$$\to_\beta (\lambda x.N_i)((\lambda x.N_i)x) \to_\beta (\lambda x.N_i)N_i \to_\beta N_i[x := N_i] = N_{i+1}$$

Note that this $\to_\beta$ rewrite sequence is not leftmost-outermost, but essentially inside-out. Now let $i \geq 1$. Then for $n = 4i$ and $M := M_i$ we find: $M = M_i = L_0 \to_\beta^n L_n = N_i$ with $|M| = |M_i| = 3(i+1) \leq 4i = n$ and $|L_n| = |N_i| = |N_i| = 2^i = 2^{n/4}$. Such an exponential depth increase contradicts the depth increase result that we will show for leftmost-outermost rewrite sequences, since, in the situation here, $|L_n| \leq |M| \cdot (n+1)$ would imply that $2^{n/4} \leq n(n+1)$ holds for infinitely many n.

The result on the linear depth increase of λ-terms along leftmost-outermost rewrite sequences will be shown here for TRS-representations of λ-terms, which will be called 'λ-TRSs'. These representations of λ-terms as orthogonal TRSs correspond to systems of 'supercombinators', which are obtained by the lambda-lifting transformation that is widely used in the implementation of functional programming languages [6]. Lambda-lifting transforms higher-order terms with binding such as λ-terms, or indeed functional programs (which can be viewed as generalized λ-terms with case and letrec constructs) into first-order terms, namely systems of combinator definitions that are called supercombinators. λ-TRSs are TRS-versions of systems of supercombinators. They are well-suited for the purpose of showing the linear depth increase result for much of the same reason why supercombinators are so useful for the evaluation of functional programs: after representing the initial term by a finite number of first-order rewrite rules (through lambda-lifting), the evaluation proceeds by repeatedly searching, and then contracting, the next leftmost-outermost redex with respect to one of these rules. This easy form of the evaluation procedure facilitates a straightforward proof of the linear depth increase invariant for steps of leftmost-outermost rewrite sequences.

While the linear depth increase statement will be shown for rewrite sequences in a TRS for simulating leftmost-outermost β-reduction, its transfer to λ-terms via a lifting theorem along lambda-lifting will only be sketched. The lifting and projection statements needed for this part are largely analogous to proofs for the correctness of fully-lazy lambda-lifting as described by Balabonski [4].

The linear depth increase property for leftmost-outermost rewrite sequences contrasts starkly with the fact that 'size explosion' can actually take place. There are λ-terms M_n of size $O(n)$ (linear size in n) such that M_n reduces in n leftmost-outermost β-reduction steps to a term of size $\Omega(2^n)$ (proper exponential size in n). As an example Accattoli and Dal Lago [1, 2] exhibit the family $\{M_n\}_n$ of $O(n)$-sized λ-terms with $M_0 = yxx$, and $M_{n+1} = (\lambda x.M_n)M_0$ for $n > 1$, which reduce in n leftmost-outermost β-reduction steps to the corresponding, $\Omega(2^n)$-sized term of the family $\{N_n\}_n$ with $N_0 = M_0$, and $N_{n+1} = N_n[x := N_0]$.

The property of the linear depth increase along leftmost-outermost rewrite sequences suggests an alternative proof of the result by Accattoli and Dal Lago, now based on a graph rewriting implementation. The crucial idea is to use (directed acyclic) graph representations of terms in λ-TRSs, which can safeguard that the implementation preserves the linear depth increase property, and to employ the power of sharing to avoid size explosion of the graph representations.[1] If additionally the overhead for the search and the simulation of the next redex contraction can be bounded polynomially in the present graph size, then leftmost-outermost rewrite sequences can be simulated efficiently (first by graph rewrite steps, which subsequently can be implemented efficiently on a reasonable machine). In Section 4 we sketch the basic idea for a graph implementation in which subterm sharing guarantees that the size increase of the graph that represents a λ-term is polynomial in the number of simulated leftmost-outermost β-reduction steps.

Notation. By $\mathbb{N} = \{0, 1, 2, \ldots\}$ we denote the natural numbers including 0. For term rewriting systems, terminology and notation from the book [8] is used, shortly summarized here. A signature Σ is a set of function symbols together with an arity function. For signature Σ we denote by $\text{Ter}(\Sigma)$ the set of terms over Σ, and by $\text{Ter}^\infty(\Sigma)$ the set of infinite terms over Σ. For a term s, $|s|$ denotes the depth of s, that is, the longest path in the syntax tree of s from the root to a leaf. A term rewriting system (TRS) is a pair $\langle \Sigma, R \rangle$ consisting of a signature Σ, and a set R of rules for terms over Σ (subject to the usual restrictions). For a TRS with rewrite relation $\to$, the many-step (zero, one or more step) rewrite relation is denoted by $\twoheadrightarrow$, and the n step rewrite relation by $\to^n$, for $n \in \mathbb{N}$. Constrasting with terms in a TRS (first-order terms), λ-terms are viewed as α-equivalence classes of pseudo-term representations with names for bound variables. For λ-terms, $\to_\beta$ denotes β-reduction, and $\to_{lo\beta}$ leftmost-outermost β-reduction.

---

[1] In the proof by Accattoli and Dal Lago the chosen representations are terms in the linear explicit substitution calculus, and size explosion is avoided by showing that linear size increase holds for 'leftmost-outermost useful' rewrite sequences.

2   Simulation of leftmost-outermost rewrite sequences

We start with the formal definition of first-order representations of λ-terms, called λ-term representations, before describing a TRS for simulating leftmost-outermost β-reduction on λ-term representations.

DEFINITION 2.1 (λ-term representations, denoted λ-terms). Let $\Sigma_\lambda := \{v_j \mid j \in \mathbb{N}\} \cup \{@\} \cup \{(\lambda v_j) \mid j \in \mathbb{N}\}$ be the signature that consists of the *variable* symbols v_j, with $j \in \mathbb{N}$, which are constants (nullary function symbols), the binary *application symbol* @, and the unary *named abstraction* symbols (λv_j), for $j \in \mathbb{N}$.

Now by a λ-*term representation* (a *(first-order) representation of a λ-term*) we mean a closed term in $\mathrm{Ter}(\Sigma_\lambda)$. A λ-term representation s denotes, by reading its symbols in the obvious way, and interpreting occurrences of variable symbols v_j that are not bound, as the variable names x_j, a unique λ-term $[\![s]\!]_\lambda$.

EXAMPLE 2.2. $(\lambda v_0)(v_0)$, $(\lambda v_1)((\lambda v_2)(v_1))$, and $(\lambda v_0)((\lambda v_1)((\lambda v_2)(@(@(v_0,v_1),@(v_1,v_2)))))$ are λ-term representations that denote the λ-terms $I = \lambda x.x$, $K = \lambda xy.x$, and $S = \lambda xyz.xz(yz)$, respectively.

The TRS below is designed to enable the simulation on λ-term representations of the evaluation of λ-terms according to the leftmost-outermost strategy. We formulate it as a motivation for a similar simulation TRS on refined λ-term representations that is introduced later in Definition 2.10, and which will be crucial for obtaining the main result. The idea behind the TRS below is as follows. Applicative terms are uncurried (by steps of rule (search$_2$)) until on the spine of the term a variable or an abstraction is encountered (detected by rules (search$_3$) or (search$_4$)). If an abstraction occurs, and the expression contains an argument for this abstraction, then a step corresponding to a β-contraction is performed (applying rule (contract)), and the procedure continues similarly from there. If there is no argument for such an abstraction, then it is part of a head normal form context, and the evaluation can descend into this abstraction (applying rule (search$_3$)) to proceed in a similar fashion. If a variable occurs on the spine (detected by rule (search$_5$)), it and the recently uncurried applications form a head normal form context, and the simulating evaluation can continue (after applying (search$_5$), (search$_6$), and repeatedly (search$_7$)), possibly in parallel, from any immediate subterm of one of the recently uncurried applications. The rules:

$$losim(x) \to losim_0(x) \qquad \text{(search}_1\text{)}$$
$$losim_n(@(x,y),y_1,\ldots,y_n) \to losim_{n+1}(x,y,y_1,\ldots,y_n) \qquad \text{(search}_2\text{)}$$
$$losim_0((\lambda v_j)(x)) \to (\lambda v_j)(losim_0(x)) \qquad \text{(search}_3\text{)}$$
$$losim_{n+1}((\lambda v_j)(x), y_1, y_2, \ldots, y_{n+1}) \to losim_n(subst(x, v_j, y_1), y_2, \ldots, y_{n+1}) \qquad \text{(contract)}$$
$$losim_0(v_j) \to v_j \qquad \text{(search}_4\text{)}$$
$$losim_{n+1}(v_j, y_1, \ldots, y_{n+1}) \to curry_{n+1}(v_j, y_1, \ldots, y_{n+1}) \qquad \text{(search}_5\text{)}$$
$$curry_1(x, y_1) \to @(x, losim_0(y_1)) \qquad \text{(search}_6\text{)}$$
$$curry_{n+2}(x, y_1, y_2, \ldots, y_{n+2}) \to curry_{n+1}(@(x, losim_0(y_1)), y_2, \ldots, y_{n+2}) \qquad \text{(search}_7\text{)}$$

have to be extended with appropriate rules for *subst* that implement capture-avoiding substitution, but which are not provided here. By $\to_{\text{subst}}$ we denote the rewrite relation induced by these rules for *subst*. By $\to_{\text{contract}}$ we mean the rewrite relation induced by the rule scheme (contract), which defines steps that initiate the simulation of a β-reduction step that proceeds with $\to_{\text{subst}}$ steps for carrying out the substitution. By $\to_{\text{search}}$ we designate the rewrite relation induced by the rules labeled with 'search', which defines steps that search for the next representation of a leftmost-outermost redex below the current position. Finally $\to_{\text{losim}}$ denotes the rewrite relation that is induced by the entire TRS.

EXAMPLE 2.3. We consider the λ-term $M = \lambda x.(\lambda y.y)((\lambda z.\lambda w.wz)x)$. Evaluating M with the leftmost-outermost rewrite strategy, symbolized by the rewrite relation $\to_{\text{lo}}$, gives rise to the rewrite sequence:

$$\lambda x.(\lambda y.y)((\lambda z.\lambda w.wz)x) \to_{\text{lo}\beta} \lambda x.(\lambda z.\lambda w.wz)x \to_{\text{lo}\beta} \lambda x.\lambda w.wx \quad (1)$$

The term $s = (\lambda v_0)(@((\lambda v_1)(v_1), @((\lambda v_2)((\lambda v_3)(@(v_3, v_2))), v_0)))$ denotes M, that is, $[\![s]\!]_\lambda = M$; other variable names are possible modulo 'α-conversion'. Simulating this leftmost-outermost rewrite sequence by means of the simulation TRS above amounts to the following $\to_{\text{losim}}$ rewrite sequence starting on $losim(s)$:

$$\begin{aligned}
losim(s) \quad &\to_{\text{search}} \quad losim_0((\lambda v_0)(@((\lambda v_1)(v_1), @((\lambda v_2)((\lambda v_3)(@(v_3, v_2))), v_0)))) \\
&\to_{\text{search}} \quad (\lambda v_0)(losim_0(@((\lambda v_1)(v_1), @((\lambda v_2)((\lambda v_3)(@(v_3, v_2))), v_0)))) \\
&\to_{\text{search}} \quad (\lambda v_0)(losim_1((\lambda v_1)(v_1), @((\lambda v_2)((\lambda v_3)(@(v_3, v_2))), v_0))) \\
&\to_{\text{contract}} \quad (\lambda v_0)(losim_0(subst(v_1, v_1, @((\lambda v_2)((\lambda v_3)(@(v_3, v_2))), v_0)))) \\
&\twoheadrightarrow_{\text{subst}} \quad (\lambda v_0)(losim_0(@((\lambda v_2)((\lambda v_3)(@(v_3, v_2))), v_0))) \\
&\to_{\text{search}} \quad (\lambda v_0)(losim_1((\lambda v_2)((\lambda v_3)(@(v_3, v_2))), v_0)) \\
&\to_{\text{contract}} \quad (\lambda v_0)(losim_0(subst((\lambda v_3)(@(v_3, v_2)), v_2, v_0))) \\
&\twoheadrightarrow_{\text{subst}} \quad (\lambda v_0)(losim_0((\lambda v_3)(@(v_3, v_0)))) \\
&\to_{\text{search}} \quad (\lambda v_0)((\lambda v_3)(losim_0(@(v_3, v_0)))) \\
&\to_{\text{search}} \quad (\lambda v_0)((\lambda v_3)(losim_1(v_3, v_0))) \\
&\to_{\text{search}} \quad (\lambda v_0)((\lambda v_3)(curry_1(v_3, v_0))) \\
&\to_{\text{search}} \quad (\lambda v_0)((\lambda v_3)(@(v_3, losim_0(v_0)))) \\
&\to_{\text{search}} \quad (\lambda v_0)((\lambda v_3)(@(v_3, v_0)))
\end{aligned}$$

Note that the $\to_{\text{contract}}$ steps indeed initiate, and the $\to_{\text{subst}}$ steps complete, the simulation of corresponding β-reduction steps in the $\to_{\text{lo}}$ rewrite sequence on λ-terms above, while $\to_{\text{search}}$ steps organize the search for the next (λ-term representation of a) leftmost-outermost β-redex. The $\to_{\text{lo}\beta}$ rewrite sequence (1) can be viewed as the projection of the $\to_{\text{losim}}$ rewrite sequence above under an extension of the denotation operation $[\![\cdot]\!]_\lambda$ on λ-term representations yielding λ-terms (which works out substitutions, and interprets uncurried application expressions $losim_n(s, t_1, \ldots, t_n)$ appropriately). Along this projection $\to_{\text{search}}$ and $\to_{\text{subst}}$ steps vanish, but $\to_{\text{contract}}$ steps project to $\to_{\text{lo}\beta}$ steps.

While the TRS above facilitates the faithful representation of leftmost-outermost rewrite sequences on λ-terms (in analogy with Lemma 2.15, see page 134), it does not lend itself well to the purpose of proving the linear depth increase result. In particular, it is not readily clear which invariant for reducts t of a term s in rewrite sequences $\sigma : s \twoheadrightarrow_{\text{losim}} t \to_{\text{losim}} u$ could make it possible to prove that the depth increase in the final step of σ is bounded by a constant d that only depends on the initial term s of the sequence (but not on t). Therefore it is desirable to consider extensions of first-order λ-term representations in which representations of leftmost-outermost β-redexes are built up from contexts that trace back to contexts in the initial term of the rewrite sequence.

λ-TRSs

We now formally define λ-TRSs as orthogonal TRSs that are able to represent λ-terms. The basic idea is that, for a λ-term M, function symbols that are called 'scope symbols' are used to represent abstractions scopes. Hereby the scope of an abstraction $\lambda x.L$ in M includes the abstraction λx and all occurrences of the bound variable x, but may leave room for subterms in L without occurrences of x bound by the abstraction. For example, the λ-term $\lambda x.zxyx$ may be denoted, with the binary scope symbol f that represents the scope of x, as the term $f(z, y)$. (In our formalization of λ-term representations the free variables z and y will be replaced by variable constants, yielding for example the term $f(\mathsf{v}_2, \mathsf{v}_1)$.) Furthermore, scopes are assumed to be strictly nested. Every scope symbol defines a rewrite rule that governs the behavior of the application of the scope to an argument. In the case of the λ-term $\lambda x.zxyx$ this leads to the first-order rewrite rule $@(f(z, y), x) \to @(@(@(z, x), y), x)$.

As mentioned earlier, λ-TRSs are TRS-representations of systems of supercombinators that are obtained by the lambda-lifting transformation. I have been familiarized with these λ-term representations by Vincent van Oostrom who pointed me to the studies of optimal reduction for weak β-reduction (β-reduction outside of abstractions or in 'maximal free' subexpressions) by Blanc, Lévy, and Maranget [5], and encouraged work by Balabonski [4] on characterizations of optimal-sharing implementations for weak β-reduction by term labellings, and on the relation with lambda-lifting.

DEFINITION 2.4 (λ-TRSs). A *λ-TRS* is a pair $\mathcal{L} = \langle \Sigma, R \rangle$, where Σ is a signature containing the binary application symbol @, and the *scope symbols* in $\Sigma^- := \Sigma \setminus \{@\}$, and where $R = \{\rho_f \mid f \in \Sigma^-\}$ consists of the *defining rules* ρ_f for scope symbols $f \in \Sigma^-$ with arity k that are of the form:

$$(\rho_f) \quad @(f(x_1, \ldots, x_k), y) \to F[x_1, \ldots, x_k, y]$$

with F a $(k+1)$-ary context that is called the *scope context* for f. For scope symbols $f, g \in \Sigma^-$ we say that f *depends on* the scope symbol g, denoted by $f \circ\!\!- g$, if g occurs in the scope context F for f. We say that $\mathcal{L}$ is *finitely nested* if the converse relation of $\circ\!\!-$, the *nested-into* relation $-\!\!\circ$, is well-founded, or equivalently (using axiom of dependent choice), if there is no infinite chain of the form $f_0 \circ\!\!- f_1 \circ\!\!- f_2 \circ\!\!- \ldots$ on scope symbols $f_0, f_1, f_2, \ldots \in \Sigma^-$.

EXAMPLE 2.5. Let $\mathcal{L} = \langle \Sigma, R \rangle$ be the λ-TRS with $\Sigma^- = \{f/2, g/0, h/0, i/1\}$, and set of rules R as follows:

$(\rho_f) \quad @(f(x_1, x_2), x) \to @(x_1, @(x_2, x))$ $\qquad$ $(\rho_h) \quad @(h, x) \to i(x)$
$(\rho_g) \quad @(g, x) \to x$ $\qquad\qquad\qquad\qquad\qquad\quad$ $(\rho_i) \quad @(i(x_1), x) \to @(x, x_1)$

This finite λ-TRS, which will facilitate to denote the λ-term M in Example 2.3, is also finitely nested, as the depends-on relation consists of merely a single link: $h \circ\!\!- i$.

In order to explain how λ-TRS terms (Definition 2.4) denote λ-term representations (Definition 2.1), we introduce, for every λ-TRS $\mathcal{L}$, an expansion TRS that makes use of the defining rules for the scope symbols in $\mathcal{L}$. Then 'denoted λ-term representations' are defined as normal forms of terms in the expansion TRS.

DEFINITION 2.6 (expansion TRS for a λ-TRS). Let $\mathcal{L} = \langle \Sigma, R \rangle$ be a λ-TRS. The *expansion TRS* $\mathcal{E}(\mathcal{L}) = \langle \Sigma_{\exp}, R_{\exp} \rangle$ *for* $\mathcal{L}$ has the signature $\Sigma_{\exp} := \Sigma \cup \Sigma_\lambda \cup \Sigma_{\text{expand}}$ with $\Sigma_{\text{expand}} := \{expand_i \mid i \in \mathbb{N}\}$ where $expand_i$ is unary for $i \in \mathbb{N}$, and $\Sigma^- \cap (\Sigma_\lambda \cup \Sigma_{\text{expand}}) = \varnothing$, and its set of rules $R_{\exp}$ consists of the rules:

$$expand_i(@(x_1, x_2)) \to @(expand_i(x_1), expand_i(x_2))$$
$$expand_i(f(x_1, \ldots, x_k)) \to (\lambda \mathsf{v}_i)(expand_{i+1}(F[x_1, \ldots, x_k, \mathsf{v}_i])) \quad \text{(where } F \text{ scope context for } f\text{)}$$
$$expand_i(\mathsf{v}_j) \to \mathsf{v}_j$$

By $\to_{\exp}$ we denote the rewrite relation of $\mathcal{E}(\mathcal{L})$.

Since expansion TRSs are orthogonal TRSs, finite or infinite normal forms are unique. Furthermore they are constructor TRSs, i.e. they have rules whose right-hand sides are guarded by constructors. This can be used to show that all terms in an expansion TRS rewrite to a unique finite or infinite normal form.

DEFINITION 2.7 (λ-term representations denoted by λ-TRS-terms). Let $\mathcal{L} = \langle \Sigma, R \rangle$ be a λ-TRS. For $s \in \text{Ter}(\Sigma)$ we denote by $[\![s]\!]^{\mathcal{L}}$ the finite or infinite $\to_{\exp}$-normal form of the term $expand_0(s)$ in $\mathcal{E}(\mathcal{L})$. If it is a λ-term representation, we say $[\![s]\!]^{\mathcal{L}}$ is the *denoted λ-term representation* of s, and write $[\![s]\!]^{\mathcal{L}}_\lambda$ for the λ-term $[\![[\![s]\!]^{\mathcal{L}}]\!]_\lambda$.

PROPOSITION 2.8. *Let $\mathcal{L}$ be a finitely nested λ-TRS. Then for every closed term s of $\mathcal{L}$, $[\![s]\!]^{\mathcal{L}}$ is a finite closed term over Σ_λ, hence a λ-term representation of the λ-term $[\![s]\!]^{\mathcal{L}}_\lambda$.*

EXAMPLE 2.9. With the λ-TRS $\mathcal{L}$ from Example 2.5 the λ-term M in Example 2.3 can be denoted as the term $f(g, h)$ expands to a λ-term representation of M (the final $\twoheadrightarrow_{\exp}$ step consists of two parallel $\to_{\exp}$ steps):

$expand_0(f(g,h)) \to_{\exp} (\lambda \mathsf{v}_0)(expand_1(@(g, @(h, \mathsf{v}_0))))$
$ \to_{\exp} (\lambda \mathsf{v}_0)(@(expand_1(g), expand_1(@(h, \mathsf{v}_0))))$
$ \to_{\exp} (\lambda \mathsf{v}_0)(@((\lambda \mathsf{v}_1)(expand_2(\mathsf{v}_1)), expand_1(@(h, \mathsf{v}_0))))$
$ \to_{\exp} (\lambda \mathsf{v}_0)(@((\lambda \mathsf{v}_1)(\mathsf{v}_1), expand_1(@(h, \mathsf{v}_0))))$
$ \to_{\exp} (\lambda \mathsf{v}_0)(@((\lambda \mathsf{v}_1)(\mathsf{v}_1), @(expand_1(h), expand_1(\mathsf{v}_0))))$
$ \to_{\exp} (\lambda \mathsf{v}_0)(@((\lambda \mathsf{v}_1)(\mathsf{v}_1), @(expand_1(h), \mathsf{v}_0)))$
$ \to_{\exp} (\lambda \mathsf{v}_0)(@((\lambda \mathsf{v}_1)(\mathsf{v}_1), @((\lambda \mathsf{v}_1)(expand_2(i(\mathsf{v}_1))), \mathsf{v}_0)))$
$ \to_{\exp} (\lambda \mathsf{v}_0)(@((\lambda \mathsf{v}_1)(\mathsf{v}_1), @((\lambda \mathsf{v}_1)((\lambda \mathsf{v}_2)(expand_3(@(\mathsf{v}_2, \mathsf{v}_1)))), \mathsf{v}_0)))$
$ \to_{\exp} (\lambda \mathsf{v}_0)(@((\lambda \mathsf{v}_1)(\mathsf{v}_1), @((\lambda \mathsf{v}_1)((\lambda \mathsf{v}_2)(@(expand_3(\mathsf{v}_2), expand_3(\mathsf{v}_1)))), \mathsf{v}_0)))$
$ \twoheadrightarrow_{\exp} (\lambda \mathsf{v}_0)(@((\lambda \mathsf{v}_1)(\mathsf{v}_1), @((\lambda \mathsf{v}_1)((\lambda \mathsf{v}_2)(@(\mathsf{v}_2, \mathsf{v}_1))), \mathsf{v}_0)))$

Hence $[\![f(g,h)]\!]^{\mathcal{L}} = (\lambda \mathsf{v}_0)(@((\lambda \mathsf{v}_1)(\mathsf{v}_1), @((\lambda \mathsf{v}_1)((\lambda \mathsf{v}_2)(@(\mathsf{v}_2, \mathsf{v}_1))), \mathsf{v}_0)))$. This λ-term representation coincides with the term s in Example 2.3 'modulo α-conversion', and $[\![f(g,h)]\!]^{\mathcal{L}}_\lambda = \lambda x.(\lambda y.y)((\lambda z.\lambda w.wz)x) = M$.

Simulation of leftmost-outermost rewrite sequences on λ-TRS-terms

We adapt the TRS for the simulation of leftmost-outermost $\to_\beta$ rewrite sequences on λ-term representations (see page 128) to 'losim-TRSs' that facilitate this simulation on terms of λ-TRSs. For every λ-TRS $\mathcal{L}$, we introduce a 'losim-TRS' with rules that are similar as before but differ for steps involving abstractions. A simulation starts on a term $losim(s)$ where s is a closed λ-TRS term. Therefore initially all abstractions are represented by scope symbols. During the simulation, abstraction representations (λv_i) are produced in stable parts of the term. The final term in the simulation of a leftmost-outermost $\to_\beta$ rewrite sequence on λ-terms will be a λ-term representation (thus with named abstraction symbols, but without scope symbols).

The altered rules concern $\to_{\text{search}}$ steps that descend into an abstraction, and $\to_{\text{contract}}$ steps that simulate the reduction of β-redexes. In both cases before the step the pertaining abstractions are represented by terms with a scope symbol at the root, and then the expansion of this scope symbol as stipulated in the expansion TRS is used. Additional substitution rules are not necessary any more, because the substitution involved in the contraction of a (represented) β-redex can now be carried out by a single first-order rewrite step. This is because such a step includes the transportation of the argument of a redex into the scope context that defines the body of the abstraction. An additional parameter i of the operation symbols $losim_{n,i}$, $curry_{n,i}$ is used to prevent that any two nested named abstractions refer to the same variable name, safeguarding that rewrite sequences denote meaningful reductions on λ-terms.

DEFINITION 2.10 (losim-TRS for λ-TRSs). Let $\mathcal{L} = \langle \Sigma, R \rangle$ be a λ-TRS. The *losim-TRS (leftmost-outermost reduction simulation TRS)* $\mathcal{LO}(\mathcal{L}) = \langle \Sigma_{\text{losim}}, R_{\text{losim}} \rangle$ *for* $\mathcal{L}$ has signature $\Sigma_{\text{losim}} := \Sigma \cup \Sigma_{\text{lored}} \cup \Sigma_\lambda$ with $\Sigma_{\text{lored}} := \{losim\} \cup \{losim_{n,i}, curry_{n,i} \mid n,i \in \mathbb{N}\}$, a signature of *operation* symbols (for simulating leftmost-outermost reduction) consisting of the unary symbol $losim$, and the symbols $losim_{n,i}$ and $curry_{n,i}$ with arity $n+1$, for $n,i \in \mathbb{N}$; the rule set R_{losim} of $\mathcal{LO}(\mathcal{L})$ consists of the following (schemes of) rules, which are indexed by scope symbols $f \in \Sigma^-$, and where F is the scope context for scope symbol f:

$$losim(x) \to losim_{0,0}(x) \qquad (\text{search})_{\text{init}}$$

$$losim_{n,i}(@(x,y), y_1, \ldots, y_n) \to losim_{n+1,i}(x, y, y_1, \ldots, y_n) \qquad (\text{search})^{@}_{n,i}$$

$$losim_{0,i}(f(x_1, \ldots, x_k)) \to (\lambda v_i)(losim_{0,i+1}(F[x_1, \ldots, x_k, v_i])) \qquad (\text{search})^{f}_{0,i}$$

$$losim_{n+1,i}(f(x_1, \ldots, x_k), y_1, y_2, \ldots, y_{n+1}) \to losim_{n,i}(F[x_1, \ldots, x_k, y_1], y_2, \ldots, y_{n+1}) \qquad (\text{contract})^{f}_{n+1,i}$$

$$losim_{0,i}(v_j) \to v_j \qquad (\text{search})^{\text{var}}_{0,i}$$

$$losim_{n+1,i}(v_j, y_1, \ldots, y_{n+1}) \to curry_{n+1,i}(v_j, y_1, \ldots, y_{n+1}) \qquad (\text{search})^{\text{var}}_{n+1,i}$$

$$curry_{1,i}(x, y_1) \to @(x, losim_{0,i}(y_1)) \qquad (\text{search})_{1,i}$$

$$curry_{n+2,i}(x, y_1, y_2, \ldots, y_{n+2}) \to curry_{n+1,i}(@(x, losim_{0,i}(y_1)), y_2, \ldots, y_{n+2}) \qquad (\text{search})_{n+2,i}$$

By $\to_{\text{losim}}$ we denote the rewrite relation of $\mathcal{LO}(\mathcal{L})$. By $\to_{\text{contract}}$ we denote the rewrite relation induced by the rule scheme $(\text{contract})^f$ where $f \in \Sigma^-$ varies among scope symbols of $\mathcal{L}$. By $\to_{\text{search}}$ we denote the rewrite relation induced by the other rules of $\mathcal{LO}(\mathcal{L})$.

EXAMPLE 2.11. For the λ-TRS $\mathcal{L}$ in Example 2.5, we reduce the term $f(g,h)$, which denotes the λ-term M in Example 2.3, in the losim-TRS $\mathcal{LO}(\mathcal{L})$ for $\mathcal{L}$:

$$\begin{aligned}
losim(f(g,h)) &\to_{\text{search}} & losim_{0,0}(f(g,h)) \\
&\to_{\text{search}} & (\lambda\mathsf{v}_0)(losim_{0,1}(@(g,@(h,\mathsf{v}_0)))) \\
&\to_{\text{search}} & (\lambda\mathsf{v}_0)(losim_{1,1}(g,@(h,\mathsf{v}_0))) \\
&\to_{\text{contract}} & (\lambda\mathsf{v}_0)(losim_{0,1}(@(h,\mathsf{v}_0))) \\
&\to_{\text{search}} & (\lambda\mathsf{v}_0)(losim_{1,1}(h,\mathsf{v}_0)) \\
&\to_{\text{contract}} & (\lambda\mathsf{v}_0)(losim_{0,1}(i(\mathsf{v}_0))) \\
&\to_{\text{search}} & (\lambda\mathsf{v}_0)((\lambda\mathsf{v}_1)(losim_{0,1}(@(\mathsf{v}_1,\mathsf{v}_0)))) \\
&\to_{\text{search}} & (\lambda\mathsf{v}_0)((\lambda\mathsf{v}_1)(losim_{1,2}(\mathsf{v}_1,\mathsf{v}_0))) \\
&\to_{\text{search}} & (\lambda\mathsf{v}_0)((\lambda\mathsf{v}_1)(curry_{1,2}(\mathsf{v}_1,\mathsf{v}_0))) \\
&\to_{\text{search}} & (\lambda\mathsf{v}_0)((\lambda\mathsf{v}_1)(@(\mathsf{v}_1,losim_{0,2}(\mathsf{v}_0)))) \\
&\to_{\text{search}} & (\lambda\mathsf{v}_0)((\lambda\mathsf{v}_1)(@(\mathsf{v}_1,\mathsf{v}_0)))
\end{aligned}$$

obtaining an 'α-equivalent' version of the λ-term representation at the end of the simulated leftmost-outermost reduction on λ-term representations in Example 2.3.

In order to define how terms in the losim-TRS denote λ-term representations we have to extend the expansion TRS from Definition 2.6 with rules that deal with operation and named-abstraction symbols.

DEFINITION 2.12 (expansion TRS for losim-TRS-terms). Let $\mathcal{L} = \langle \Sigma, R \rangle$ be a λ-TRS. The *expansion TRS* $\mathcal{E}_{\text{losim}}(\mathcal{L}) = \langle \Sigma_{\text{losim}} \cup \Sigma_{\text{expand}}, R_{\text{exp}} \cup R_{\text{exp}'} \rangle$ *for losim-TRS-terms* has as its signature the union of the signature Σ_{losim} of $\mathcal{LO}(\mathcal{L})$ and the signature Σ_{expand} of $\mathcal{E}(\mathcal{L})$, and as rules the rules R_{exp} of $\mathcal{E}(\mathcal{L})$ together with the set of rules $R_{\text{exp}'}$ that consists of:

$$\begin{aligned}
expand_i((\lambda\mathsf{v}_j)(x)) &\to (\lambda\mathsf{v}_j)(expand_{\max\{i,j\}+1}(x)) \\
expand_i(losim(x)) &\to expand_i(x) \\
expand_i(losim_{0,j}(x)) &\to expand_{\max\{i,j\}}(x) \\
\left.\begin{array}{l} expand_i(losim_{n+1,j}(x,y_1,\ldots,y_{n+1})) \\ expand_i(curry_{n+1,j}(x,y_1,\ldots,y_{n+1})) \end{array}\right\} &\to @(\cdots @(expand_{i'}(x), expand_{i'}(y_1))\ldots, expand_{i'}(y_{n+1}))
\end{aligned}$$
$$\text{(where } i' = \max\{i,j\}\text{)}$$

The rewrite relation of $\mathcal{E}_{\text{losim}}(\mathcal{L})$ will again be denoted by $\to_{\text{exp}}$.

DEFINITION 2.13 (denoted λ-term (representation), extended to losim-TRS-terms). Let $\mathcal{L} = \langle \Sigma, R \rangle$ be a λ-TRS. For terms $s \in \text{Ter}(\Sigma_{\text{losim}})$ in $\mathcal{LO}(\mathcal{L})$, we also denote by $[\![s]\!]^{\mathcal{L}}$ the finite or infinite $\to_{\text{exp}}$-normal form of the term $expand_0(s)$. If it is a λ-term representation, then we say that $[\![s]\!]^{\mathcal{L}}$ is the *denoted λ-term representation* of s, and we again write $[\![s]\!]^{\mathcal{L}}_\lambda$ for the λ-term $[\![[\![s]\!]^{\mathcal{L}}]\!]_\lambda$.

We now sketch the relationship between rewrite sequences in losim-TRSs with β-reduction rewrite sequences on the denoted λ-terms. For this we formulate statements about the projections of $\to_{\text{losim}}$ steps to steps on λ-terms, and about the lifting of leftmost-outermost β-reduction rewrite sequences to leftmost-outermost rewrite sequences in losim-TRSs. These statements can be illustrated by means of the running example. We do not prove these statements here, as they are closely analogous to the correctness statement for fully-lazy lambda-lifting, and in particular, to the correspondence between weak β-reduction steps on λ-terms and combinator reduction steps on supercombinator representations obtained by fully-lazy lambda-lifting. The latter result was formulated and proved by by Balabonski in [4]. The statements below can be established in a very similar manner.

The first statement concerns the projection of $\rightarrow_{\text{losim}}$ steps to $\rightarrow_\beta$ or empty steps on λ-terms.

LEMMA 2.14 (Projection of $\rightarrow_{\text{losim}}$ steps via $\llbracket \cdot \rrbracket_\lambda^{\mathcal{L}}$). *Let $\mathcal{L} = \langle \Sigma, R \rangle$ be a λ-TRS. Let $s \in \text{Ter}(\Sigma_{\text{losim}})$ be a closed term in $\mathcal{LO}(\mathcal{L})$ such that $\llbracket s \rrbracket_\lambda^{\mathcal{L}} = M$ for a λ-term M. Then the following statements hold concerning the projection of $\rightarrow_{\text{losim}}$ steps via $\llbracket \cdot \rrbracket_\lambda^{\mathcal{L}}$ to steps on λ-terms, for all $s_1 \in \text{Ter}(\Sigma_{\text{losim}})$:*

(i) *If $s \rightarrow_{\text{search}} s_1$, then $\llbracket s \rrbracket_\lambda^{\mathcal{L}} = \llbracket s_1 \rrbracket_\lambda^{\mathcal{L}}$, that is, the projection of a $\rightarrow_{\text{search}}$ step via $\llbracket \cdot \rrbracket_\lambda^{\mathcal{L}}$ vanishes.*

(ii) *If $s \rightarrow_{\text{contract}} s_1$, then $\llbracket s \rrbracket_\lambda^{\mathcal{L}} \rightarrow_\beta \llbracket s_1 \rrbracket_\lambda^{\mathcal{L}}$, that is, every $\rightarrow_{\text{contract}}$ step projects via $\llbracket \cdot \rrbracket_\lambda^{\mathcal{L}}$ to a $\rightarrow_\beta$ step.*

(iii) *If $s \rightarrow_{\text{contract}} s_1$ is actually a leftmost-outermost step, then $\llbracket s \rrbracket_\lambda^{\mathcal{L}} \rightarrow_{\text{lo}\beta} \llbracket s_1 \rrbracket_\lambda^{\mathcal{L}}$ holds, that is, leftmost-outermost $\rightarrow_{\text{contract}}$ steps project to leftmost-outermost β-reduction steps.*

The next lemma states that every leftmost-outermost β-reduction step $M \rightarrow_{\text{lo}\beta} M_1$ can be lifted to a leftmost-outermost many-step $s \twoheadrightarrow_{\text{search}} \cdot \rightarrow_{\text{contract}} s_1$ in a losim-TRS, provided that s denotes M, and s has been obtained by the simulation of a $\rightarrow_{\text{lo}\beta}$ rewrite sequence.

LEMMA 2.15 (Lifting of $\rightarrow_{\text{lo}\beta}$ steps to $\twoheadrightarrow_{\text{search}} \cdot \rightarrow_{\text{contract}}$ steps w.r.t. $\llbracket \cdot \rrbracket_\lambda^{\mathcal{L}}$). *Let $\mathcal{L} = \langle \Sigma, R \rangle$ be a λ-TRS. Let $s \in \text{Ter}(\Sigma)$ be a closed term such that $\llbracket s \rrbracket_\lambda^{\mathcal{L}} = M_0$ for a λ-term M_0. Furthermore let $u \in \text{Ter}(\Sigma_{\text{losim}})$ with $\llbracket u \rrbracket_\lambda^{\mathcal{L}} = M$ for a λ-term M be the final term of a leftmost-outermost rewrite sequence $losim(s) \twoheadrightarrow_{\text{losim}} u$.*

Then for a $\rightarrow_{\text{lo}\beta}$ step $\rho : \llbracket u \rrbracket_\lambda^{\mathcal{L}} = M \rightarrow_{\text{lo}\beta} M_1$ with λ-term M_1 as target there are terms $u', u_1 \in \text{Ter}(\Sigma_{\text{losim}})$ and a leftmost-outermost $\rightarrow_{\text{losim}}$ rewrite sequence $\hat{\rho} : u \twoheadrightarrow_{\text{search}} u' \rightarrow_{\text{contract}} u_1$ whose projection via $\llbracket \cdot \rrbracket_\lambda^{\mathcal{L}}$ amounts to the step ρ, and hence, $\llbracket u' \rrbracket_\lambda^{\mathcal{L}} = M$, and $\llbracket u_1 \rrbracket_\lambda^{\mathcal{L}} = M_1$.

Now by using this lemma in a proof by induction on the length of a $\rightarrow_{\text{lo}\beta}$ rewrite sequence the theorem below can be obtained. It justifies the use of losim-TRSs for the simulation of $\rightarrow_{\text{lo}\beta}$ rewrite sequences.

THEOREM 2.16 (Lifting of $\rightarrow_{\text{lo}\beta}$ to leftmost-outermost $\rightarrow_{\text{losim}}$ rewrite sequences). *Let $\mathcal{L} = \langle \Sigma, R \rangle$ be a λ-TRS. Let $s \in \text{Ter}(\Sigma)$ be a closed term with $\llbracket s \rrbracket_\lambda^{\mathcal{L}} = M$ for a λ-term M. Then every $\rightarrow_{\text{lo}\beta}$ rewrite sequence:*

$$\sigma : M = L_0 \rightarrow_{\text{lo}\beta} L_1 \rightarrow_{\text{lo}\beta} \cdots \rightarrow_{\text{lo}\beta} L_k \; (\rightarrow_{\text{lo}\beta} L_{k+1} \rightarrow_{\text{lo}\beta} \cdots)$$

of finite or infinite length l lifts via $\llbracket \cdot \rrbracket_\lambda^{\mathcal{L}}$ to a leftmost-outermost $\rightarrow_{\text{losim}}$ rewrite sequence:

$$\hat{\sigma} : losim(s) = u_0 \twoheadrightarrow_{\text{search}} \cdot \rightarrow_{\text{contract}} u_1 \twoheadrightarrow_{\text{search}} \cdots$$
$$\cdots \rightarrow_{\text{contract}} u_k \; (\twoheadrightarrow_{\text{search}} \cdot \rightarrow_{\text{contract}} u_{k+1} \twoheadrightarrow_{\text{search}} \cdots)$$

with precisely l $\rightarrow_{\text{contract}}$ steps such that furthermore $\llbracket u_i \rrbracket_\lambda^{\mathcal{L}} = L_i$ holds for all $i \in \{0, 1, \ldots, l\}$.

3 Linear depth increase

In this section we establish the main result by first deriving bounds for the depth increase of the denoted λ-terms in $\to_{\text{losim}}$ rewrite sequences in losim-TRSs. In order to reason directly on terms of the losim-TRS, we define the notion of 'λ-term depth' for these terms as the depth of the denoted λ-term representations.

DEFINITION 3.1 (λ-term depth). Let $\mathcal{L} = \langle \Sigma, R \rangle$ be a λ-TRS, and let $\mathcal{LO}(\mathcal{L})$ be the losim-TRS for $\mathcal{L}$. For terms s in $\mathcal{LO}(\mathcal{L})$, the λ-*term (representation) depth* $|s|_\lambda$ of s is defined as the depth of the λ-term representation denoted by s, giving rise to the function $|\cdot|_\lambda : \text{Ter}(\Sigma_{\text{lo}}) \to \mathbb{N} \cup \{\infty\}$, $s \mapsto |s|_\lambda := |[\![s]\!]^{\mathcal{L}}|$.

Since a λ-term representation s and the λ-term $[\![s]\!]_\lambda$ denoted by it have the same depth, the λ-term depth of a term s that denotes a λ-term M is the depth of M.

PROPOSITION 3.2. *Let $\mathcal{L} = \langle \Sigma, R \rangle$ be a λ-TRS, and let $\mathcal{LO}(\mathcal{L})$ be the losim-TRS for $\mathcal{L}$. If for a term s in $\mathcal{LO}(\mathcal{L})$ it holds that $[\![s]\!]_\lambda^{\mathcal{L}} = M$ for a λ-term M, then $|s|_\lambda = |[\![s]\!]^{\mathcal{L}}| = |[\![s]\!]_\lambda^{\mathcal{L}}| = |M|$.*

The following proposition formulates clauses for the λ-term depth depending on the outermost symbol of a term in a losim-TRS. For finitely nested λ-TRSs, these clauses can be read as an inductive definition. They can be proved in a straightforward manner by making use of the definition via the expansion TRS of the λ-term representations $[\![s]\!]^{\mathcal{L}}$ for terms s of the losim-TRS for a λ-TRS $\mathcal{L}$.

PROPOSITION 3.3. *Let $\mathcal{L} = \langle \Sigma, R \rangle$ be a λ-TRS, and let $\mathcal{LO}(\mathcal{L})$ be the losim-TRS for $\mathcal{L}$. The λ-term depth of terms in $\mathcal{LO}(\mathcal{L})$ satisfies the following clauses, for $i, j, n \in \mathbb{N}$, terms $x, t, t_1, t_2, s_1, \ldots s_k$, and $f \in \Sigma^-$:*

$$
\begin{aligned}
|x|_\lambda &= 0 & (x \text{ variable}) \\
|@(t_1, t_2)|_\lambda &= 1 + \max\{|t_1|_\lambda, |t_2|_\lambda\} \\
|f(s_1, \ldots, s_k)|_\lambda &= |(\lambda\mathsf{v}_j)(F[s_1, \ldots, s_k, \mathsf{v}_j])|_\lambda \quad (f \in \Sigma^-, F \text{ as in the rule } \rho_f, \mathsf{v}_j \text{ fresh}) \\
|\mathsf{v}_j|_\lambda &= 0 & (j \in \mathbb{N}) \\
|(\lambda\mathsf{v}_j)(t)|_\lambda &= 1 + |t|_\lambda \\
|losim(t)|_\lambda &= |t|_\lambda \\
\left. \begin{array}{l} |losim_{n,i}(s, t_1, \ldots, t_n)|_\lambda \\ |curry_{n,i}(s, t_1, \ldots, t_n)|_\lambda \end{array} \right\} &= |@(\cdots @(s, t_1) \ldots, t_n)|_\lambda \\
&= \max\{|s|_\lambda + n, |t_1|_\lambda + n, \ldots, |t_n|_\lambda + 1\}
\end{aligned}
$$

PROPOSITION 3.4. *Let $\mathcal{L}$ be a finitely nested λ-TRS, and let $\mathcal{LO}(\mathcal{L})$ be the losim-TRS for $\mathcal{L}$. Then every term $t \in \text{Ter}(\Sigma_{\text{lo}})$ has finite λ-term depth $|t|_\lambda \in \mathbb{N}$, and hence the λ-term depth function on terms of $\mathcal{LO}(\mathcal{L})$ is well-defined of type $|\cdot|_\lambda : \text{Ter}(\Sigma_{\text{lo}}) \to \mathbb{N}$.*

We extend the concept of λ-term depth also to scope symbols. Let $\mathcal{L}$ be a λ-TRS. The λ-*term depth* $|f|_\lambda$ of a scope symbol $f \in \Sigma^-$ with arity k is defined as $|f(x_1, \ldots, x_k)|_\lambda \in \mathbb{N} \cup \{\infty\}$, the λ-term depth of the term $f(x_1, \ldots, x_k)$. Note that if $\mathcal{L}$ is finitely nested, then Proposition 3.4 entails $|f|_\lambda = |f(x_1, \ldots, x_k)|_\lambda \in \mathbb{N}$. We also define $|\mathcal{L}|_\lambda := \max\{|f|_\lambda \mid f \in \Sigma^-\} \in \mathbb{N} \cup \{\infty\}$, the *maximal λ-term depth* of a scope symbol in $\mathcal{L}$. Note that if, in addition to being finitely nested, $\mathcal{L}$ is also finite, then it holds that $|\mathcal{L}|_\lambda < \infty$.

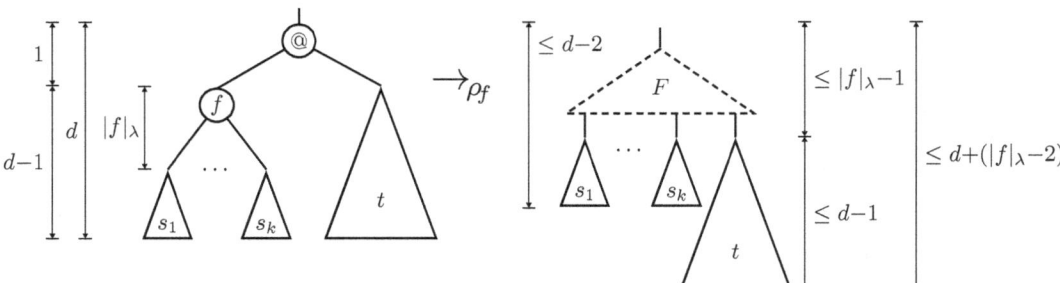

Figure 3.1. Illustration of the λ-term depth increase that is caused by the simulation of a $\to_\beta$ step at the root of a λ-term M on a λ-TRS-term that denotes M: the depth increase in a step $@(f(s_1,\ldots,s_k),t) \to F[s_1,\ldots,s_k,t]$ according to the defining rule ρ_f for the scope symbol f is at most $|f|_\lambda - 2$. The subterm t could be duplicated in the step and occur several times below F, but only one such occurrence is displayed.

PROPOSITION 3.5. *Let $\mathcal{L} = \langle \Sigma, R \rangle$ be a λ-TRS, and let $\mathcal{LO}(\mathcal{L})$ be the losim-TRS for $\mathcal{L}$. If for a term s in $\mathcal{LO}(\mathcal{L})$ it holds that $[\![s]\!]_\lambda^{\mathcal{L}} = M$ for a λ-term M, then $|\mathcal{L}|_\lambda \le |M|$.*

For analyzing the depth increase of steps in losim-TRSs, the following two lemmas will be instrumental. They relate the λ-term depth of contexts filled with terms to the λ-term depths of occurring terms.

LEMMA 3.6. *Let $\mathcal{L} = \langle \Sigma, R \rangle$ be a finitely nested λ-TRS. For all unary contexts C over Σ, terms $s,t \in \text{Ter}(\Sigma)$, and $d \in \mathbb{N}$ the following statements hold:*

$$|s|_\lambda \le |t|_\lambda + d \;\Rightarrow\; |C[s]|_\lambda \le |C[t]|_\lambda + d, \qquad (2)$$
$$|s|_\lambda = |t|_\lambda \;\Rightarrow\; |C[s]|_\lambda = |C[t]|_\lambda. \qquad (3)$$

Proof. Statement (2) can be established by straightforward induction on the structure of the context C, using the clauses concerning the λ-term depth from Proposition 3.3. Statement (3) is an easy consequence. ⊣

LEMMA 3.7. *Let $\mathcal{L} = \langle \Sigma, R \rangle$ be a finitely nested λ-TRS. Then for all $(k+1)$-ary contexts C over Σ, where $k \in \mathbb{N}$, and for all terms $s_1,\ldots,s_k,t \in \text{Ter}(\Sigma)$ it holds:*

$$|C[s_1,\ldots,s_k,t]|_\lambda \le \max\{|C[s_1,\ldots,s_k,x]|_\lambda, |C[x_1,\ldots,x_{k+1}]|_\lambda + |t|_\lambda\}.$$

Proof. By a straightforward induction on the structure of the context C. ⊣

Now we can formulate, and prove, a crucial lemma (Lemma 3.8). Its central statement is that the depth increase in a $\to_{\text{contract}}$ step (with respect to a losim-TRS) at the root of a term is bounded by the depth of the scope context of the scope symbol that is involved in the step. See Figure 3.1 for an illustration of the underlying intuition for the analogous case of a step according to the defining rule of a scope symbol. Then we obtain a lemma (Lemma 3.9) concerning the depth increase in general $\to_{\text{contract}}$ and $\to_{\text{search}}$ steps.

LEMMA 3.8. Let $\mathcal{L} = \langle \Sigma, R \rangle$ be a finitely nested λ-TRS. Then for every scope symbol $f \in \Sigma^-$ with arity k and scope context F, and for all terms $s_1, \ldots, s_k, t \in \text{Ter}(\Sigma)$, and all $i \in \mathbb{N}$, it holds:

(i) $|F[s_1, \ldots, s_k, t]|_\lambda \leq |@(f(s_1, \ldots, s_k), t)|_\lambda + |f|_\lambda - 2$.

(ii) $|losim_{n,i}(F[s_1, \ldots, s_k, t_1], t_2, \ldots, t_{n+1})|_\lambda \leq$
$\leq |losim_{n+1,i}(f(s_1, \ldots, s_k), t_1, \ldots, t_{n+1})|_\lambda + |f|_\lambda - 2$.

Proof. For all f, F, $s_1, \ldots, s_k, t$, and i as assumed in the statement of the lemma, we find:

$$\begin{aligned}
|@(f(s_1, \ldots, s_k), t)|_\lambda - 1 &= \max\{|f(s_1, \ldots, s_k)|_\lambda, |t|_\lambda\} \\
&= \max\{|(\lambda v_j)(F[s_1, \ldots, s_k, v_j])|_\lambda, |t|_\lambda\} \\
&= \max\{1 + |F[s_1, \ldots, s_k, v_j]|_\lambda, |t|_\lambda\}
\end{aligned} \quad (4)$$

$$|F[x_1, \ldots, x_{k+1}]|_\lambda = |(\lambda v_j)(F[s_1, \ldots, s_k, v_j])|_\lambda - 1 = |f(x_1, \ldots, x_k)|_\lambda - 1 = |f|_\lambda - 1 \quad (5)$$

by using clauses from Proposition 3.3. By applying this inequality, we obtain the statement in item (i):

$$\begin{aligned}
|F[s_1, \ldots, s_k, t]|_\lambda &\leq \max\{|F[s_1, \ldots, s_k, x_{k+1}]|_\lambda, |F[x_1, \ldots, x_k, x_{k+1}]|_\lambda + |t|_\lambda\} \\
&\leq \max\{|F[s_1, \ldots, s_k, x_{k+1}]|_\lambda, |t|_\lambda\} + |F[x_1, \ldots, x_k, x_{k+1}]|_\lambda \\
&\leq \max\{1 + |F[s_1, \ldots, s_k, x_{k+1}]|_\lambda, |t|_\lambda\} + |F[x_1, \ldots, x_k, x_{k+1}]|_\lambda \\
&= |@(f(s_1, \ldots, s_k), t)|_\lambda - 1 + |f|_\lambda - 1 \\
&= |@(f(s_1, \ldots, s_k), t)|_\lambda + (|f|_\lambda - 2)
\end{aligned}$$

where the first step is justified by Lemma 3.7, and the forth step by (4) and (5). For the statement in item (ii) we first note, using Proposition 3.3 again, that:

$$\begin{aligned}
&|losim_{n+1,i}(f(s_1, \ldots, s_k), t_1, t_2, \ldots, t_{n+1})|_\lambda \\
&= |@(\ldots @(@(f(s_1, \ldots, s_k), t_1), t_2) \ldots, t_{n+1})|_\lambda \\
&= |losim_{n,i}(@(f(s_1, \ldots, s_k), t_1), t_2, \ldots, t_{n+1})|_\lambda
\end{aligned} \quad (6)$$

holds due to the definition of the λ-term depth via $\to_{\exp}$ steps. With (6) the statement in (ii) follows by using Lemma 3.6 with item (i), and context $C := losim_{n,i}(\Box, t_2, \ldots, t_{n+1})$, and constant $d := |f|_\lambda - 2$. ⊣

LEMMA 3.9. Let $\mathcal{L} = \langle \Sigma, R \rangle$ be a finite, and finitely nested λ-TRS. Then the following statements hold concerning the preservation or the increase of the λ-term depth in steps of the losim-TRS $\mathcal{LO}(\mathcal{L})$ for $\mathcal{L}$ between terms $t_1, t_2 \in \text{Ter}_{\text{lo-red}}(\Sigma_{\text{lo}})$:

(i) If $t_1 \to_{\text{contract}} t_2$, then $|t_2|_\lambda \leq |t_1|_\lambda + (|f|_\lambda - 2)$, where f is the scope symbol involved in the step.

(ii) If $t_1 \to_{\text{search}} t_2$, then $|t_1|_\lambda = |t_2|_\lambda$.

Proof. We first establish the inequality in item (i). For $\twoheadrightarrow_{\text{contract}}$ steps at the root, which are of the form:

$$losim_{n+1,i}(f(s_1,\ldots,s_k),t_1,\ldots,t_{n+1}) \twoheadrightarrow_{\text{contract}} losim_{n,i}(F[s_1,\ldots,s_k,t_1],t_2,\ldots,t_{n+1})$$

the desired inequality follows by using $|F[x_1,\ldots,x_{k+1}]|_\lambda = |f(x_1,\ldots,x_{k+1})|_\lambda - 1 \leq D - 1$ from Lemma 3.8, (ii). For non-root $\twoheadrightarrow_{\text{contract}}$ steps this inequality is lifted into a rewriting context by appealing to Lemma 3.6.

We continue with showing item (ii). By means of the clauses of Proposition 3.3 it is straightforward to check that $\twoheadrightarrow_{\text{search}}$ steps at the root preserve the λ-term depth. This can be extended to all $\twoheadrightarrow_{\text{search}}$ steps by Lemma 3.6, (3). ⊣

Using this lemma we now can obtain, quite directly, our main result concerning the depth increase of terms in $\twoheadrightarrow_{\text{losim}}$ rewrite sequences.

THEOREM 3.10. *Let $\mathcal{L} = \langle \Sigma, R \rangle$ be a finite, and finitely nested λ-TRS, and let $D := |\mathcal{L}|_\lambda$. Then for all finite or infinite $\twoheadrightarrow_{\text{losim}}$ rewrite sequences σ with initial term s and length $l \in \mathbb{N} \cup \{\infty\}$, which can be construed as:*

$$\sigma : s = u_0 \twoheadrightarrow_{\text{search}} u_0' \twoheadrightarrow_{\text{contract}} u_1 \twoheadrightarrow_{\text{search}} \cdots$$
$$\cdots \twoheadrightarrow_{\text{search}} u_{k-1}' \twoheadrightarrow_{\text{contract}} u_k \, (\twoheadrightarrow_{\text{search}} u_k' \twoheadrightarrow_{\text{contract}} \cdots),$$

the following statements hold for all $n \in \mathbb{N}$ with $n \leq l$:

(i) $|u_n|_\lambda = |u_n'|_\lambda$, *and* $|u_{n+1}|_\lambda \leq |u_n'|_\lambda + (D - 2)$ *if $n + 1 \leq l$, that is more precisely, the λ-term depth remains the same in the $\twoheadrightarrow_{\text{search}}$ steps, and it increases by at most $D - 2$ in the $\twoheadrightarrow_{\text{contract}}$ steps.*

(ii) $|u_n|_\lambda, |u_n'|_\lambda \leq |s|_\lambda + (D - 2) \cdot n$, *that is, the increase of the λ-term depth along σ is linear in the number of $\twoheadrightarrow_{\text{contract}}$ steps performed, with $(D - 2)$ as multiplicative constant.*

Proof. Statement (i) follows directly from Lemma 3.9, (i) and (ii). Statement (ii) is obtained by adding up the uniform bound D on the λ-term depth increase of each of the n $\twoheadrightarrow_{\text{contract}}$ steps in the rewrite sequence σ. ⊣

From this theorem we obtain our main theorem, the linear depth increase result for leftmost-outermost β-reduction rewrite sequences, by making use of three auxiliary results we have obtained. Namely first that $\twoheadrightarrow_{\text{lo}\beta}$ rewrite sequences lift to $\twoheadrightarrow_{\text{losim}}$ rewrite sequences (Proposition 2.16); second, that λ-term and λ-term representation depths coincide (Proposition 3.2), and third, that the depth of a λ-TRS that represents a λ-term M is bounded by the depth of M (Proposition 3.5).

THEOREM 3.11 (Linear depth increase in $\twoheadrightarrow_{\text{lo}\beta}$ rewrite sequences). *Let M be a λ-term. Then for every finite or infinite leftmost-outermost rewrite sequence $\sigma : M = L_0 \twoheadrightarrow_{\text{lo}\beta} L_1 \twoheadrightarrow_{\text{lo}\beta} \cdots \twoheadrightarrow_{\text{lo}\beta} L_k \, (\twoheadrightarrow_{\text{lo}\beta} L_{k+1} \twoheadrightarrow_{\text{lo}\beta} \cdots)$ from M with length $l \in \mathbb{N} \cup \{\infty\}$ it holds:*

(i) $|L_{n+1}| \leq |L_n| + |M|$ *for all $n \in \mathbb{N}$ with $n + 1 \leq l$, that is, the depth increase in each step of σ is uniformly bounded by $|M|$.*

(ii) $|L_n| \leq |M| + n \cdot |M| = (n+1) \cdot |M|$, *and hence $|L_n| - |M| \in O(n)$, for all $n \in \mathbb{N}$ with $n \leq l$, that is, the depth increase along σ to the n-th reduct is linear in n, with $|M|$ as multiplicative constant.*

4 Idea for a Graph Rewriting Implementation

The linear depth increase result suggests a directed-acyclic-graph implementation of leftmost-outermost β-reduction that keeps subterms shared as much as possible, particularly in the search for the representation of the next leftmost-outermost redex. The idea is that steps used in the search for the next leftmost-outermost redex do not perform any unsharing, but only use markers to organize the search and keep track of its progress. All such search steps together only increase the size of the graph by at most a constant multiple. The number of search steps necessary for finding the next leftmost-outermost redex is linear in the size of the current graph. Unsharing of the graph only takes place once the next (representation of the) leftmost-outermost redex is found: then the part of the graph between this redex and the root is unshared (copied), and subsequently the (represented) redex is contracted.

Since by Theorem 3.11 the depth of the term L_n after $n \to_{\text{contract}}$ steps in a $\to_{\text{losim}}$ rewrite sequence (and hence after n already performed simulated $\to_{\text{lo}\beta}$ steps) is bounded linearly in n, this also holds for the directed acyclic graph that represents L_n after n simulated $\to_{\text{lo}\beta}$ steps on the sharing graphs. Hence unsharing work necessary for the simulation of the $(n+1)$-th $\to_{\text{lo}\beta}$ step is linear in n. This can be used to show that the size increase of the graph after n contractions of (represented) leftmost-outermost redexes is at most quadratic in n. Since consequently the work for searching and contracting the n-th leftmost-outermost redex is also quadratic in n, such an implementation can make it possible to simulate n leftmost-outermost β-reduction steps on sharing graphs in time that is cubic in n.

Acknowledgement. I want to sincerely thank: Vincent van Oostrom, for familiarizing me with TRS-representations of λ-terms, and with the simulation of weak-β reduction by orthogonal TRSs; Dimitri Hendriks, for his many comments on my drafts, and for his questions about it that always helped me; and Jörg Endrullis, for his suggestion about how to typeset Figure 3.1 with TikZ.

BIBLIOGRAPHY

[1] Beniamino Accattoli and Ugo Dal Lago. Beta Reduction is Invariant, Indeed. In *Proceedings of the joint conference CSL-LICS '14*, pages 8:1–8:10, New York, NY, USA, 2014. ACM.

[2] Beniamino Accattoli and Ugo Dal Lago. Beta Reduction is Invariant, Indeed (Long Version). Technical report, arXiv.org, 2014. http://arxiv.org/abs/1405.3311.

[3] Andrea Asperti and Jean-Jacques Levy. The Cost of Usage in the λ-Calculus. In *Proceedings of LICS 2013*, LICS '13, pages 293–300, Washington, DC, USA, 2013. IEEE Computer Society.

[4] Thibaut Balabonski. A Unified Approach to Fully Lazy Sharing. In *Proceedings of the Symposium POPL '12*, pages 469–480, New York, NY, USA, 2012. ACM.

[5] Tomasz Blanc, Jean-Jacques Lévy, and Luc Maranget. Sharing in the Weak Lambda-Calculus. In *Processes, Terms and Cycles: Steps on the Road to Infinity. Essays dedicated to Jan Willem Klop*, number 3838 in LNCS. Springer, 2005.

[6] John Hughes. Graph Reduction with Supercombinators. Technical Report PRG28, Oxford University Computing Laboratory, June 1982.

[7] Charles Stewart. Reducibility between Classes of Port Graph Grammar. *Journal of Computer and System Sciences*, 65(2):169 – 223, 2002.

[8] Terese. *Term Rewriting Systems*, volume 55 of *Cambridge Tracts in Theoretical Computer Science*. Cambridge University Press, 2003.

Labelled Tableaux for Interpretability Logics

TUOMAS HAKONIEMI AND JOOST J. JOOSTEN

Abstract

In this paper we present a labelled tableau proof system that serves a wide class of interpretability logics. The system is proved sound and complete for any interpretability logic characterised by a frame condition given by a set of universal strict first order Horn sentences. As such, the current paper adds to a better proof-theoretical understanding of interpretability logics.

1 Introduction

Provability logics like the Gödel-Löb logic **GL** describe the structural behaviour of formalized provability in a simple propositional modal language. Interpretability logics are natural extensions of provability logics: they describe the structural behaviour of relative interpretability.

Essentially since Solovay's landmark paper [17] we know that any Σ_1 sound theory that extends elementary arithmetic has the same provability logic **GL**. The situation is very different for interpretability logics. Basically, for two different kind of theories we know the corresponding interpretability logics.

On the one hand Shavrukov [15] and independently Berarducci [1] determined the interpretability logic of any sound and essentially reflexive theory like Peano Arithmetic to be **ILM**. On the other hand, Visser has proven in [18] that the interpretability logic of any sound and finitely axiomatised theory that proves the totality of super-exponentiation –like $I\Sigma_1$– is **ILP**.

In case the base theory is neither finitely axiomatizable nor essentially reflexive, the situation turns out to be much more difficult and actually, to determine the interpretability logics in those situations remain open problems. Some partial results are known in the case of Primitive Recursive Arithmetic ([2]) or in the case when we consider those modal principles that are provable in any reasonable arithmetical theory [8, 9, 11]. In this sense, interpretability logics are in need of more study compared to provability logics.

A miracle happens writes Albert Visser as the first line of [19]: Whereas provability is a Σ_1 complete predicate, the logic **GL** that governs its structural behaviour is nice, well behaved and simple[1]. The situation with interpretability seems even more extreme since Shavrukov has shown in [16] that interpretability is Σ_3 complete and again, the modal logic describing the structural behaviour is nice.

But again here we see a discrepancy between provability and interpretability logics. In the case of interpretability logics we actually know to a much lesser extent *how* nice they are. In particular,

---

[1] "Only" of complexity PSPACE.

apart from some observations on the closed fragment ([3, 10]), close to nothing is known about the computational complexity of interpretability logics. Also, very little is known about well-behaved proof systems for interpretability logics with the sole exception of some work by Sasaki such as [14]. The current paper is intended to add to the proof-theoretic understanding of interpretability logics by studying labelled tableaux proof systems for them.

Tableaux proof systems are tightly related to sequent proof systems and are dual to them in many aspects. Rules in sequent proof systems typically have possibly multiple antecedents and single conclusions/succedents. Moreover, sequent-style proofs generally are trees that have the root at the bottom and are based on validity: the consequence of a rule is valid if (and often only if) all of the antecedents are valid.

On the other hand, rules in tableaux systems typically have single antecedent and possibly multiple succedents. Moreover, Tableaux proofs generally are trees that have the root at the top and are based on satisfiability: the antecedent of a rule is satisfiable if and only if some of the 'consequents'/succedents is satisfiable.

Labelled tableaux introduce extra devices to the syntax that aim to represent the accessibility relation in the corresponding Kripke-style semantics. This extra syntax allows us to give tableaux proof systems for many logics lacking a traditional one, where nodes of the tableaux carry only (sets of) formulas. A precursor for this idea of bringing a bit of semantics into the syntax appears already in [12], and labelled tableaux as they are now known were introduced prominently by Fitting in [5]. Standard references here are [6] and [7]. For more on the history and development of tableaux systems for modal logics see e.g. [7]. Naturally, labels have also been incorporated into sequent calculi. We refer the reader to [13] for details on labelled sequent calculi for modal logics.

Outline of the paper. After introducing the necessary preliminaries in Section 2, we use Section 3 to introduce the *labelled tableaux system* for all interpretability logics **ILX** characterised by a set of first order Horn formulas. It is shown how a *systematic tableau* can be assigned to a finite set Γ of formulas so that the tableau contains all the necessary information as to the satisfiability of Γ.

In Sections 4 and 5 we show that the tableau proofs are sound and complete with respect to **ILX**-validity. In the last section we remark that our results concern most of the interpretability logics encountered in the literature, but not all.

2 Preliminaries

Interpretability logics are propositional modal logics with a unary modality $\Box$ whose dual modality $\Diamond$ is defined as $\Diamond := \neg\Box\neg$ corresponding to provability and consistency respectively, and a binary modality $\rhd$ corresponding to relative interpretability.

In this paper we shall work with the Boolean connectives $\neg$ and $\rightarrow$. Thus, with Prop a countable set of propositional variables, the formulas $\mathcal{F}$ of interpretability logic are defined as

$$\mathcal{F} := \mathsf{Prop} \mid (\neg \mathcal{F}) \mid (\mathcal{F} \rightarrow \mathcal{F}) \mid (\Box \mathcal{F}) \mid (\mathcal{F} \rhd \mathcal{F}).$$

As always we will use the other connectives and Boolean constants freely since they can be defined from $\neg$ and $\rightarrow$. In order to use less parentheses we omit outer parentheses and shall say that $\neg, \Box$ and $\Diamond$ bind strongest, followed by the equally strong binding $\vee$ and $\wedge$ who bind stronger than

$\triangleright$ which in turn binds stronger than $\rightarrow$. Thus, for example,

$$p \triangleright q \rightarrow p \wedge \Box r \triangleright q \wedge \Box r$$

is short for

$$\left[(p \triangleright q) \rightarrow \left((p \wedge (\Box r)) \triangleright (q \wedge (\Box r)) \right) \right].$$

DEFINITION 2.1. *The axioms of the basic interpretability logic* **IL** *are, apart from all substitution instances (in the language of interpretability) of all propositional tautologies, given by the following axiom schemata*

L1 $\Box(A \rightarrow B) \rightarrow (\Box A \rightarrow \Box B)$;

L2 $\Box(\Box A \rightarrow A) \rightarrow \Box A$;

J1 $\Box(A \rightarrow B) \rightarrow A \triangleright B$;

J2 $(A \triangleright B) \wedge (B \triangleright C) \rightarrow A \triangleright C$;

J3 $(A \triangleright C) \wedge (B \triangleright C) \rightarrow A \vee B \triangleright C$;

J4 $A \triangleright B \rightarrow (\Diamond A \rightarrow \Diamond B)$;

J5 $\Diamond A \triangleright A$.

The rules are Modus Ponens and Necessitation: $A/\Box A$.

The following lemma collects two easily obtainable and well-known properties of **IL** that will play prominent role in our tableaux systems.

LEMMA 2.2.

1. **IL** $\vdash \Diamond A \rightarrow \Diamond(A \wedge \Box \neg A)$;
2. **IL** $\vdash B \triangleright B \wedge \Box \neg B$.

The logic **GL** is the fragment of **IL** where the modal language is restricted to $\Box$. We shall consider various extensions of **IL**. By **ILX** we denote the logic that arises by adding the axiom scheme(s) X to **IL**. The extensions of **IL** obtained by the following axiom schemes play a prominent role in the literature.

P: $A \triangleright B \rightarrow \Box(A \triangleright B)$;

M: $A \triangleright B \rightarrow A \wedge \Box C \triangleright B \wedge \Box C$.

Interpretability logics allow for a relational semantics very much in the sense as **GL** does.

DEFINITION 2.3. *An* **IL**-*frame is a triple* $\langle W, R, S \rangle$ *where* W *is a non-empty domain set, whose members are often called* worlds, *and* R *is a binary relation on* W *that is transitive and Noetherian (no infinite chains* $x_0 R x_1 R x_2 \ldots$). S *is a ternary relation on* W *that is often considered as a collection* $\{S_x\}_{x \in W}$ *of binary relations by fixing the first argument* x *of the ternary* S. *It is required* that each S_x is a transitive and reflexive binary relation on $\{y \in W : xRy\}$ satisfying the following property:

$$\text{if } xRyRz, \text{ then } yS_xz.$$

An **IL**-model is a quadruple $\langle W, R, S, V \rangle$, where $\langle W, R, S \rangle$ is an **IL**-frame and V is a function assigning a collection $V(p)$ of worlds to a propositional variable p. Given an **IL**-model $\langle W, R, S, V \rangle$ we define a forcing relation $\Vdash$ between worlds and formulas as usual:

- $M, x \Vdash p \Leftrightarrow x \in V(p)$;

- $M, x \Vdash \neg A \Leftrightarrow M, x \nVdash A$;

- $M, x \Vdash A \to B \Leftrightarrow M, x \nVdash A \text{ or } M, x \Vdash B$;

- $M, x \Vdash \Box A \Leftrightarrow \forall y (xRy \Rightarrow M, y \Vdash A)$;

- $M, x \Vdash A \rhd B \Leftrightarrow \forall y \big(xRy \wedge M, y \Vdash A \Rightarrow \exists z (yS_xz \wedge M, z \Vdash B)\big)$.

We shall write $x \in M$ whenever $M = \langle W, R, S, V \rangle$ with $x \in W$ and likewise for frames. We write $M \vDash A$ to denote that $M, x \Vdash A$ for all $x \in M$. The above defined semantics is good in that one can prove completeness for **IL** as was first done in [4]:

$$\mathbf{IL} \vdash A \iff \forall M \; M \vDash A.$$

An extension **L** of **IL** can be specified either axiomatically or semantically by restricting the class of models for example by specifying so called *frame conditions*. We say that a frame $F := \langle W, R, S \rangle$ validates A and we write $F \vDash A$ whenever for all valuations V on F we have $\langle W, R, S, V \rangle \vDash A$.

A set of first or higher order sentences $\mathcal{C}$ in the language $\{\overline{R}, \overline{S}\}$ with $\overline{R}$ a binary and $\overline{S}$ a ternary first-order relation symbol is called a *frame condition* for a logic **L** extending **IL** whenever we have

$$\langle W, R, S \rangle \vDash \mathbf{L} \iff \langle W, R, S \rangle \vDash_{\mathsf{fo/ho}} \mathcal{C},$$

where in the right-hand side the interpretations of $\overline{R}$ and $\overline{S}$ are R and S, respectively. Then we also say that $\mathcal{C}$ characterises the logic **L**. As always, $\langle W, R, S \rangle \vDash \mathbf{L}$ denotes that $\langle W, R, S \rangle \vDash A$ for any theorem A of **L** and we use a similar convention for models. From now on we will use the same symbol R for $\overline{R}$ and its interpretation and likewise for S.

In case an axiomatic extension **ILX** of **IL** is characterised by a set of strict universal Horn sentences in the language $\{R, S\}$ we say that **ILX** is a *Horn* logic. By a strict universal Horn sentence we mean a first order formula of the form

$$\forall \ldots \forall (\varphi_1 \wedge \ldots \wedge \varphi_n \to \psi),$$

where $n \geq 0$, $\varphi_1, \ldots, \varphi_n$ and ψ are atomic formulas and $\forall \ldots \forall$ denotes the universal closure. In case that **ILX** is a Horn logic, we shall denote the corresponding frame condition by $\mathcal{C}_X$ and call an **IL**-frame satisfying $\mathcal{C}_X$ an **ILX**-frame

For example, the logic **ILP** is characterised by the (universal closure of the) first order formula

$$xRy \wedge yRz \wedge zS_x u \to zS_y u,$$

and **IL** is characterised by the empty frame condition (or $\forall x \, (xRx \to xRx)$ for that matter).

In this paper, we shall – given a frame – reduce the binary modality $\triangleright$ to a series of unary ones. We will do so, so that the corresponding tableaux rules become more amenable. Thus, given an **IL**-frame $\mathcal{F} = \langle W, R, S \rangle$, we introduce new unary modal operators $\Box_x$ for each $x \in W$ and give the following truth definition for the operators in a model M on the frame $\mathcal{F}$

$$M, y \Vdash \Box_x A \Leftrightarrow \forall z \, (yS_x z \Rightarrow M, z \Vdash A).$$

Now it is easy to verify that for any **IL**-model

$$M, x \Vdash A \triangleright B \quad \Leftrightarrow \quad M, x \Vdash \Box(A \to \neg\Box_x \neg B). \tag{1}$$

3 Tableaux for Horn interpretability logics

In this section we define a tableau proof method for interpretability logics which are Horn. Moreover, we will give a systematic tableau procedure for such **ILX** that yields a canonical tableau given a finite set of formulas.

As always, our tableaux will be downward growing trees. Each node of the tree carries a labelled formula. A labelled formula is a pair with a label and a formula. The label corresponds to a possible world where the formula is to be satisfied. We will show the unsatisfiability of a finite set of formulas in case all branches in the systematic tableau close (precise definition follow). In case the systematic tableau contains an open branch, that branch will carry information about a satisfying model.

DEFINITION 3.1. *Labels are strings composed of non-negative integers and letters R and S. The set of all labels is defined recursively as follows:*

- *0 is a label;*

- *If σ is a label, then $\sigma R n$ is a label for all $n \in \mathbb{N}$;*

- *If σ and ρ are labels and ρ is a strict non-empty prefix of σ, then $\sigma S_\rho n$ is a label for all $n \in \mathbb{N}$.*

Now that we have a sufficiently large set of labels we will describe how we generically build (almost) frames from them.

DEFINITION 3.2 (**ILX**-label structure). *Given a Horn logic **ILX** and a set of labels Λ, we define relations $\mathbf{R}_{\mathbf{ILX}}^\Lambda$ and $\mathbf{S}_{\mathbf{ILX}}^\Lambda$ on the set Λ as the least relations on Λ such that:*

1. *If $\sigma, \sigma R n \in \Lambda$, then $\langle \sigma, \sigma R n \rangle \in \mathbf{R}_{\mathbf{ILX}}^\Lambda$ for all labels σ and $n \in \mathbb{N}$;*

2. If $\langle \sigma, \tau \rangle \in \mathbf{R}_{\mathbf{ILX}}^{\Lambda}$ and $\langle \tau, \rho \rangle \in \mathbf{R}_{\mathbf{ILX}}^{\Lambda}$, then $\langle \sigma, \rho \rangle \in \mathbf{R}_{\mathbf{ILX}}^{\Lambda}$;

3. If $\sigma, \rho, \sigma S_\rho n \in \Lambda$, then $\langle \rho, \sigma, \sigma S_\rho n \rangle \in \mathbf{S}_{\mathbf{ILX}}^{\Lambda}$ for all labels σ and ρ and all $n \in \mathbb{N}$;

4. If $\langle \sigma, \tau \rangle \in \mathbf{R}_{\mathbf{ILX}}^{\Lambda}$, then $\langle \sigma, \tau, \tau \rangle \in \mathbf{S}_{\mathbf{ILX}}^{\Lambda}$;

5. If $\langle \rho, \sigma \rangle \in \mathbf{R}_{\mathbf{ILX}}^{\Lambda}$ and $\langle \sigma, \tau \rangle \in \mathbf{R}_{\mathbf{ILX}}^{\Lambda}$, then $\langle \rho, \sigma, \tau \rangle \in \mathbf{S}_{\mathbf{ILX}}^{\Lambda}$;

6. If $\langle \rho, \sigma, \tau \rangle \in \mathbf{S}_{\mathbf{ILX}}^{\Lambda}$ and $\langle \rho, \tau, \upsilon \rangle \in \mathbf{S}_{\mathbf{ILX}}^{\Lambda}$, then $\langle \rho, \sigma, \upsilon \rangle \in \mathbf{S}_{\mathbf{ILX}}^{\Lambda}$;

7. If $\langle \rho, \sigma, \tau \rangle \in \mathbf{S}_{\mathbf{ILX}}^{\Lambda}$, then $\langle \rho, \sigma \rangle \in \mathbf{R}_{\mathbf{ILX}}^{\Lambda}$ and $\langle \rho, \tau \rangle \in \mathbf{R}_{\mathbf{ILX}}^{\Lambda}$;

8. $\langle \Lambda, \mathbf{R}_{\mathbf{ILX}}^{\Lambda}, \mathbf{S}_{\mathbf{ILX}}^{\Lambda} \rangle \vDash_{\mathsf{fo}} \mathcal{C}_{\mathsf{X}}$.

Note that the least relations exist since $\mathcal{C}_{\mathsf{X}}$ is a set of strict first order Horn sentences. If the context allows us so, we will drop both the sub- and the superscripts in $\mathbf{R}_{\mathbf{ILX}}^{\Lambda}$ and $\mathbf{S}_{\mathbf{ILX}}^{\Lambda}$. Moreover, when $\langle \rho, \sigma, \tau \rangle \in \mathbf{S}$ we will denote this by $\sigma \mathbf{S}_\rho \tau$ and likewise for $\mathbf{R}$.

Note that $\mathbf{R}$ is irreflexive in case $\mathbf{ILX}$ is consistent and Λ sufficiently nice. Moreover, apart from $\mathbf{R}$ being Noetherian, all the other properties of $\mathbf{ILX}$-frames are satisfied: $\mathbf{R}$ is transitive; $\mathbf{S}_\rho$ is a relation on $\{\sigma \in \Lambda \mid \rho \mathbf{R} \sigma\}$ that is transitive and reflexive so that $\rho \mathbf{R} \sigma \mathbf{R} \tau \Rightarrow \sigma \mathbf{S}_\rho \tau$.

We will now define the generating rules for tableaux for Horn interpretability logics. As mentioned, the nodes carry *labelled formulas* which consist of a pair $\sigma :: A$, where σ is a label and A is a formula. Recall that the idea of the labels is, that they will correspond to worlds in a model where the corresponding formula will be satisfied if satisfiable.

The rules that we present are not entirely local since, for example, we have to guarantee that new labels have not yet been used in relevant parts of the tableau so far. Thus, we define the rules relative to a set of labels.

DEFINITION 3.3 (Tableau rules). *Let* $\mathbf{ILX}$ *be a Horn interpretability logic and let* Λ *be a set of labels. The* $\mathbf{ILX}$-*tableau rules with respect to* Λ *are as follows:*

Propositional rules:

$$(\neg) \; \frac{\sigma :: \neg\neg A}{\sigma :: A} \; ;$$

$$(\rightarrow) \; \frac{\sigma :: A \rightarrow B}{\sigma :: \neg A \quad | \quad \sigma :: B} \; ;$$

$$(\neg \rightarrow) \frac{\sigma :: \neg(A \rightarrow B)}{\substack{\sigma :: A \\ \sigma :: \neg B}}.$$

(ν)-rules:

$$(\nu_\Box, \Lambda) \frac{\sigma :: \Box A}{\tau :: A}, \text{ when } \sigma \mathbf{R} \tau;$$

$$(\nu_S, \Lambda) \frac{\sigma :: \Box_\rho A}{\tau :: A}, \text{ when } \sigma \mathbf{S}_\rho \tau;$$

$$(\nu_\rhd, \Lambda) \frac{\sigma :: A \rhd B}{\tau :: \neg A \quad | \quad \tau :: \neg\Box_\sigma \neg B}, \text{ when } \sigma \mathbf{R} \tau.$$

(π)-rules:

$$(\pi_\Box, \Lambda) \frac{\sigma :: \neg \Box A}{\substack{\sigma R n :: \neg A \\ \sigma R n :: \Box A}}, \text{ where } n \in \mathbb{N} \text{ is such that } \sigma R n \notin \Lambda;$$

$$(\pi_S, \Lambda) \frac{\sigma :: \neg \Box_\rho A}{\substack{\sigma S_\rho n :: \neg A \\ \sigma S_\rho n :: \Box A}}, \text{ where } n \in \mathbb{N} \text{ is such that } \sigma S_\rho n \notin \Lambda;$$

$$(\pi_\rhd, \Lambda) \frac{\sigma :: \neg(A \rhd B)}{\substack{\sigma R n :: A \\ \sigma R n :: \Box_c \neg B \\ \sigma R n :: \Box \neg A}}, \text{ where } n \in \mathbb{N} \text{ is such that } \sigma R n \notin \Lambda.$$

We call the labelled formula above the line in the rules above the *antecedent* and the labelled formula(s) under the line *succedent(s)*.

Some clarifying remarks on the tableau rules seem in order. First we note that we use the symbol "|" in the rules ($\to$), and ($\nu_\triangleright, \Lambda$) to denote branching in proof-trees. Next, we observe that various non-branching rules have multiple succedents such as the rules ($\neg \to$), ($\pi_\square, \Lambda$) and ($\pi_\triangleright, \Lambda$). These succedents are to be understood as different nodes one placed under the other. Lemma 2.2.1 is reflected in the rules ($\pi_\square, \Lambda$) and ($\pi_\triangleright, \Lambda$) and Lemma 2.2.2 is reflected in the (π_S, Λ) rule.

Another non-local feature of the tableaux proof system will be that we will allow to apply rules to any node $\sigma :: A$ in a branch, not necessarily only to bottom-nodes. Upon application of the rule, the succedent(s) with possible branching can be appended to the bottom of *any* branch passing through $\sigma :: A$. If $\mathcal{B}$ is a branch in a tree whose nodes are labelled formulas, by $\mathrm{lab}(\mathcal{B})$ we denote the collection of labels that occur in $\mathcal{B}$.

DEFINITION 3.4 (**ILX**-Tableaux, open and closed). *Given a Horn logic* **ILX** *and a finite set Γ of formulas, an* **ILX**-*tableau for Γ is a binary irreflexive directed downward growing tree with nodes carrying labelled formulas defined inductively as follows:*

- *A single node tree T with $0 :: A$ as the sole node for some formula $A \in \Gamma$ is an* **ILX**-*tableau for Γ.*

- *If T is an* **ILX**-*tableau for Γ, then a tree T' obtained by extending (appending below) any branches of T with $0 :: A$ for some formula $A \in \Gamma$ is an* **ILX**-*tableau for Γ.*

- *Let T be an* **ILX**-*tableau for Γ, $\mathcal{B}$ be a branch of T, and let (ρ) be a rule w.r.t.* $\mathrm{lab}(\mathcal{B})$. *If some labelled formula $\sigma :: A$ that occurs in $\mathcal{B}$ is the antecedent of an instance of (ρ), then the tree T' obtained by extending $\mathcal{B}$ with the appropriate succedents of (ρ) in any particular ordering (with possible branching) is an* **ILX**-*tableau for Γ.*

A branch $\mathcal{B}$ of an **ILX**-*tableau T for Γ is called* closed *if there is σ and A such that $\sigma :: A \in \mathcal{B}$ and $\sigma :: \neg A \in \mathcal{B}$. Otherwise the branch is* open. *An* **ILX**-*tableau T for Γ is closed if all of its branches are closed. Otherwise T is open.*

Given a Horn logic **ILX**, we are now ready to assign to a finite set of formulas Γ what we call a *systematic* **ILX**-*tableau for Γ* which will contain all the information as to the satisfiability of Γ. The systematic tableau method given below follows closely the procedure given in [7].

DEFINITION 3.5 (Systematic **ILX**-tableau). *For a Horn logic* **ILX**, *a systematic* **ILX**-*tableau for a finite set Γ of formulas is constructed in stages. Throughout the stages, the nodes in the tree T_i will be marked with exactly one of* awake, asleep *or* finished. *The marked version of T_i will be denoted by $\mu(T_i)$.*
Stage 0: *Form the initial tableau T_0 with $0 :: A$ for all $A \in \Gamma$ in some order on top of each other and mark them all awake.*
Stage n+1: *Look for an awake $\sigma :: A$ in $\mu(T_n)$ closest to the root of the tableau; if there are several with the same distance, choose the leftmost one. If $A = p$ or $A = \neg p$ for some propositional variable p, then T_{n+1} and $\mu(T_{n+1})$ are as T_n and $\mu(T_n)$ respectively except that we mark the node $\sigma :: A$ as finished and we end Stage n+1.*
Otherwise we obtain T_{n+1} and $\mu(T_{n+1})$ as follows:

- *If $A = \neg\neg B$ for some B, for every open branch $\mathcal{B}$ that passes through $\sigma :: A$, extend $\mathcal{B}$ with $\sigma :: B$ marking it awake and marking $\sigma :: A$ as finished. Here and below 'extending $\mathcal{B}$' means 'appending new nodes to the bottom of $\mathcal{B}$'.*

- If $A = (B \to C)$ for some B and C, for every open branch $\mathcal{B}$ that passes through $\sigma :: A$, split the end of $\mathcal{B}$ and extend the left fork with $\sigma :: \neg B$ and the right fork with $\sigma :: C$. Both new nodes will be marked awake and $\sigma :: A$ will be marked as finished.

- If $A = \neg(B \to C)$ for some B and C, for every open branch $\mathcal{B}$ that passes through $\sigma :: A$, extend $\mathcal{B}$ with $\sigma :: B$ and $\sigma :: \neg C$ in whatever order. Both new nodes will be marked awake and $\sigma :: A$ will be marked as finished.

- If $A = \Box B$ for some B, for every open branch $\mathcal{B}$ that passes through $\sigma :: A$ and for all $\tau \in \mathrm{lab}(\mathcal{B})$, if $\sigma \mathbf{R} \tau$, then extend $\mathcal{B}$ with $\tau :: B$. These new nodes will be marked awake and $\sigma :: A$ will be marked as asleep.

- If $A = \neg \Box B$ for some B, for every open branch $\mathcal{B}$ that passes through $\sigma :: A$, extend $\mathcal{B}$ with $\sigma Rn :: \neg B$ and $\sigma Rn :: \Box B$, where $n \in \mathbb{N}$ is the least number such that $\sigma Rn \notin \mathrm{lab}(\mathcal{B})$. Mark both $\sigma Rn :: \neg B$ and $\sigma Rn :: \Box B$ awake and $\sigma :: A$ finished. Moreover, mark as awake every $\tau :: \Box B \in \mathcal{B}$ and $\tau :: B \triangleright C \in \mathcal{B}$ whenever $\tau \mathbf{R} \sigma Rn$ and mark awake every $\tau :: \Box_\rho B \in \mathcal{B}$ whenever $\tau \mathbf{S}_\rho \sigma Rn$.

- If $A = \Box_\rho B$ for some ρ and B, for every branch $\mathcal{B}$ that passes through $\sigma :: A$ and for all $\tau \in \mathrm{lab}(\mathcal{B})$, if $\sigma \mathbf{S}_\rho \tau$, then extend $\mathcal{B}$ with $\tau :: B$. Mark $\tau :: B$ awake and $\sigma :: A$ asleep.

- If $A = \neg \Box_\rho B$ for some ρ and B, for every open branch $\mathcal{B}$ that passes through $\sigma :: A$, extend $\mathcal{B}$ with $\sigma S_\rho n :: \neg B$ and $\sigma S_\rho n :: \Box B$, where $n \in \mathbb{N}$ is the least number such that $\sigma S_\rho n \notin \mathrm{lab}(\mathcal{B})$. Mark both $\sigma S_\rho n :: \neg B$ and $\sigma S_\rho n :: \Box B$ awake and mark $\sigma :: A$ finished. Moreover, mark awake every $\tau :: \Box B \in \mathcal{B}$ and $\tau :: B \triangleright C \in \mathcal{B}$ such that $\tau \mathbf{R} \sigma S_\rho n$ and every $\tau :: \Box_\rho B \in \mathcal{B}$ such that $\tau \mathbf{S}_\rho \sigma S_\rho n$.

- If $A = B \triangleright C$ for some B and C, for every open branch $\mathcal{B}$ that passes through $\sigma :: A$ and every $\tau \in \mathrm{lab}(\mathcal{B})$, if $\sigma \mathbf{R} \tau$, split the end of $\mathcal{B}$ and extend the left fork with $\tau :: \neg B$ and the right fork with $\tau :: \neg \Box_\sigma \neg C$. Both new nodes are marked awake and $\sigma :: A$ will be marked asleep.

- If $A = \neg(B \triangleright C)$ for some B and C, for every open branch $\mathcal{B}$ that passes through x, pick the smallest $n \in \mathbb{N}$ such that $\sigma Rn \notin \mathrm{lab}(\mathcal{B})$ and extend $\mathcal{B}$ with $\sigma Rn :: B$, with $\sigma Rn :: \Box_\sigma \neg C$ and with $\sigma Rn :: \Box \neg B$ in whatever order you like. All new nodes are marked awake and $\sigma :: A$ finished. Moreover, mark awake every $\tau :: \Box B \in \mathcal{B}$ and $\tau :: B \triangleright C \in \mathcal{B}$ such that $\tau \mathbf{R} \sigma Rn$ and every $\tau :: \Box_\rho B \in \mathcal{B}$ such that $\tau \mathbf{S}_\rho \sigma Rn$.

By this procedure we construct a chain $\langle T_i : i \in \omega \rangle$ of **ILX**-tableaux for Γ. We call $\bigcup_{i \in \omega} T_i$ a systematic **ILX**-tableau for X.

REMARK 3.6. *A systematic **ILX**-tableau T for a finite set Γ of formulas is not in general[2] an **ILX**-tableau in the sense of Definition 3.4. However, if T is finite it is an **ILX**-tableau. In particular, if T closes, then T is an **ILX**-tableau. Moreover, if there is a closed **IL**-tableau T' for Γ, then there is a closed systematic tableau for Γ.*

---

[2] An example is when $\Gamma = \{\Diamond p, p \triangleright q, q \triangleright p\}$: The $\Diamond p$ yields a world where p holds, and since we do not reuse worlds the circular $p \triangleright q, q \triangleright p$ keep on creating new worlds.

LEMMA 3.7 (Fairness). *If $\sigma :: A$ is awake at stage n+1, the systematic **ILX**-tableau procedure visits $\sigma :: A$ at some later stage.*

Proof. Straightforward and similar to Lemma 6.4.4 in [7]. ⊣

4 Soundness

As usual, for a Horn logic **ILX**, we call a formula A **ILX**-tableau provable whenever the systematic **ILX**-tableaux for $\{\neg A\}$ closes. In this section we shall show that this notion of provability is sound with respect to **ILX** frames.

DEFINITION 4.1. *A set $\mathcal{X}$ of labelled formulas is **ILX**-satisfiable if there exists an **ILX**-model $M = \langle W, R, S, V \rangle$ and an interpretation $I : \mathrm{lab}(\mathcal{X}) \to W$ such that*

(i) *If $\sigma, \tau \in \mathrm{lab}(\mathcal{X})$ and $\sigma \mathbf{R} \tau$, then $I(\sigma) R I(\tau)$;*

(ii) *If $\rho, \sigma, \tau \in \mathrm{lab}(\mathcal{X})$ and $\sigma \mathbf{S}_\rho \tau$, then $I(\sigma) S_{I(\rho)} I(\tau)$;*

(iii) $M, I(\sigma) \Vdash A$ *for all $\sigma :: A \in \mathcal{X}$;*

*An **ILX**-tableau T is **ILX**-satisfiable, if there is a branch $\mathcal{B}$ of T such that $\mathcal{B}$ is **ILX**-satisfiable*

The next lemma tells us that satisfiable **ILX**-tableaux are closed under applying the rules to them for Horn logics **ILX**.

LEMMA 4.2. *Let **ILX** be a Horn logic and let T be a satisfiable **ILX**-tableau. Then for any rule, the tableau T' obtained by the application of the rule is also **ILX**-satisfiable.*

Proof. Suppose $\mathcal{B}$ is an **ILX**-satisfiable branch of T. We show that if we apply some rule to a some labelled formula in $\mathcal{B}$, we obtain a branch that is **ILX**-satisfiable. The cases for the **propositional rules** are trivial.

Suppose $\sigma :: \Box A \in \mathcal{B}$ and consider the branch $\mathcal{C}$ obtained by an **application of the $(\nu_\Box)$-rule** with $\tau :: A$ added to the branch for some $\tau \in \mathrm{lab}(\mathcal{B})$ such that $\sigma \mathbf{R} \tau$. By assumption there is an **ILX**-model $M = \langle W, R, S, V \rangle$ and an interpretation $I : \mathrm{lab}(\mathcal{B}) \to W$ such that $M, I(\sigma) \Vdash \Box A$. Now $\tau \in \mathrm{lab}(\mathcal{B})$ and $\sigma \mathbf{R} \tau$. Hence $I(\sigma) R I(\tau)$ and so $M, I(\tau) \Vdash A$.

Suppose $\sigma :: \Box_\rho A \in \mathcal{B}$ and consider the branch $\mathcal{C}$ obtained by an **application of the (ν_S) rule** with $\tau :: A$ added to the branch for some $\tau \in \mathrm{lab}(\mathcal{B})$ such that $\sigma \mathbf{S}_\rho \tau$. By assumption there is an **ILX**-model $M = \langle W, R, S, V \rangle$ and an interpretation $I : \mathrm{lab}(\mathcal{B}) \to W$ such that $M, I(\sigma) \Vdash \Box_{I(\rho)} A$. Now since $\tau, \rho \in \mathrm{lab}(\mathcal{B})$ and $\sigma \mathbf{S}_\rho \tau$, we also have that $I(\sigma) S_{I(\rho)} I(\tau)$. Hence $M, I(\tau) \Vdash A$.

Suppose $\sigma :: A \rhd B \in \mathcal{B}$ and consider the two branches obtained by an **application of the $(\nu_\rhd)$ rule** with $\tau :: \neg A$ in the left branch and $\tau :: \neg \Box_\sigma \neg B$ in the right branch for some $\tau \in \mathrm{lab}(\mathcal{B})$ such that $\sigma \mathbf{R} \tau$. Now by assumption there is an **ILX**-model $M = \langle W, R, S, V \rangle$ and an interpretation $I : \mathrm{lab}(\mathcal{B}) \to W$ such that $M, I(\sigma) \Vdash A \rhd B$. Now since $\tau \in \mathrm{lab}(\mathcal{B})$ and $\sigma \mathbf{R} \tau$, we have that $I(\sigma) R I(\tau)$. If $M, I(\tau) \Vdash \neg A$, then the left branch is satisfiable with I. If on the other hand $M, I(\tau) \Vdash A$, then there exists some $x \in W$ such that $I(\tau) S_{I(\sigma)} x$ and $M, x \Vdash B$. Hence $M, I(\tau) \Vdash \neg \Box_{I(\sigma)} \neg B$.

Suppose $\sigma :: \neg \Box A \in \mathcal{B}$ and consider the branch $\mathcal{C}$ obtained by **an application of the $(\pi_\Box)$ rule** with $\sigma R n :: \neg A$ and $\sigma R n :: \Box A$ added to the branch for some $n \in \mathbb{N}$ such that $\sigma R n \notin \mathrm{lab}(\mathcal{B})$.

By assumption there is an **ILX**-model $M = \langle W, R, S, V \rangle$ and an interpretation $I\colon \mathrm{lab}(\mathcal{B}) \to W$ such that $M, I(\sigma) \Vdash \neg \Box A$. Hence there is some $x \in W$ such that $I(\sigma)Rx$ and $M, x \Vdash \neg A$ and $M, x \Vdash \Box A$. Now, since $\sigma Rn \notin \mathrm{lab}(\mathcal{B})$, we can extend I to I' by putting $I'(\sigma Rn) = x$. Now define $\mathbf{R}_{I'}$ and $\mathbf{S}_{I'}$ on $\mathrm{lab}(\mathcal{B}) \cup \{\sigma Rn\}$ by

$$\langle \tau, \rho \rangle \in \mathbf{R}_{I'} \Leftrightarrow I'(\tau) R I'(\rho),$$

$$\langle \upsilon, \tau, \rho \rangle \in \mathbf{S}_{I'} \Leftrightarrow I'(\tau) S_{I'(\upsilon)} I'(\rho).$$

Now $\mathbf{R}_{I'}$ and $\mathbf{S}_{I'}$ satisfy conditions (1)-(8) in Definition 3.2. Hence $\mathbf{R} \subseteq \mathbf{R}_{I'}$ and $\mathbf{S} \subseteq \mathbf{S}_{I'}$, and so I' is an interpretation from $\mathcal{C}$ to M.

Suppose $\sigma :: \neg \Box_\rho A \in \mathcal{B}$ and consider the branch $\mathcal{C}$ obtained by **an application of the** (π_S) **rule** with $\sigma S_\rho n :: \neg A$ and $\sigma S_\rho n :: \Box A$ added to the branch $\mathcal{B}$ for some $n \in \mathbb{N}$. By assumption there is an **ILX**-model $M = \langle W, R, S, V \rangle$ and an interpretation $I\colon \mathrm{lab}(\mathcal{B}) \to W$ such that $M, I(\sigma) \Vdash \neg \Box_{I(\rho)} A$. Now there is some $x \in W$ such that $I(\sigma) S_{I(\rho)} x$ and $M, x \Vdash \neg A$. Now if $M, x \Vdash \Box A$, we extend I to I' by putting $I'(\sigma S_\rho n) = x$. On the other hand if $M, x \Vdash \neg \Box A$, then there is $y \in W$ such that xRy and $M, y \Vdash \neg A$ and $M, y \Vdash \Box A$. Now $I(\rho) R x R y$ and so $x S_{I(\rho)} y$. Hence $I(\sigma) S_{I(\rho)} y$. Now extend I to I' by putting $I'(\sigma S_\rho n) = y$. Now I' is again an interpretation from $\mathcal{C}$ to M.

Suppose finally $\sigma :: \neg (A \rhd B) \in \mathcal{B}$ and consider the branch obtained by **an application of the** $(\pi_\rhd)$ **rule** with $\sigma Rn :: A$, $\sigma Rn :: \Box_\sigma \neg B$ and $\sigma Rn :: \Box \neg A$ added to the branch for some $n \in \mathbb{N}$. By assumption there is an **ILX**-model $M = \langle W, R, S, V \rangle$ and an interpretation $I\colon \mathrm{lab}(\mathcal{B}) \to W$ such that $M, I(\sigma) \Vdash \neg (A \rhd B)$. Hence there is $x \in W$ such that $I(\sigma) Rx$, $M, x \Vdash A$ and $M, x \Vdash \Box_{I(\sigma)} \neg B$. If $M, x \Vdash \Box \neg A$, we may extend I to I' by putting $I'(\sigma Rn) = x$. On the other hand if $M, x \Vdash \neg \Box \neg A$, then there is $y \in W$ such that $M, y \Vdash A$ and $M, y \Vdash \Box \neg A$. But now since $I(\sigma) R x R y$, we have that $x S_{I(\sigma)} y$ and so $M, y \Vdash \Box_{I(\sigma)} \neg B$. So now we may extend I to I' by putting $I'(\sigma Rn) = y$. Again, I' is an interpretation from $\mathcal{C}$ to M. ⊣

THEOREM 4.3. *Let* **ILX** *be a Horn logic. If a systematic* **ILX**-*tableau for a set of formulas* Γ *closes, then* Γ *is* **ILX**-*unsatisfiable.*

Proof. Suppose a systematic tableau T for Γ is closed, but that there is an **ILX**-model $M = \langle W, R, S, V \rangle$ and $x \in W$ such that $M, x \Vdash \Gamma$. Now consider the initial tableau T_0 for Γ with $0 :: A$ for all $A \in \Gamma$.

Now, by assumption, $I = \{\langle 0, x \rangle\}$ is an interpretation from $\mathrm{lab}(T_0)$ to W. By the above lemma, every tableau obtained from the initial tableau is **ILX**-satisfiable. In particular, the closed tableau T obtained by the systematic procedure is **ILX**-satisfiable. A contradiction since, for any branch $\mathcal{B}$ of T there is σ and A such that $\sigma :: A \in \mathcal{B}$ and $\sigma :: \neg A \in \mathcal{B}$. ⊣

For the sake of being explicit let us formulate the soundness of our tableaux as an immediate corollary.

COROLLARY 4.4. *Let* **ILX** *be a Horn logic. If a systematic* **ILX**-*tableau for* $\{\neg A\}$ *closes, then* A *is* **ILX**-*valid.*

5 Completeness

In this section we shall show that our proof system is also complete w.r.t. **ILX**- frames.

DEFINITION 5.1. *A set $\mathcal{X}$ of labelled formulas is a **ILX**-Hintikka set if the following hold:*

(i) *There is no σ and A such that $\sigma :: A \in \mathcal{X}$ and $\sigma :: \neg A \in \mathcal{X}$;*

(ii) *If $\sigma :: \neg\neg A \in \mathcal{X}$, then $\sigma :: A \in \mathcal{X}$;*

(iii) *If $\sigma :: A \to B \in \mathcal{X}$, then $\sigma :: \neg A \in \mathcal{X}$ or $\sigma :: B \in \mathcal{X}$;*

(iv) *If $\sigma :: \neg(A \to B) \in \mathcal{X}$, then $\sigma :: A \in \mathcal{X}$ and $\sigma :: \neg B \in \mathcal{X}$;*

(v) *If $\sigma :: A \rhd B \in \mathcal{X}$, then $\tau :: \neg A \in \mathcal{X}$ or $\tau :: \neg\Box_\sigma \neg B \in \mathcal{X}$ for all $\tau \in \mathrm{lab}(\mathcal{X})$ such that $\sigma \mathbf{R} \tau$;*

(vi) *If $\sigma :: \neg(A \rhd B) \in \mathcal{X}$, then there is $\tau \in \mathrm{lab}(\mathcal{X})$ such that $\tau :: A \in \mathcal{X}$, $\tau :: \Box_\sigma \neg B \in \mathcal{X}$ and $\sigma \mathbf{R} \tau$;*

(vii) *If $\sigma :: \Box A \in \mathcal{X}$, then $\tau :: A \in \mathcal{X}$ for each $\tau \in \mathrm{lab}(\mathcal{X})$ such that $\sigma \mathbf{R} \tau$;*

(viii) *If $\sigma :: \neg\Box A \in \mathcal{X}$, then there is $\tau \in \mathrm{lab}(\mathcal{X})$ such that $\tau :: \neg A \in \mathcal{X}$ and $\sigma \mathbf{R} \tau$;*

(ix) *If $\sigma :: \Box_\rho A \in \mathcal{X}$, then $\tau :: A \in \mathcal{X}$ for each $\tau \in \mathrm{lab}(\mathcal{X})$ such that $\sigma \mathbf{S}_\rho \tau$;*

(x) *If $\sigma :: \neg\Box_\rho A \in \mathcal{X}$, then there is $\tau \in \mathrm{lab}(\mathcal{X})$ such that $\tau :: \neg A \in \mathcal{X}$ and $\sigma \mathbf{S}_\rho \tau$.*

Hintikka sets contain all the needed information to extract a model from them. This is clearly manifested in the proof of the following lemma.

LEMMA 5.2 (Truth Lemma). *Let **ILX** be a Horn logic. If $\mathcal{X}$ is an **ILX**-Hintikka set and $\mathbf{R}_{\mathbf{ILX}}^{\mathrm{lab}(\mathcal{X})}$ is Noetherian, then $\mathcal{X}$ is **ILX**-satisfiable.*

Proof. As mentioned before, we will omit various sub and superscripts. Thus, if $\mathbf{R}$ is Noetherian, then $\mathcal{F} = \langle \mathrm{lab}(\mathcal{X}), \mathbf{R}, \mathbf{S} \rangle$ is clearly an **ILX**-frame. Define a valuation V on $\mathrm{lab}(\mathcal{X})$ by putting

$$V(p) = \{\sigma \in \mathrm{lab}(\mathcal{X}) : \sigma :: p \in \mathcal{X}\} \text{ for all propositional variables } p.$$

Now we can prove by an easy induction on the complexity of formulas that for all σ and A:

$$\sigma :: A \in \mathcal{X} \Rightarrow \langle \mathcal{F}, V \rangle, \sigma \Vdash A \text{ and}$$
$$\sigma :: \neg A \in \mathcal{X} \Rightarrow \langle \mathcal{F}, V \rangle, \sigma \nVdash A.$$

Hence $\langle \mathcal{F}, V \rangle$ satisfies $\mathcal{X}$ with the identity interpretation. ⊣

LEMMA 5.3. *If $\mathcal{B}$ is an open branch in a systematic **ILX**-tableau for a finite set Γ, then $\mathcal{B}$ is a Hintikka set and $\mathbf{R}$ is Noetherian.*

Proof. That $\mathcal{B}$ is a Hintikka set follows easily from the fairness of the systematic **ILX**-tableau procedure.

Notice that if $\sigma :: \Box A \in \mathcal{B}$ for some σ and A, then A is either a subformula of a formula from Γ or the negation of a subformula of a formula from Γ. Let $\tilde{\Gamma} = \mathrm{sub}[\Gamma] \cup \{\neg A : A \in \mathrm{sub}[\Gamma]\}$.

Now suppose towards a contradiction that there is an ascending $\mathbf{R}$-chain $\langle \sigma_i : i \in \omega \rangle$ in $\mathrm{lab}(\mathcal{B})$. Without loss of generality we may assume that $\sigma_0 = 0$ and $\sigma_i \neq 0$ for all $i > 0$.

Now we show that for any $i \in \omega$ there is A_i such that $\sigma_{i+1} :: \Box A_i \in \mathcal{B}$, but $\sigma_j :: \Box A_i \notin \mathcal{B}$ for all $j \leq i$.

If $\sigma_{i+1} = \tau R n$ for some $\tau \in \mathrm{lab}(\mathcal{B})$ and $n \in \mathbb{N}$, then σ_{i+1} is introduced either with a $(\pi_\Box)$-rule applied to some $\tau :: \neg \Box A \in \mathcal{B}$ or by a $(\pi_\rhd)$-rule applied to some $\tau :: \neg(A \rhd B) \in \mathcal{B}$. In the first case $\sigma_{i+1} :: \Box A \in \mathcal{B}$, but $\sigma_j :: \Box A \notin \mathcal{B}$ for all $j \leq i$. In the second case $\sigma_{i+1} :: \Box \neg A \in \mathcal{B}$, but $\sigma_j :: \Box \neg A \notin \mathcal{B}$ for all $j \leq i$.

If $\sigma_{i+1} = \tau S_\rho n$ for some $\tau, \rho \in \mathrm{lab}(\mathcal{B})$ and $n \in \mathbb{N}$, then σ_{i+1} is introduced with a (π_S)-rule applied to some $\tau :: \neg \Box_\rho A$. Now $\sigma_{i+1} :: \Box A \in \mathcal{B}$, but $\sigma_j :: \Box A \notin \mathcal{B}$ for all $j \leq i$.

Now for large enough m,

$$|\{A : A \in \tilde{\Gamma} \text{ and } \sigma_m :: A \in \mathcal{B}\}| > |\tilde{\Gamma}|.$$

⊣

We have the following corollaries from this lemma.

COROLLARY 5.4. *Let **ILX** be a Horn logic. If a systematic **ILX**-tableau for a finite Γ has an open branch, then Γ is **ILX**-satisfiable.*

In particular, we can formulate completeness of our proof systems.

COROLLARY 5.5. *Let **ILX** be a Horn logic. If A is **ILX**-valid, then any systematic **ILX**-tableau for $\{\neg A\}$ closes.*

6 Scope of our results

Our results apply to all interpretability logics extending **IL** that are Horn. In particular the results apply to the most important systems **ILM** and **ILP**. At first sight, the restriction of the logic being Horn might seem quite severe. However, most logics that occur in the literature turn out to be Horn. In particular also the logics based on

R: $A \rhd B \to \neg(A \rhd \neg C) \rhd B \land \Box C$;

and the two series of generalizations of this principle as presented in [9] are Horn logics. An important logic that falls out of the scope of this paper is **ILW** since the corresponding frame condition is second order.

Acknowledgement

We would like to thank Rajeev Goré for his comments on a draft version of this paper. Further thanks go to Volodya Shavrukov and an anonymous referee for helping to improve the paper. The second author was supported by the Generalitat de Catalunya under grant number 2014 SGR 437 and from the Spanish Ministry of Science and Education under grant number MTM2014-59178-P.

BIBLIOGRAPHY

[1] A. Berarducci. The interpretability logic of peano arithmetic. *Journal of Symbolic Logic*, 55(3):1059–1089, 1990.
[2] M. Bilkova, D. de Jongh, and J. J. Joosten. Interpretability in PRA. *Annals of Pure and Applied Logic*, 161(2):128–138, 2009.
[3] F. Bou and J. J. Joosten. The closed fragment of IL is PSPACE hard. *Electronical Notes Theoretical Computer Science*, 278:47–54, 2011.
[4] D.H.J. de Jongh and F. Veltman. Provability logics for relative interpretability. In P.P. Petkov, editor, *Mathematical Logic, Proceedings of the Heyting 1988 summer school in Varna, Bulgaria*, pages 31–42. Plenum Press, Boston, New York, 1990.
[5] M. Fitting. Tableau methods of proof for modal logics. *Notre Dame Journal of Formal Logic*, 13(2):237–247, 1972.
[6] M. Fitting. *Proof Methods for Modal and Intuitionistic Logics*, volume 169 of *Synthese Library*. Springer Netherlands, 1983.
[7] R. Goré. Tableau methods for modal and temporal logics. In M. D'Agostino, D. M. Gabbay, R. Hähnle, and J. Posegga, editors, *Handbook of Tableau Methods*, pages 297–396. Springer Netherlands, 1999.
[8] E. Goris and J. J. Joosten. A new principle in the interpretability logic of all reasonable arithmetical theories. *Logic Journal of the IGPL*, 19(1):14–17, 2011.
[9] E. Goris and J. J. Joosten. Two series of formalized interpretability principles for weak systems of arithmetic. *arXiv:1503.09130 [math.LO]*, 2015.
[10] P. Hájek and V. Švejdar. A note on the normal form of closed formulas of interpretability logic. *Studia Logica*, 50(1):25–28, 1991.
[11] J. J. Joosten and A. Visser. The interpretability logic of all reasonable arithmetical theories. The new conjecture. *Erkenntnis*, 53(1-2):3–26, 2000.
[12] S. Kanger. *Provability in Logic*. Stockholm Studies in Philosophy, University of Stockholm, Almqvist and Wiksell, Sweden, 1957.
[13] S. Negri. Proof analysis in modal logic. *Journal of Philosophical Logic*, 34(5-6):507–544, October 2005.
[14] Katsumi Sasaki. A cut-free sequent system for the smallest interpretability logic. *Studia Logica*, 70(3):353–372, 2002.
[15] V. Y. Shavrukov. The logic of relative interpretability over Peano arithmetic. Preprint, Steklov Mathematical Institute, Moscow, 1988. In Russian.
[16] V. Y. Shavrukov. Interpreting reflexive theories in finitely many axioms. *Fundamenta Mathematicae*, 152:99–116, 1997.
[17] R. M. Solovay. Provability interpretations of modal logic. *Israel Journal of Mathematics*, 28:33–71, 1976.
[18] A. Visser. Interpretability logic. In P.P. Petkov, editor, *Mathematical Logic, Proceedings of the Heyting 1988 summer school in Varna, Bulgaria*, pages 175–209. Plenum Press, Boston, New York, 1990.
[19] A. Visser. An overview of interpretability logic. In M. Kracht, M. de Rijke, and H. Wansing, editors, *Advances in modal logic '96*, pages 307–359. CSLI Publications, Stanford, CA, 1997.

The Root of Evil – A Self-Referential Play in One Act

VOLKER HALBACH

A teaching room in an Oxbridge college. Fred Philonous, a tutor of the college, sits in front of a large desk. At the other end of the room two armchairs and a sofa are grouped around a fragile looking small table. In the background, close to the window, is a large whiteboard. Through the window perpendicular buildings can be seen. Fred is typing slowly on his computer. After a short time a student, Sophia, peeks in through one of the window panes. She disappears again and after a few seconds a knocking can be heard.[1]

FRED Come in.

SOPHIA (*entering the room.*) Hi.

FRED (*While speaking, he gets up from his chair and walks slowly to an armchair.*) Hello. I got your essay. It was on paradoxes I thought we had agreed that you write an essay on 'Should we define truth in terms of correspondence or coherence?'

SOPHIA No, you just said I should write an essay on truth.

FRED But the question on the correspondence and coherence theory was the next question on the reading list.

SOPHIA Yes, I skipped it. The next question was also on truth.

FRED The week before we talked about definitions of knowledge and the Gettier problem. There is a reason why I put the question on the correspondence theory as the next question on the list. In epistemology we use a correspondence definition of truth. Almost all epistemologists subscribe to a correspondence theory of truth and define truth in terms of correspondence with reality. It would have made sense to write on the correspondence theory because it would have been an analysis of the truth condition in definitions of knowledge.

SOPHIA I started reading Russell. But then they said in a survey article that Tarski showed that truth cannot be defined, and I thought that there is a general problem for definitions of truth in terms of correspondence or whatever, if Tarski is right.

FRED Yes, there may be a problem, but believe me: In an exam it's much easier to write an essay on the correspondence or coherence theory of truth than on Tarski's account of truth.

SOPHIA I am sorry, I answered the question 'What is the source of semantic paradox?'

FRED (*sinking slowly into the armchair.*) Ok, so what is the root of evil?

---

[1] In this play I have used some motives from [HV14a] and [HV14b]. My thanks go to all those who helped us with these two papers. Of course the present paper is my very personal take on some of the topics touched upon in the two joint papers. I am grateful to Susanne Bobzien for comments on an earlier version.

SOPHIA (*sitting down on the sofa*) Some people say that self-reference is the source of all semantic paradoxes. Others say that it's the truth predicate that can apply to sentences containing the very same truth predicate.

FRED Well, clearly, if we distinguish between object and metalanguage, the paradoxes disappear. That's how mathematical logicians solve the liar paradox. But distinguishing between object and metalanguage doesn't really help to solve the liar paradox in natural language. In natural language we don't observe this distinction; we apply the truth predicate to all sentences whether they contain the truth predicate or not.

SOPHIA Even that isn't clear to me. There's this paradox by Visser.[2] It seems to show that we can still get into trouble even when we have only typed truth predicates. He has a hierarchy of languages as Tarski did; but he still gets a paradox.

FRED I think you are confused. Many students here become confused about these logical matters because they learn logic from a confusing textbook. (*He gets up and paces slowly up and down.*) If you have truth predicates each of which apply only to sentences containing only truth predicates lower in the hierarchy, then all paradoxes are blocked. The liar paradox relies on a truth predicate that is applied to a sentence containing the same truth predicate. (*He looks out of the window and continues after a pause.*) Or maybe you have two truth predicates $true_1$ and $true_2$ and two sentences. Now you write on one side of a postcard 'The sentence on the other side is not $true_1$' and on the other side 'The sentence on the other side is $true_2$'. When you think about it, the truth predicate '$true_1$' is only applied to the sentence on the other side and this sentence contains only the truth predicate '$true_2$'. Conversely, the sentence on the other page says that the sentence on the first page is $true_2$. So, the sentence on the second page is only applied to a sentence containing '$true_1$' but not '$true_2$' itself. So none of the sentences talks about a sentence that contains the same truth predicate. But still, both sentences are self-referential. Both sentences refer to themselves. Their self-reference is indirect. The source of paradox is self-reference. (*He stops walking behind his armchair.*)

SOPHIA I have come across the postcard paradox. But I mean you can get a paradox without any circularity of this kind.

FRED Um, I remember. Yablo's paradox is supposed to show that there is semantic paradox without self-reference.[3]

SOPHIA (*interrupting* FRED) No, I don't mean Yablo's paradox. Yablo used a truth predicate that is applied to sentences that contain the same truth predicate. In Visser's paradox there are many truth predicates linearly ordered in a hierarchy. Each of these predicates applies only to sentences with predicates that are strictly lower in the hierarchy.

FRED But didn't Tarski prove that such a hierarchy blocks the paradox?

SOPHIA The weird feature of Visser's hierarchy is that it is illfounded.

FRED What does that mean?

SOPHIA We have truth predicates '$true_n$' for every natural number n. For a given n, '$true_n$' is a truth predicate for sentences that contain only truth predicates with an index bigger than n.

---

[2] See [Vis89]. Drafts of this paper were circulated in the early 1980s.
[3] See [Yab85, p. 340], [Yab93] and the recent monograph [Coo14].

This means that the predicate '$true_0$' is the topmost predicate; it applies to sentences possibly containing '$true_1$', '$true_2$' and so on. The predicate '$true_1$' is one level lower in the hierarchy; it applies to sentences possibly containing '$true_2$', '$true_3$' and so on, but not to sentences containing '$true_0$' or '$true_1$'.

FRED (*sinking slowly back into his armchair*) Sounds a little weird. But it's still a hierarchy. What's paradoxical about it?

SOPHIA (*jumps up from the sofa and walks to the whiteboard.*) Unlike Yablo, Visser used the Gödel fixed-point theorem for defining the problematic sentences, but we can write them just as a list, in the way Yablo did. (*After trying out some pens, Sophia finds one with some ink left in it and writes the following lines on the board:*)

>No sentence below this one is $true_0$.
>No sentence below this one is $true_1$.
>No sentence below this one is $true_2$.
>No sentence below this one is $true_3$.
>⋮

This list shows that we can have paradox without a global truth. The truth predicates apply only to sentences containing truth predicates lower in the hierarchy.

FRED (*interrupting*) What? If we assume the first sentence at the top, the sentence in the second line isn't $true_0$. Therefore some sentence below the second line must be $true_1$, thereby contradicting the first line. So, the first line can't hold.

SOPHIA Exactly! Therefore there must be a sentence below the topmost sentence that is $true_0$.

FRED (*sounding bored*) But assuming one of the sentences lower down leads to a contradiction in the same way that assuming the first line led to a contradiction. That looks very much like Yablo's paradox.

SOPHIA (*erasing the type indices with a very dirty sponge*) Yes, you get Yablo's list by erasing the indices. (*This leaves the following result on the whiteboard:*)

>No sentence below this one is true.
>No sentence below this one is true.
>No sentence below this one is true.
>No sentence below this one is true.
>⋮

FRED So, why did you bring up Visser's paradox at all?

SOPHIA Because you claimed that the semantic paradoxes disappear if we apply the object language versus metalanguage distinction. I think Visser's paradox shows that you can get a paradox even if you observe the distinction and you admit only truth predicates that apply to sentences with truth predicates lower in the hierarchy. Yablo's paradox, in contrast, is merely another proof of the inconsistency of the full T-sentences. We knew that before. But we didn't know that even the hierarchical approach can lead to paradox. After all, it was Kripke who mentioned the illfounded hierarchies.[4] There is no hint that Kripke realized that there was

---

[4] See [Kri75, p. 697].

a paradox. Let me see. (*She takes her computer out of her back, opens and continues after searching on the computer.*). Here. Kripke writes: ' Even if unrestricted truth definitions are in question, standard theorems easily allow us to construct a *descending* chain of first-order languages $L_0, L_1, L_2, \ldots$, such that L_i contains a truth predicate for L_{i+1}.' That doesn't sound as if he had been aware that such a hierarchy doesn't have a standard model.

FRED Ok, so Visser gave a proof of a new result, and Yablo gave a new proof of an old result.

SOPHIA Yes, and Visser showed that strict hierarchies of truth predicates don't save us from paradox. He showed that we can have paradox without such a global predicate. For his paradox we use truth predicates that apply only to sentences containing truth predicates lower in the hierarchy. In this sense Visser's paradox is much stronger than Yablo's paradox.

FRED Well, which axioms does Visser need to get the inconsistency? What does Yablo need?

SOPHIA (*sitting down again*) There are different ways to formalize these paradoxes. But usually you get only an an ω-inconsistency. For Yablo's paradox you need something like the unrestricted T-sentences. For Visser's paradox you need the T-sentences for instances of lower levels. Thus his T-sentences obey Tarski's object language versus metalanguage distinction.[5]

FRED (*staring out of the window*) Ok, then the problem isn't the unrestricted truth predicate. The root of the evil must be self-reference. Hasn't Priest shown that a proper definition of the Yablo list needs the fixed-point lemma or the recursion theorem or something that involves self-reference?[6] Doesn't Visser need that too?

SOPHIA Yes, Visser used the fixed-point lemma. When you formulate all this in arithmetic, you would normally use the fixed-point lemma or the recursion theorem.

FRED Why 'normally'?

SOPHIA Priest and others pointed at the use of the fixed-point lemma in the proofs of the inconsistency or ω-inconsistency. You can also use the recursion theorem. That's even nicer because you define a function that corresponds to the sentences in the Yablo or Visser lists. But I don't know a proof that the use of the fixed-point lemma or the recursion theorem cannot be avoided and other proofs cannot be given. It don't even know what exactly it means to say that Yablo's or Visser's list cannot be defined without the fixed-point lemma or the recursion theorem. You could always change the formulation of the fixed-point lemma a little to show that you don't have to go exactly through a certain lemma.

FRED Yes, there are variants, but in the end it all comes down to the old Gödel fixed point lemma or a trivial variation thereof. And that means in the end that the root of paradox is self-reference. Gödel himself already described his sentence as a sentence that states its own unprovability. Gödel produced a sentence that is equivalent to the sentence saying that the sentence isn't provable. (*He gets up and writes in the corner of the whiteboard:*)

G iff A('G')

If this equivalence is provable, then A says about itself that it has the property expressed by the formula $A(x)$, whether $A(x)$ expresses non-provability, falsity, or still something else.

SOPHIA Really? Truth predicates are supposed to have many fixed points. Probably you want

---

[5]Cf. [Lei01] for a comparison of Yablo's with Visser's and other paradoxes.
[6]See [Pri94].

>   G iff true('G')
>
for many sentences, not just for sentences G that are constructed using Gödel's trick? For instance, you want to have it for 'snow is white' as an instance. 'Snow is white' is a fixed point of the truth predicate. But do you really want to claim that it's self-referential?

FRED (*sinking into his armchair with a sigh*) Err, no. To be self-referential, the sentence must have been obtained via the Gödel's construction. And vice versa, if a sentence has been obtained via that construction, it's self-referential.

SOPHIA That's strange. I read that Gödel's diagonal construction is a trick. Now you seem to claim that it's an analysis of self-reference and the only way to get self-reference.

> *From time to time some people pass the window. While* FRED *is speaking, a lady passes with two plastic bags full of scripts. One of the scripts slips out, hovers for a second in the air and then sinks slowly down to the floor.*

FRED Well, in arithmetic Gödel's construction is the only way to get self-reference. In English there are more options. Think of the notorious sentence 'I am not true'. It sounds a little odd, because the personal pronoun usually refers to the speaker, not to the sentence that is uttered. We can also get self-reference via description. For instance, we can express a liar sentence by saying: (*whispering*) The whispered sentence isn't true.

SOPHIA Sentences with a first-order personal pronoun are used when people make claims about themselves. People usually don't use descriptions to make claims about themselves. It would be very strange if I referred to myself by a definite description. I say 'I will write an essay on truth for next week' and not 'The person sitting on the sofa will write an essay on truth for next week.' Personal pronouns seem to be the most direct way to make claims about oneself. Of course pretending that a sentence can make claims about itself using a first-person personal pronoun is weird. But this entire way of talking about making claims about itself and so on is just a metaphor that treats sentences like persons that can make claims about others and themselves.

> *The lady with the plastic bag has returned and can be seen through the window again. She picks up the lost scripts and continues her way again, while she is being watched through the window by* FRED *and* SOPHIA.

FRED Why?

SOPHIA Assume that a few seconds ago the woman spotted the script on the floor. When she thought 'I lost that piece of paper,' she had a belief about herself. If she had just thought that the last person who passed the window lost the script, without realizing that it was her, she wouldn't have come back. She thought that she *herself* lost the script.

FRED Even if she had thought that the last person who passed the window lost the script, it would still be a belief about herself, because she was that last person. Using the personal pronoun is just shorter than a definite description.

SOPHIA Is it really a belief about herself in a strict sense? The belief 'The last person who passed by lost the script' is only accidentally about herself, and she needn't know that it's about herself. If she thinks 'I lost that script', then it's not accidental – and only the latter belief prompts her to pick it up again.

FRED Ok, by using the pronoun 'I' she would refer to herself without relying on any description. She would have a *de re* belief about herself that she lost it, as far as it is ok to say that she is a *res*. But in arithmetic these differences play no role whatsoever. 'The last person who passed the window' refers to whoever passed the window, while the first-person pronoun 'I' rigidly always refers to the speaker. You remember what we said when we talked about *Naming and Necessity*?[7]

SOPHIA Of course.

FRED In arithmetic the differences don't matter, because in arithmetic nothing is contingent or a posteriori. Whatever satisfies a definite description in arithmetic does so by necessity and a priori. So all designators are rigid.

SOPHIA What I have in mind isn't the *de re/de dicto* distinction. I mean something that is stronger than *de re*. It's more what Lewis called *de se*.[8] The lady could think: 'Somebody must have been there, in front of the window. That person must have lost the script.' She would believe something about this person in a *de re* way, referring to him or her using the phrase 'that person'; but she need not believe *de se* that she herself lost the paper.

FRED I think that you confuse two topics here. We don't talk about propositional attitudes here. That's a topic for a later week.

SOPHIA (*getting up again*) Perhaps this really hasn't anything to do with the liar paradox and self-reference in logic. But *de se* beliefs are the paradigmatic self-referential beliefs.

FRED Why does this matter?

SOPHIA (*wiping the whiteboard again*) Because whether a sentence is true or not can depend on how the sentence refers to itself when we don't use pronouns like 'I'. I am not sure.

FRED The main point is that we have a term in the language that refers to the sentence itself – or its code.

SOPHIA Consider the following sentence (*writing on the whiteboard and reading out the following sentence:*)

$$(A) = (A) \tag{A}$$

This sentence says about (A), which happens to be the very sentence, that it's identical to (A). You would say that the sentence says about itself that it's identical to (A), don't you?

FRED Of course.

SOPHIA Now consider another sentence (*again writing on the whiteboard and reading out:*)

$$(B) = (A) \tag{B}$$

This sentence says the same about itself, namely that it is identical to (A). They say the same thing about themselves, according to you. Both sentences claim about themselves that they are identical to (A). Although they say the same thing about themselves, (A) is true while (B) is false.

---

[7] Fred refers to [Kri72].
[8] See [Lew79].

FRED Actually I retract my claim that (A) says about itself that it is identical with (A). It says about itself that it is self-identical. I mean, (A) states that it is identical with itself. If we allow sentences to refer to themselves in the first person, then the sentence (A) expresses the same proposition as the sentence 'I am identical with myself.'

SOPHIA Now you resort to the pronouns used to express *de se*-beliefs. But neither (A) nor (B) contain a personal pronoun. They don't express anything *de se*. You understand the label '(A)' as a first-person pronoun. The sentence 'I am identical to myself' says only one thing of itself. It only says that it is self-identical. But as soon as you use some other designator such as the name '(A)', that is no longer the case. Our sentence (A) says about itself that it is self-identical, but also that it is identical to (A) and that (A) is identical to it. I could have used another example with a relation that isn't symmetric. I could have written (*writing and speaking the following sentence*):

        (C) comes earlier in the alphabetical order than (C)   (C)

By 'coming earlier in the alphabet' I mean 'coming earlier if we ordered all sentences in the usual alphabetical order'. Sentence (C) says at least three things about itself; or, if you prefer, it ascribes three properties to itself. First, it says that it comes earlier in the alphabetical order than itself. But it also says about itself that (C) comes earlier than it – and that it comes earlier than (C). These are three different things. We can also construct sentences that ascribe to themselves only one of those two last properties: (*continues writing*):

        (C) comes earlier in the alphabetical order than (D)   (D)

and also

        (E) comes earlier in the alphabetical order than (C)   (E)

The sentence (C) is false, (D) is true, because C precedes D in the alphabet; (E) is false again.

FRED That's confusing.

SOPHIA All the confusion disappears once you use first-order pronouns.

FRED I think they add extra complications.

SOPHIA Actually they simplify the situation. The sentence 'I am identical to myself' ascribes to itself only the property of self-identity; it only says that it is identical to itself. You read (A) in this way (*pointing at* (A)). You think that (A) says the same as 'I am identical to myself'. But (A) ascribes at least two properties to itself, while the sentence 'I am identical to myself' ascribes only one property to itself.

FRED (*frowning*) What?

SOPHIA We have the sentence 'I am identical to myself'. It says about itself that it is self-identical. And we have (A). That sentence makes probably three claims about itself. That's why first-person pronouns and reflexive pronouns simplify everything. When we say that a sentence says something about itself, e.g., that it's not true, not provable, true, self-identical, and so on, we use the reflexive pronoun 'itself' in the same way that we said about the woman that she thought about herself that she had lost that script.

FRED Sorry, how is that related to her?

SOPHIA If the woman says 'I lost that script', she ascribes to herself the property of having lost that script. It's the same with the sentence 'I am identical to myself'. It ascribes to itself the property of being self-identical.

FRED Sure.

SOPHIA If the woman thinks 'The last person to pass by lost the script', then it's far less clear whether she ascribes any property to herself. Probably she does ascribe to herself a property, but that's contingent on the assumption that she is the last person to pass the window. She doesn't ascribe to herself the property *de se*. It's only second-rate self-reference. First-rate self-reference is obtained by using first-person pronouns.

FRED How is all this relevant to self-reference in formal languages where we get self-reference via the diagonal lemma?

SOPHIA We describe the Gödel sentence as the sentence as the sentence that says: 'I am not provable', which we take to mean that the sentence says of itself that it's not provable. Similarly, we describe the liar as the sentence that says: 'I am not true.' And we describe the truth teller as the sentence that says: 'I am true.' Then we study these sentences in formal languages that don't have personal pronouns or a similar device. In these languages we have only definite descriptions. We study sentences corresponding to the claim 'The last person who passed the window lost the script.' We don't study sentences of the form 'I lost the script.' We don't have first-person pronouns in arithmetic. There we have only definite descriptions of the sort 'the result of substituting the numeral of ...' The allegedly self-referential sentences in formal languages such as the language of arithmetic are not *de se*.

FRED (*impatiently*) We don't have much time left. Why is that relevant to the liar paradox?

SOPHIA You said that the source of the paradox is self-reference. Then you also claimed that the use of Gödel's diagonal lemma shows that self-reference is involved. But it may be misleading. The Gödel and liar sentences shouldn't be compared with a person saying: 'I lost the script'; it should be compared with somebody saying: 'The last person lost the script.' If we describe the self-referentiality of the liar sentence by seeing it as a sentence that says about itself that it is not true, then we ascribe a kind of self-referentiality to it that it doesn't have.

FRED So we don't have self-reference in formal languages?

SOPHIA I didn't say that.

FRED But we get paradoxes using the weaker form of self-reference without personal pronouns.

SOPHIA Yes, but I don't fully understand what self-reference is. Not any fixed-point sentence is self-referential. In the literature people tend to describe sentences obtained via Gödel's diagonal construction as sentences saying: 'I am X.' But, as I just said, that strikes me as unjustified.

FRED (*looking at the clock on the wall*) Ok, forget about the personal pronouns. Is Yablo's paradox self-referential?

SOPHIA I don't know.

FRED We better stop here. I still think you should write an essay on the correspondence theory. So the topic for next week is: 'Should we define truth in terms of correspondence or coherence?'

SOPHIA (*gasps with her mouth and eyes wide open*) Ok.

FRED (*getting up and searching for a script on his desk*) Bye.

SOPHIA See you next week.

SOPHIA (*leaves the room. Two new students come in.*)

FRED (*nodding to the two students*) David, George, your essays on the coherence and correspondence theories were brilliant.

BIBLIOGRAPHY

[Coo14] Roy Cook. *The Yablo Paradox: An Essay on Circularity*. Oxford University Press, Oxford, 2014.
[HV14a] Volker Halbach and Albert Visser. Self-Reference in Arithmetic I. *Review of Symbolic Logic*, 7:671–691, 2014.
[HV14b] Volker Halbach and Albert Visser. Self-Reference in Arithmetic II. *Review of Symbolic Logic*, 7:692–712, 2014.
[Kri72] Saul Kripke. Naming and Necessity. In Donald Davidson and Gilbert Harman, editors, *The Semantics of Natural Languages*, pages 253–355. Reidel, Dordrecht, 1972.
[Kri75] Saul Kripke. Outline of a Theory of Truth. *Journal of Philosophy*, 72:690–712, 1975. reprinted in [Mar84].
[Lei01] Hannes Leitgeb. Theories of Truth which have no Standard Models. *Studia Logica*, 21:69–87, 2001.
[Lew79] David Lewis. Attitudes De Dicto and De Se. *The Philosophical Review*, 88:513–543, 1979.
[Mar84] Robert L. Martin, editor. *Recent Essays on Truth and the Liar Paradox*. Clarendon Press and Oxford University Press, Oxford and New York, 1984.
[Pri94] Graham Priest. The Structure of the Paradoxes of Self-Reference. *Mind*, 103:25–34, 1994.
[Vis89] Albert Visser. Semantics and the liar paradox. In Dov Gabbay and Franz Günthner, editors, *Handbook of Philosophical Logic*, volume 4, pages 617–706. Reidel, Dordrecht, 1989.
[Yab85] Stephen Yablo. Truth and Reflection. *Journal of Philosophical Logic*, 14:297–349, 1985.
[Yab93] Stephen Yablo. Paradox Without Self-Reference. *Analysis*, 53:251–252, 1993.

Reasoning in Circles

ROSALIE IEMHOFF

For Albert, on the occasion of his 65th birthday

Abstract

Circular proofs, introduced by Daniyar Shamkanov, are proofs in which assumptions are allowed that are not axioms but do appear at least twice along a branch. Shamkanov has shown that a formula belongs to the provability logic GL exactly if it has a circular proof in the modal logic K4. Shamkanov uses Tait style proof systems and infinitary proofs. In this paper we prove the same result but then for sequent calculi and without the detour via infinitary systems. We also obtain a mild generalisation of the result, implying that its intuitionistic analogue holds as well.

1 Introduction

In the spring of 2015 Lev Beklemishev told Albert Visser and me about a theorem that his then student Daniyar Shamkanov had proved, a theorem stating that when a certain form of circularity is allowed in proofs in the modal logic K4, the resulting proof system is sound and complete with respect to the provability logic GL. Later that week Albert told me: "Such a theorem makes me happy for the whole day". For this Liber Amicorum, in honor of his 65th birthday and retirement, I take a closer look at this source of happiness.

In [Shamkanov(2014)] develops a notion of *circular proof*, that extends the standard notion of proof in a given Gentzen or Tait calculus by allowing derivations in which the leafs are not axioms but equal to a sequent below that leaf. For references to earlier occurrences of notions of circularity in proofs, see [Brotherston(2006)]. Clearly, proofs are special instances of circular proofs, namely those in which all leafs are axioms. Shamkanov shows that being provable in GL is equal to having a circular proof in K4. For example, the following is a circular proof of Löb's principle in the well-known sequent calculus for K4 as given in Section 2.

$$\frac{\dfrac{\Box(\Box\varphi \to \varphi) \Rightarrow \Box\varphi \quad \Box(\Box\varphi \to \varphi), \varphi \Rightarrow \varphi}{\boxdot(\Box\varphi \to \varphi) \Rightarrow \varphi} L\to}{\boxdot(\Box\varphi \to \varphi) \Rightarrow \Box\varphi} R_{\mathsf{K4}}$$

When I read Shamkanov's clever paper after Lev's visit to Utrecht I wondered whether his approach, which uses Tait style calculi, could be adapted for (two–sided) sequent calculi and whether it could be generalized to other logics. The answer to the first question is *yes*, and to the second question *I am not so sure*. In this paper I will explain why.

The key ideas in the paper are certainly Shamkanov's, but I do present certain facts in a different way. Most importantly, I do not use infinitary proof systems as an intermediary step between standard and circular proof systems, as is done in [Shamkanov(2014)].

In trying to establish whether Shamkanov's Theorem could be generalized to other logics, I have tried, in this paper, to generalize the assumptions under which the main theorem holds. And in doing so, I discovered that the results actually seem to very much depend on particular properties of the logic GL. The only immediate corollary from the generalization is the insight that an analogue of the main theorem holds for the intuitionistic versions of the modal logics (Theorem 5.4).

Structure of the paper

We proceed as follows. If $\vdash_G S$ denotes that S has a proof in Gentzen calculus G and $\vdash_G^\circ S$ denotes that S has a circular proof in G, then my translation of Shamkanov's Theorem in terms of the standard sequent calculi G3K4 and G3GL for K4 and GL, can be expressed as

$$\vdash_{\mathsf{G3GL}} S \text{ if and only if } \vdash_{\mathsf{G3K4}}^\circ S.$$

In this paper I generalize this to Theorem 5.1: For every extension G of G3p by ordered box rules that are closed under weakening and contraction and for any slim box rules R_1 and R_2 such that GR_2 is the circular companion of GR_1:

$$\vdash_{\mathsf{GR}_1} S \text{ if and only if } \vdash_{\mathsf{GR}_2}^\circ S.$$

The technical terms will be explained in the next sections, but let me mention here that all requirements are met by many sequent calculi for modal logics, except for the last requirement about modal companion. That G3K4 is the circular companion of G3GL seems, as we will see, to depend strongly on the properties of GL.

Theorem 5.1 consists of two directions, the one from left to right is Lemma 3.1 and the other direction is Lemma 4.4. Section 5 contains the main result and the application to intuitionistic modal logics. It also shows why the obvious generalization to Grzegorczyk logic does not work.

I thank two anonymous referees for useful comments on an earlier version of this paper.

2 Logics and sequent calculi

The logics we consider are modal propositional logics, formulated in a language $\mathcal{L}$ that contains constants $\top$ and $\bot$, propositional variables or atoms $p, q, r, \dots$ and the connectives $\wedge, \vee, \neg, \to$ and the modal operator $\Box$. The expression $\boxdot \varphi$ stands for $\varphi \wedge \Box \varphi$. All logics that we consider are extensions of classical propositional logics, but we do not assume them to be normal.

We will mainly work with sequents, which are expression $\Gamma \Rightarrow \Delta$, where Γ and Δ are finite multisets of formulas in $\mathcal{L}$, that are interpreted as $I(\Gamma \Rightarrow \Delta) = (\bigwedge \Gamma \to \bigvee \Delta)$. We denote finite multisets by $\Gamma, \Pi, \Delta, \Sigma$. We also define ($a$ for antecedent, s for succedent):

$$(\Gamma \Rightarrow \Delta)^a \equiv_{df} \Gamma \quad (\Gamma \Rightarrow \Delta)^s \equiv_{df} \Delta.$$

When sequents are used in the setting of formulas, we often write S for $I(S)$, such as in $\vdash S$, which thus means $\vdash I(S)$. Multiplication of sequents is defined as

$$S_1 \cdot S_2 \equiv_{df} (S_1^a \cup S_2^a \Rightarrow S_1^s \cup S_2^s).$$

Given a multiset Γ, we write $\Box\Gamma$ for the multiset obtained by putting a box in front of every formula in Γ, and similarly for $\boxdot\Gamma$. For a sequent S we write $\Box S$ for the sequent $\Box S^a \Rightarrow \Box S^s$, and similarly for $\boxdot S$. For example, $\Box(p \Rightarrow)$ denotes $(\Box p \Rightarrow)$.

We will be interested in multisets in which the repetition of formulas occurs for certain formulas only. Given a multiset Γ we denote by $\Gamma_{\boxdot}$ the largest set $\boxdot\Pi$ such that $\boxdot\Pi \subseteq \Gamma$, and the multiset $\Gamma_{\boxdot} \cup \{\varphi \in \Gamma \mid \varphi \notin \Gamma_{\boxdot}\}$ by Γ^*. Thus $\Gamma^* \backslash \Gamma_{\boxdot}$ is a set. With every sequent $S = (\Gamma \Rightarrow \Delta)$ the *set–sequent* $S^* = (\Gamma^* \Rightarrow \Delta^*)$ is associated. A sequent S is a *set–sequent* if it is of the form S_0^* for some sequent S_0. Two sequents are *set–equivalent* if their set–sequents are equal.

Given formulas $\chi(p)$ and φ, $\chi(\varphi)$ denotes the result of replacing p everywhere by φ in χ. For a multiset Γ, we use $\chi(\Gamma)$ as abbreviation for the set of formulas $\{\chi(\varphi) \mid \varphi \in \Gamma\}$ and $\chi(S)$ for $\chi(I(S))$. For example, if $\chi = \Box p$, then $\chi(\{\varphi, \psi\}) = \{\Box\varphi, \Box\psi\}$.

The complexity of formulas is defined as usual, where connectives and modal operators increase the complexity by 1. We define a partial order $\preccurlyeq$ on sequents, based on the Dershowitz-Manna well-ordering $\preccurlyeq_{dm}$ on multisets, in the usual way: $S_1 \preccurlyeq S_2$ exactly if $S_1^a \cup S_1^s \preccurlyeq_{dm} S_2^a \cup S_2^s$. Here $\preccurlyeq_{dm}$ is the reflexive transitive closure of the ordering $\prec^-_{dm}$ between multisets, where $\Gamma \prec^-_{dm} \Pi$ precisely if Γ is the result of replacing a formula in Π by finitely many formulas of lower complexity than that formula. Furthermore, $S_1 \prec S_2$ precisely if $S_1 \preccurlyeq S_2$ and S_1 and S_2 are not equal as sequents.

Gentzen calculi

A rule R is an expression of the form

$$\frac{S_1 \quad \ldots \quad S_n}{S_0} \, R$$

where the S_i are sequents. R_c denotes the formula $I(S_0)$ corresponding to the conclusion, and R^a denotes the formula $\bigwedge_i I(S_i)$ corresponding to the conjunction of the premises. An *axiom* is a rule with no premises, thus in our view, axioms are rules.

Given an extension G of G3p, to be defined below, and a rule R, we denote the calculus $G + R$ by GR. In the case of G3p, we leave out the "p" and write G3R instead of G3pR.

Rule R is a *box rule* if it satisfies:

○ The conclusion of R is of the form $\Box S \cdot (\Gamma \Rightarrow \Sigma)$ for some sequent S and two multisets Γ, Σ not occurring, as multiset symbols, in S nor in the premises of R.

○ All premises of R consist of subformulas of formulas in S.

○ If an instance of R is of the form

$$\frac{S_1 \quad \ldots \quad S_n}{(\Gamma \Rightarrow \Sigma) \cdot \Box(S_0 \cdot S'_0 \cdot S'_0)}$$

there are sequents S'_i and S''_i such that $S_i = S'_i \cdot S''_i \cdot S''_i$ and

$$\frac{S'_1 \cdot S''_1 \quad \ldots \quad S'_n \cdot S''_n}{(\Gamma \Rightarrow \Sigma) \cdot \Box(S_0 \cdot S'_0)}$$

is an instance of R as well.

The last requirement guarantees that when box rules are added to a sequent calculus, closure under weakening and contraction is preserved, as will be proved in Lemma 2.4.

Examples of a box rule (left) and a rule that is not a box rule (right):

$$\frac{p \Rightarrow}{\Pi, \Box p \Rightarrow \Delta} \qquad \frac{\Pi, p \Rightarrow \Delta}{\Pi, \Box p \Rightarrow \Delta}$$

Important in this paper are the box rules R_{K4}, R_{GL} and R_{Grz}, which are, respectively,

$$\frac{\Box\Gamma \Rightarrow \varphi}{\Pi, \Box\Gamma \Rightarrow \Box\varphi, \Delta} \; R_{\mathsf{K4}} \qquad \frac{\Box\Gamma, \Box\varphi \Rightarrow \varphi}{\Pi, \Box\Gamma \Rightarrow \Box\varphi, \Delta} \; R_{\mathsf{GL}} \qquad \frac{\Box\Gamma, \Box(\varphi \to \Box\varphi) \Rightarrow \varphi}{\Pi, \Box\Gamma \Rightarrow \Box\varphi, \Delta} \; R_{\mathsf{Grz}}$$

A *Gentzen calculus* or a *sequent calculus* is a finite set of rules. In this paper we only consider Gentzen calculi of the form G3p + $\mathcal{R}$ for some set of box rules $\mathcal{R}$, where G3p is given as follows.

The Gentzen calculus G3p

$$\frac{}{\Gamma, p \Rightarrow p, \Delta} \; Ax \; (p \text{ an atom}) \qquad \frac{}{\Gamma, \bot \Rightarrow \Delta} \; L\bot$$

$$\frac{\Gamma, \varphi, \psi \Rightarrow \Delta}{\Gamma, \varphi \wedge \psi \Rightarrow \Delta} \; L\wedge \qquad \frac{\Gamma \Rightarrow \varphi, \Delta \quad \Gamma \Rightarrow \psi, \Delta}{\Gamma \Rightarrow \varphi \wedge \psi, \Delta} \; R\wedge$$

$$\frac{\Gamma, \varphi \Rightarrow \Delta \quad \Gamma, \psi \Rightarrow \Delta}{\Gamma, \varphi \vee \psi \Rightarrow \Delta} \; L\vee \qquad \frac{\Gamma \Rightarrow \varphi, \psi, \Delta}{\Gamma \Rightarrow \varphi \vee \psi, \Delta} \; R\vee$$

$$\frac{\Gamma \Rightarrow \varphi, \Delta \quad \Gamma, \psi \Rightarrow \Delta}{\Gamma, \varphi \to \psi \Rightarrow \Delta} \; L\to \qquad \frac{\Gamma, \varphi \Rightarrow \psi, \Delta}{\Gamma \Rightarrow \varphi \to \psi, \Delta} \; R\to$$

A *derivation tree* for S in a calculus G is a finite tree labelled with sequents, where the root is labelled with S, and every inner node (not a leaf) with all its parent(s) forms an instance of a rule in G. A *derivation* or *(standard) proof* of S in G is a derivation tree for which all the leafs are axioms. We write $\vdash_\mathsf{G} S$ if sequent S has a derivation in G, and when G is clear from the context we write $\vdash$ instead of $\vdash_\mathsf{G}$. We write $\vdash_d S$ if S has a proof of depth (length of the longest branch of the derivation tree) at most d.

THEOREM 2.1. *[Avron(1984)]* $\vdash_\mathsf{G3GL} S$ *if and only if* $\vdash_\mathsf{GL} I(S)$.

A *substitution* σ is a map from formulas in $\mathcal{L}$ to formulas in $\mathcal{L}$ that commutes with the connectives and the modal operator. σS denotes the sequent $\{\sigma\varphi \mid \varphi \in S^a\} \Rightarrow \{\sigma\varphi \mid \varphi \in S^s\}$. We say that φ *admissibly derives* ψ in a logic L, notation $\varphi \mathrel{\vert\!\sim}_\mathsf{L} \psi$, if for every substitution σ, if $\vdash_\mathsf{L} \sigma\varphi$, then $\vdash_\mathsf{L} \sigma\psi$. For sequents S and S', S *admissibly derives* S' in a calculus G, notation $S \mathrel{\vert\!\sim}_\mathsf{G} S'$, if for every substitution σ, if $\vdash_\mathsf{G} \sigma S$, then $\vdash_\mathsf{G} \sigma S'$. Lemmas 2.2 below provides a typical example of admissibility. Clearly, if $\varphi \vdash_\mathsf{L} \psi$, then $\varphi \mathrel{\vert\!\sim}_\mathsf{L} \psi$. But the converse is not always the case, more on this topic can be found in [Jeřábek(2005)].

A leaf with label S for which there is a node at its branch properly below it with the same label S is *circular*. A *circular derivation* or *circular proof* of S in a calculus G is a derivation tree for S in which every leaf either is an axiom of G or is circular. We write $\vdash_\mathsf{G}^\circ S$ if sequent S has a circular derivation in G. A circular derivation is in particular a derivation tree.

Clearly, $\vdash_G S$ implies $\vdash_G^\circ S$, but not vice versa, as the following circular proof of the sequent version of *Löb's principle* shows.

$$\frac{\dfrac{\Box(\Box\varphi \to \varphi) \Rightarrow \Box\varphi \quad \Box(\Box\varphi \to \varphi), \varphi \Rightarrow \varphi}{\dfrac{\boxdot(\Box\varphi \to \varphi) \Rightarrow \varphi}{\Box(\Box\varphi \to \varphi) \Rightarrow \Box\varphi} R_{\mathsf{K4}}} L{\to}}$$

Thus we can conclude that the sequent version of Löb's principle has a circular proof in K4: $\vdash_{\mathsf{K4}}^\circ$ $\Box(\Box\varphi \to \varphi) \Rightarrow \Box\varphi$. As the principle is not provable in K4 this shows that $\vdash_{\mathsf{K4}}^\circ$ is strictly stronger than $\vdash_{\mathsf{K4}}$. One of the corollaries of the main theorem of this note is that, actually, a sequent has a proof in GL if and only if it has a circular proof in K4.

If we weaken the requirement of circular leafs to: there is a node at its branch properly below it with a label that has the same set-sequent as S, the system is no longer sound, as the following circular proof shows.

$$\frac{\varphi \wedge \psi, \varphi, \psi, \varphi, \psi \Rightarrow}{\varphi \wedge \psi, \varphi \wedge \psi, \varphi, \psi \Rightarrow} L\wedge$$

LEMMA 2.2. $\Box\Pi, \boxdot\Sigma, \Box\varphi \Rightarrow \varphi \vdash_{\mathsf{G3GL}} \Box\Pi, \boxdot\Sigma \Rightarrow \varphi$.

Proof. The following steps prove the lemma, using in the third step that $S \vdash_{\mathsf{G3GL}} \Box S$ for any S.

$$\begin{array}{ll}
\Box\Pi, \boxdot\Sigma, \Box\varphi \Rightarrow \varphi & \vdash_{\mathsf{G3GL}} \\
\Box\Pi, \boxdot\Sigma \Rightarrow \Box\varphi \to \varphi & \vdash_{\mathsf{G3GL}} \\
\Box\Pi, \boxdot\Sigma \Rightarrow \Box(\Box\varphi \to \varphi) & \vdash_{\mathsf{G3GL}} \\
\Box\Pi, \boxdot\Sigma \Rightarrow \Box\varphi & \vdash_{\mathsf{G3GL}} \text{ (using first line)} \\
\Box\Pi, \boxdot\Sigma \Rightarrow \varphi. &
\end{array}$$

⊣

Weakening and contraction

LEMMA 2.3. *(Inversion Lemma) If $\mathcal{R}$ is a set of box rules, then in* G3p + $\mathcal{R}$ *the following holds.*

1. $\vdash_d \Gamma, \varphi \wedge \psi \Rightarrow \Delta$ *implies* $\vdash_d \Gamma, \varphi, \psi \Rightarrow \Delta$.

2. $\vdash_d \Gamma, \varphi_0 \vee \varphi_1 \Rightarrow \Delta$ *implies* $\vdash_d \Gamma, \varphi_i \Rightarrow \Delta$ *for* $i = 0, 1$.

3. $\vdash_d \Gamma, \varphi \to \psi \Rightarrow \Delta$ *implies* $\vdash_d \Gamma, \psi \Rightarrow \Delta$ *and* $\vdash_d \Gamma \Rightarrow \varphi, \Delta$.

4. $\vdash_d \Gamma \Rightarrow \varphi_0 \wedge \varphi_1, \Delta$ *implies* $\vdash_d \Gamma \Rightarrow \psi_i, \Delta$ *for* $i = 0, 1$.

5. $\vdash_d \Gamma \Rightarrow \varphi \vee \psi, \Delta$ *implies* $\vdash_d \Gamma \Rightarrow \varphi, \psi, \Delta$.

6. $\vdash_d \Gamma \Rightarrow \varphi \to \psi, \Delta$ *implies* $\vdash_d \Gamma, \varphi \Rightarrow \psi, \Delta$.

Proof. Analogues to the proof of Lemma 5.1.6 in [Troelstra and Schwichtenberg(1996)]. With induction to d. The case that $d = 1$ is straightforward. In the induction step we consider the last inference of the derivation and distinguish by cases. For inferences that are instances of rules in G3p we reason as in [Troelstra and Schwichtenberg(1996)]. For an instance $S_1 \ldots S_n / (\Pi \Rightarrow \Sigma) \cdot \Box S_0$ of a box rule R in $\mathcal{R}$, it follows that any formula φ in the conclusion that is not boxed can be replaced by any formula ψ and still have a valid proof, thus proving that also in this case all six properties in the lemma hold. ⊣

LEMMA 2.4. *For any set of box rules $\mathcal{R}$ weakening and contraction are depth preserving admissible in the calculus* G3p $+ \mathcal{R}$: *In* G3p $+ \mathcal{R}$, *for any sequents S and S' the following holds.*

- *If* $\vdash_d S$, *then* $\vdash_d S' \cdot S$.
- *If* $\vdash_d S' \cdot S' \cdot S$, *then* $\vdash_d S' \cdot S$.

Proof. We prove the lemma with induction to d. The proof for weakening is straightforward and therefore left to the reader. The key ingredient is the observation that for any instance

$$\frac{S_1 \quad \ldots \quad S_n}{S_0}$$

of a box rule R and any sequent S,

$$\frac{S_1 \quad \ldots \quad S_n}{S \cdot S_0}$$

is an instance of R as well.

We turn to contraction. Suppose G3p $+ \mathcal{R}$ derives $S' \cdot S' \cdot S$. Clearly, it suffices to treat the case that S' consist of a single formula, say φ. We treat the case that $S' = (\varphi \Rightarrow)$, the other case being analogous. If $d = 1$, then $S' \cdot S' \cdot S$ is an instance of an axiom. If it is an axiom of G3p, inspection of the possible axioms shows that whence $S' \cdot S$ is an instance of that axiom too. If the axiom belongs to $\mathcal{R}$, then the third requirement in the definition of box rules implies that $S' \cdot S$ is an instance of the axiom too.

If $d > 1$, consider the last inference

$$\frac{S_1 \quad \ldots \quad S_n}{S' \cdot S' \cdot S} \tag{1}$$

of the derivation. If it is an instance of a rule R in $\mathcal{R}$, $S' \cdot S' \cdot S = (\Gamma \Rightarrow \Sigma) \cdot \Box S_0$ for some Γ, Σ and S_0. There are several cases to consider: (1) φ occurs twice in $\Box S_0^a$ or (2) φ occurs twice in Γ or (3) φ occurs in S_0^a and Γ.

In case (1) the third requirement in the definition of box rules implies that there exist S_i' and S_i'' such that $S_i = S_i' \cdot S_i' \cdot S_i''$ and

$$\frac{S_1' \cdot S_1'' \quad \ldots \quad S_n' \cdot S_n''}{S' \cdot S}$$

is an instance of R. By the induction hypothesis, the $S_i' \cdot S_i''$ have proofs of depth smaller than d, which proofs that $S' \cdot S$ has proof of depth at most d. In cases (2) and (3) it follows that

$$\frac{S_1 \quad \ldots \quad S_n}{S' \cdot S}$$

is an instance of R, and we are done immediately.

If (1) is an instance of a rule of G3p, we have to distinguish by cases. We treat the left implication rule. Therefore assume (1) is of the form

$$\frac{\Gamma \Rightarrow \varphi, \Delta \quad \Gamma, \psi \Rightarrow \Delta}{\Gamma, \varphi \to \psi \Rightarrow \Delta} \; R$$

where either Δ or Γ contains a formula twice or Γ contains $\varphi \to \psi$. In the first two cases the induction hypothesis immediately applies. In the last case, by applying Lemma 2.3 to the two premises, it follows that $\Gamma\backslash\{\varphi \to \psi\} \Rightarrow \varphi, \varphi, \Delta$ and $\Gamma\backslash\{\varphi \to \psi\}, \psi, \psi \Rightarrow \Delta$ have proofs of depth $< d$. Hence so do $\Gamma\backslash\{\varphi \to \psi\} \Rightarrow \varphi, \Delta$ and $\Gamma\backslash\{\varphi \to \psi\}, \psi \Rightarrow \Delta$ by the induction hypothesis. An application of $L \to$ gives $\Gamma\backslash\{\varphi \to \psi\}, \varphi \to \psi \Rightarrow \Delta$.

⊣

COROLLARY 2.5. *Weakening and contraction are admissible in* G3K4 *and* G3GL.

In this paper we do not need the admissibility of cut in G3K4 and G3GL, but it is worth mentioning that the rule is indeed admissible. For a proof, see, for example, [Avron(1984)].

Ordered rules and proofs

A rule is *ordered* if all its premises are $\prec$–lower than its conclusion and consist solely of subformulas of formulas in the conclusion.

An instance of a rule is a *set*–instance if the premises are set–sequents. Given a set of rules $\mathcal{R}$, a proof is $\mathcal{R}$–*set* if every instance in the proof of a rule in $\mathcal{R}$ is a set-instance.

Given a calculus G, denote by $\mathcal{R}_G$ the set of those rules in G in which the premises are not $\prec$–lower than the conclusion. A calculus G is *ordered* if every provable sequent has a proof that is $\mathcal{R}_G$-set.

A rule R is *slim* if for every instance $S_1 \ldots S_n / S$ of it, $S_1^* \ldots S_n^* / S$ is an instance of R as well. Observe that both R_{GL} and R_{K4} are slim rules.

LEMMA 2.6. *For every set $\mathcal{R}$ of box rules that are slim or ordered: for any sequent S provable in* G3p $+ \mathcal{R}$, *there is a finite set of sequents $\mathcal{S}$ such that in any proof of S in* G3p $+ \mathcal{R}$ *that is $\mathcal{R}$–set, only sequents in $\mathcal{S}$ occur.*

Proof. Let $\mathcal{S}'$ consist of all set–sequents that consist of subformulas of formulas in S. $\mathcal{S}$ denote the $\mathcal{S}'$ union all sequents that are $\prec$–lower than a sequent in $\mathcal{S}'$. Because of the subformula property that box rules as well as rules in G3p satisfy, every $\mathcal{R}$–set proof of S contains only sequents in $\mathcal{S}$. ⊣

LEMMA 2.7. *For every extension* G *of* G3p *by ordered box rules and for every set $\mathcal{R}$ of slim box rules: for every proof in* G $+ \mathcal{R}$, *there exists an $\mathcal{R}$–set proof in* G $+ \mathcal{R}$ *of the same endsequent of depth no greater than the original proof.*

Proof. Consider a proof in $G' = G + \mathcal{R}$. With induction on the depth $d(\mathcal{D})$ of the lowest inferences that violate that $\mathcal{D}$ is $\mathcal{R}$–set, with a subinduction to the number $m(\mathcal{D})$ of those lowest inferences that violate that $\mathcal{D}$ is $\mathcal{R}$–set. If $d(\mathcal{D}) = 0$, then $\mathcal{D}$ is $\mathcal{R}$–set and there is nothing to prove.

If $d(\mathcal{D}) > 0$, consider an inference

$$\frac{S_1 \quad \ldots \quad S_n}{S}$$

at depth $d(\mathcal{D})$ which is an application of a rule $R \in \mathcal{R}$ such that not all S_i are set–sequents. As R is closed under contraction, Lemma 2.4 implies that the sequent S_i^* has a proof of the same or lower depth than the proof of S_i. Since R is a set–rule, this implies S has a proof of depth $\leq d(\mathcal{D})$ in which the last inference is a set-instance of R. Replacing the subproof of S in $\mathcal{D}$ by this proof results in a proof $\mathcal{D}'$ with the same endsequent as $\mathcal{D}$ for which either $d(\mathcal{D}') < d(\mathcal{D})$, or $d(\mathcal{D}') = d(\mathcal{D})$ and $m(\mathcal{D}') < m(\mathcal{D})$. In both cases the induction hypothesis applies and we obtain a proof of the endsequent of $\mathcal{D}$ that is $\mathcal{R}$–set. ⊣

3 From standard proofs to circular proofs

LEMMA 3.1. *For every extension* G *of* G3p *by ordered box rules: if* R_1, R_2 *are slim box rules such that* $R_1^a \vdash_{\mathsf{GR}_1} R_2^a$ *and* $R_1^c = R_2^c$, *then* $\vdash_{\mathsf{GR}_1} S$ *implies* $\vdash_{\mathsf{GR}_2}^\circ S$.

Proof. First we need to introduce some terminology. Given a derivation $\mathcal{D}$, let $h_\mathcal{D}^R$ denote the height of the lowest application of R in $\mathcal{D}$, where the height on a application of a rule R is the number of nodes from the root of the tree to the conclusion of that application. If $\mathcal{D}$ does not contain applications of R we put $h_\mathcal{D}^R = 0$. With $n_\mathcal{D}^R$ we denote the number of applications of R at height $h_\mathcal{D}^R$ in $\mathcal{D}$.

Let G be G3p extended by R_1 and R_2. Suppose $\vdash_{\mathsf{GR}_1} S$ and let $\mathcal{D}_0$ be an $\{R_1, R_2\}$–set proof of S in GR_1, which exists by the previous lemma. We construct a sequence $\mathcal{D}_0, \mathcal{D}_1, \mathcal{D}_2, \ldots$ of $\{R_1, R_2\}$–set proofs in G with the following properties, where $h_i = h_{\mathcal{D}_i}^{R_1}$ and $n_i = n_{\mathcal{D}_i}^{R_1}$. For every i either $n_i = 0$, or $h_{i+1} = h_i$ and $n_{i+1} < n_i$, or $h_{i+1} > h_i$. In no $\mathcal{D}_i$ there are applications of R_2 above applications of R_1. In other words, subproofs that end in an application of R_1, are proofs in GR_1.

If $\mathcal{D}_i$ contains no application of R_1, then the sequence stops at $\mathcal{D}_i$ with $n_i = 0$. Otherwise consider the leftmost application of R_1 at height h_i and let S_0 and S_1 be its conclusion and its premiss, respectively. The subproof of S_0 therefore is a proof in GR_1. Since $S_1 = R_1^a \vdash_{\mathsf{GR}_1} R_2^a$, there exists a proof in GR_1 of R_2^a. As R_1 is a slim rule, Lemmas 2.4 and 2.7 imply that there is an $\{R_1\}$–set proof of $(R_2^a)^*$ in GR_1. Let $\mathcal{D}$ denote this proof followed by an application of R_2. Thus $\mathcal{D}$ is an $\{R_1, R_2\}$–set proof of S_0. Let $\mathcal{D}_{i+1}$ be the result of replacing the considered subproof of S_0 by $\mathcal{D}$. We show that it has the required properties.

That there is no application of R_2 above applications of R_1 is clear. If $n_i > 1$, then $n_{i+1} = n_i - 1 < n_i$ and $h_i = h_{i+1}$. If, on the other hand, $n_i = 1$, then $h_{i+1} > h_i$ or $n_{i+1} = 0$. This proves that a sequence of proofs as described above can be constructed.

Since all $\mathcal{D}_i$ are $\{R_1, R_2\}$–set proofs in G it follows from Lemma 2.6 that there exists a finite set of sequents $\mathcal{S}$ such that every sequent that occurs in some $\mathcal{D}_i$ belongs to $\mathcal{S}$. There are two possibilities: the sequence of the $\mathcal{D}_i$ is finite or it is infinite. It follows from the construction that in the first case the last proof in the sequence does not contain applications of R_1. Thus it is a proof in GR_2. Hence $\vdash_{\mathsf{GR}_2} S$ and therefore $\vdash_{\mathsf{GR}_2}^\circ S$. If the sequence is infinite, Consider $\mathcal{D}_i$ for an i for which h_i is larger than the number of sequents in $\mathcal{S}$. The length of any branch in $\mathcal{D}_i$ is either greater than h_i or at most h_i. In the last case, it cannot contain applications of R_1. In the first case, the sequent at height h_{i+1} has to occur at that branch at a height lower than h_i as well. Therefore, if we cut away all nodes at height h_{i+2} and higher we obtain a circular proof of S in GR_2. ⊣

4 From circular proofs to standard proofs

LEMMA 4.1. *For every extension* G *of* G3p *by ordered box rules, if in a proof of a sequent in* G + R *there is a branch with two nodes with the same label, then there is an application of R between these two occurrences along the branch.*

Proof. In all rules in G the premises are $\prec$–lower than the conclusion. ⊣

Given a calculus G and two rules R_1 and R_2, calculus GR_2 is the *circular companion* of calculus GR_1 if there exist formulas $\chi(p)$ and $\eta(p)$ such that for any instance $S_1 \ldots S_n/S_0$ of R_2 and for all multisets Π and Σ (recall that $\chi(S)$ stands for $\chi(I(S))$ and $\chi(\Gamma)$ for $\{\chi(\varphi) \mid \varphi \in \Gamma\}$, and likewise for η):

- $R_1^a \vdash_{GR_1} R_2^a$ and $R_1^c = R_2^c$;
- $\vdash_{GR_1} \eta(\varphi) \to \varphi$ for all formulas φ;
- $\{\chi(\Pi), \eta(\Sigma), S_i^a \Rightarrow S_i^s \mid 1 \le i \le n\} \vdash_{GR_1} \chi(\Pi \cup \Sigma), S_0^a \Rightarrow S_0^s$;
- $\chi(\Pi), \eta(\Sigma), \chi(S), S^a \Rightarrow S^s \vdash_{GR_1} \chi(\Pi), \eta(\Sigma), S^a \Rightarrow S^s$ for any sequent S;
- for every instance $S_1' \ldots S_n'/S_0'$ of a rule in G, $S' \cdot S_1' \ldots S' \cdot S_n'/S' \cdot S_0'$ is an instance as well, for S' of the form $(\chi(\Pi), \eta(\Sigma) \Rightarrow)$.

REMARK 4.2. *The last two requirements in the definition of circular companions imply that for such companions also holds:*

$$\chi(\Pi), \eta(\Sigma), \eta(S_0), S_i^a \Rightarrow S_i^s \mid 1 \le i \le n\} \vdash_{GR_1} \chi(\Pi \cup \Sigma), S_0^a \Rightarrow S_0^s.$$

REMARK 4.3. G3K4 *is the circular companion of* G3GL *by taking* $\eta(p) = \Box p$ *and* $\chi(p) = \Box p$. *In fact, for any extension* G *of* G3p, $G + R_{K4}$ *is the circular companion of* $G + R_{GL}$ *for the same* η *and* χ. *That the second requirement holds is trivial. For the third one the following observations suffice, recalling that* $\Box(\Box\Gamma \Rightarrow \Box\varphi)$ *denotes* $\Box(\bigwedge \Box\Gamma \to \Box\varphi)$.

$$\Box\Pi, \Box\Sigma, \Box\Gamma \Rightarrow \varphi \qquad \vdash_{G3GL}$$
$$\Box\Pi, \Box\Sigma, \Box\Gamma, \Box\varphi \Rightarrow \varphi \qquad \vdash_{G3GL}$$
$$\Box\Pi, \Box\Sigma, \Box\Gamma \Rightarrow \Box\varphi.$$

The fourth requirement follows from Lemma 2.2 with $\varphi = I(S)$, *and the first requirement is left to the reader.*

LEMMA 4.4. *For every extension* G *of* G3p *by ordered box rules: if* R_1, R_2 *are slim box rules such that* GR_2 *is the circular companion of* GR_1, *then* $\vdash^\circ_{GR_2} S$ *implies* $\vdash_{GR_1} S$.

Proof. Given a derivation tree $\mathcal{D}$ in GR_2, a leaf labelled with sequent S is an *assumption leaf* if it is not circular and S is not an axiom. Denote by $ap_\mathcal{D}$ and $nap_\mathcal{D}$ the sets of formulas of the form $I(S)$, where S is the label of an assumption leaf that has, respectively does not have, an application of R_2 along its branch.

Suppose GR_2 is the circular companion of GR_1 and let (χ, η) be the witness of it. We prove with induction to the height of a circular derivation tree $\mathcal{D}$ in GR_2 with root S:

$$\vdash_{\mathsf{GR}_1} \chi(ap_{\mathcal{D}}), \eta(nap_{\mathcal{D}}), S^a \Rightarrow S^s. \tag{2}$$

Since for a circular proof in GR_2, both $ap_{\mathcal{D}}$ and $nap_{\mathcal{D}}$ are empty, this will prove the lemma.

If $\mathcal{D}$ consists of one sequent only, it is either an axiom of GR_2, in which case (2) clearly holds, or it is an assumption leaf with no application of R_2 along its branch, which also implies (2) because $\eta(\varphi)$ implies φ for all formulas φ.

Suppose the height of $\mathcal{D}$ is greater than one and suppose the last inference of $\mathcal{D}$ is an application of a rule R and let $S_1, \ldots, S_m$ be its premisses. The induction hypothesis and the fact that GR_1 is closed under weakening gives for every i:

$$\vdash_{\mathsf{GR}_1} \chi(\bigcup_{i=1}^{m} ap_{\mathcal{D}_i}), \eta(\bigcup_{i=1}^{m} nap_{\mathcal{D}_i}), S_i^a \Rightarrow S_i^s, \tag{3}$$

We distinguish the cases that any leaf in $\mathcal{D}$ that is circular is circular in one of the $\mathcal{D}_i$, and the opposite case. In the first case, if R is one of the rules of G, then $ap_{\mathcal{D}}$ is equal to $\bigcup \{ap_{\mathcal{D}_i} \mid i \leq m\}$, and similarly for $nap_{\mathcal{D}}$. Therefore (2) follows from the last requirement of circular companions and an application of R to (3). If $R = R_2$, then $nap_{\mathcal{D}}$ is empty and

$$ap_{\mathcal{D}} = \bigcup_{i=1}^{m} (ap_{\mathcal{D}_i} \cup nap_{\mathcal{D}_i}).$$

As GR_2 is the circular companion of GR_1, it follows that $\vdash_{\mathsf{GR}_1} \chi(ap_{\mathcal{D}}), S^a \Rightarrow S^s$, which implies (2).

Next, consider the case that in $\mathcal{D}$ there is a circular leaf that is not circular in any of the $\mathcal{D}_i$. Note that all such leafs are labelled with the same sequent as the endsequent of the proof, S, and that they may become assumption leafs in the $\mathcal{D}_i$. Hence

$$ap_{\mathcal{D}} \cup nap_{\mathcal{D}} = (\bigcup_{i=1}^{m} (ap_{\mathcal{D}_i} \cup nap_{\mathcal{D}_i})) \setminus \{S\}.$$

First consider the case that R is one of the rules of G. By Lemma 4.1 it follows that there is an application of R_2 along branches that have leaf S, which means that if S occurs in $\bigcup_{i=1}^{m}(ap_{\mathcal{D}_i} \cup nap_{\mathcal{D}_i})$, it occurs in $\bigcup_{i=1}^{m} ap_{\mathcal{D}_i}$, and therefore as $\chi(S)$ in (3). An application of R to (3) gives

$$\vdash_{\mathsf{GR}_1} \chi(ap_{\mathcal{D}}), \eta(nap_{\mathcal{D}}), \chi(S), S^a \Rightarrow S^s.$$

The fact that GR_2 is the circular companion of GR_1 now implies (2).

If, on the other hand, $R = R_2$, then $nap_{\mathcal{D}}$ is empty and for any i, S may be in either $ap_{\mathcal{D}_i}$ or $nap_{\mathcal{D}_i}$. And thus appear as $\chi(S)$ or $\eta(S)$ in (3). Then the fact that GR_2 is the circular companion of GR_1 and Remark 4.2 imply (2). ⊣

5 Standard proofs versus circular proofs

The results in the previous section lead to a sufficient condition for being the circular companion of a logic, from which Shamkanov's results follow.

THEOREM 5.1. *For every extension G of G3p by ordered box rules and for any slim box rules R_1 and R_2 such that GR_2 is the circular companion of GR_1:*

$$\vdash_{GR_1} S \text{ if and only if } \vdash^{\circ}_{GR_2} S.$$

THEOREM 5.2. *For every extension G of G3p by ordered box rules:*

$$\vdash_{G+R_{GL}} S \text{ if and only if } \vdash^{\circ}_{G+R_{K4}} S.$$

Proof. Use Theorem 5.1 and Remark 4.3 with $R_1 = R_{GL}$ and $R_2 = R_{K4}$. ⊣

These theorems immediately give us Shamkanov's Theorem:

COROLLARY 5.3. $\vdash_{G3GL} S$ *if and only if* $\vdash^{\circ}_{G3K4} S$.

Intuitionistic modal logics

Inspection of the proofs of the theorems above show that they also hold when G3p is replaced by one of the standard single-conclusion Gentzen calculi for intuitionistic logic without structural rules, such as the propositional part of G3i from [Troelstra and Schwichtenberg(1996)], or Dyckhoff's calculus [Dyckhoff(1992)]. If iG3GL and iG3K4 denotes the extension of one of Dyckhoff's calculus by the single conclusion versions of the rules R_{GL} and R_{GL}, respectively, we can conclude the following.

THEOREM 5.4. $\vdash_{iG3GL} S$ *if and only if* $\vdash^{\circ}_{iG3K4} S$.

Grzegorczyk logic

Recall that there is a cut–free sequent calculus for S4, which consists of G3p plus R_{K4} and R_T, where R_T is the rule

$$\frac{\Gamma, \varphi \Rightarrow \Delta}{\Gamma, \Box\varphi \Rightarrow \Delta} R_T.$$

In [Avron(1984)] it is shown that the calculus $G3p + R_{Grz} + R_T$ has cut-elimination. In fact, it is shown that a variant of $G + R_{Grz} + R_T$ with explicit weakening has cut-elimination, but it is not hard to see that this implies the former result.

Note that R_T is an ordered rule. It is not a box rule, but it is not hard to see that the reasoning in the previous proofs about box rules applies to this rule as well. We therefore have the following.

COROLLARY 5.5. $\vdash_{G3p+R_{Grz}+R_T} S$ *implies* $\vdash^{\circ}_{G3S4} S$.

Proof. Proved in a similar way as Lemma 3.1 with $G = G3p + R_T$, $R_1 = R_{Grz}$ and $R_2 = R_{K4}$. ⊣

The converse, however, does not hold, since Löb's principle has a circular proof in K4, as we saw, but is not provable in Grzegorczyk logic.

BIBLIOGRAPHY

[Avron(1984)] Avron, A. On Modal Systems Having Arithmetical Interpretations. *Journal of Symbolic Logic* 49 (3): 935–942 (1984)

[Brotherston(2006)] Brotherston, J. *Sequent calculus proof systems for inductive definitions.* PhD thesis, University of Edinburgh (2006)

[Dershowitz and Manna(1979)] Dershowitz, N. and Manna, Z. Proving termination with multiset orderings. *Communications of the ACM* 22: 465–476 (1979)

[Dyckhoff(1992)] Dyckhoff, R. Contraction-Free Sequent Calculi for Intuitionistic Logic. *Journal of Symbolic Logic* 57 (3): 795–807 (1992)

[Jeřábek(2005)] E. Jeřábek, Admissible rules of modal logics, *Journal of Logic and Computation* 15(4), 2005, pp.411-431.

[Leivant(1981)] Leivant, D. On the Proof Theory of the Modal Logic for Arithmetic Provability. *Journal of Symbolic Logic* 46 (3): 531–538 (1981)

[Shamkanov(2014)] Shamkanov, D. S. Circular proofs for the Gödel–Löb provability logic *Mathematical Notes* 96 (4): 575–585 (2014)

[Troelstra and Schwichtenberg(1996)] Troelstra, A.S. and Schwichtenberg, H. *Basic Proof Theory.* Cambridge Tracts in Theoretical Computer Science 43, Cambridge University Press (1996)

[Visser(1996a)] Visser, A. Bisimulations, Model Descriptions and Propositional Quantifiers. *Logic Group Preprint Series* 161, Utrecht University (1996)

[Visser(1996b)] Visser, A. Uniform interpolation and layered bisimulation. *Lecture Notes in Logic* 6: 139–164 (1996)

NNIL Axioms Have the Finite Model Property

JULIA ILIN, DICK DE JONGH, FAN YANG

Abstract

Albert Visser introduced NNIL-formulas in 1983-1984 in a study of Σ_1-subsitutions in Heyting Arithmetic. NNIL-formulas are propositional formulas that do not allow nesting of implication to the left. The main results about these formulas were obtained in a paper of 1995 with Albert Visser as first author where it was shown that NNIL-formulas are those formulas that are preserved under taking submodels of Kripke models. In the present paper a different characterization by N. Bezhanishvili and D. de Jongh of NNIL-formulas as the formulas which are backwards preserved by monotonic maps of Kripke models is used to give a new proof of the fact that intermediate logics axiomatized by NNIL-formulas have the finite model property.

To Albert at 65

1 Introduction

NNIL-formulas are formulas with *no* *n*esting of *i*mplications to the *l*eft. These formulas are very expressive but considerably easier to handle than the class of all formulas in the language of the intuitionistic propositional calculus IPC. The study of these formulas is one of the many topics introduced by Albert Visser. He introduced them in 1983-1984 when working on Σ_1 substitutions of propositional formulas in Heyting Arithmetic [10], an investigation that was continued with D. de Jongh [6]. He instigated the research on the purely propositional properties of these formulas which was undertaken in [11], and in which, among other things, it is shown that these formulas are (up to provable equivalence) the ones that are preserved under taking submodels of Kripke models. It was remarked in [2] and later in [12] that this implies that these formulas are also preserved under taking subframes. They axiomatize so-called subframe logics. Modal subframe logics were first introduced by Fine [7]. Intermediate subframe logics were defined by Zakharyaschev [13] (see [4, 11.3]) who also proved the finite model property of these logics. Generally, these logics are not thought of as being axiomatized by NNIL-formulas. For example, [1] used $[\wedge, \rightarrow]$-formulas to axiomatize these logics and to prove their finite model property. To obtain NNIL-axiomatizations, in [2] (see also [3]), for each finite rooted frame $\mathfrak{F}$ a NNIL-formula is constructed from a model $\mathfrak{M}$ on that frame that fails on a descriptive frame $\mathfrak{G}$ iff $\mathfrak{F}$ is a p-morphic image of a subframe of $\mathfrak{G}$. Using ideas from [3] we refine this here to show that such formulas fail on a descriptive frame $\mathfrak{G}$ iff $\mathfrak{M}$ can be mapped into $\mathfrak{G}$ by a monotonic function (which satisfies an additional condition defined by us: color-consistency). This result is then used in the proof of the finite model property.

2 Subframe logics and NNIL-formulas

Let us start by fixing the basic notations to be used in this paper. We will have to be somewhat sketchy. A more extensive treatment with all the relevant issues treated can be found in [3]. For the intuitionistic propositional calculus IPC (see e.g. [4], [5], [2]) we have the usual Kripke semantics with intuitionistic frames $\mathfrak{F} = (W, R)$, and intuitionistic models $\mathfrak{M} = (\mathfrak{F}, V)$ as well as descriptive intuitionistic frames $\mathfrak{F} = (W, R, \mathcal{P})$. In these frames and models W is a nonempty set, R is a partial order on W, V is a persistent valuation and $\mathcal{P}$ is a Heyting subalgebra of the set of upward closed sets in W. All frames we consider in the paper are assumed to be *rooted*. If $W' \subseteq W$ the subframe on W' is the frame $(W', R \upharpoonright W')$. Descriptive subframes of descriptive frames need to satisfy the additional *topo-subframe* condition (see e.g. [3]).

Our models will always be n-models, i.e., Kripke models for n propositional variables $p_1, \ldots, p_n$. We will write $V(\varphi)$ for $\{w \in W \mid w \models \varphi\}$, and we write $\mathfrak{M}_w$ for the *submodel of $\mathfrak{M}$ generated by* w, i.e. the model with the domain $R(w) = \{w' \mid wRw'\}$. wR^+w' will stand for $wRw' \land w \neq w'$, i.e., w' is a proper successor of w. A point w is called a *maximal point* if it has no proper successor, i.e., wRv implies $w = v$. The *depth* of a point w in a finite model $\mathfrak{M}$ will be the maximal length k of a chain $w = w_1 R^+ \ldots R^+ w_k$ in $\mathfrak{M}$, in particular, the depth of a maximal point is 1. The *depth of a finite model* $\mathfrak{M}$ will be the maximal depth of the points in the model.

Next we will introduce subframe logics.

DEFINITION 2.1. *An (intermediate) subframe logic is a logic with a class of descriptive frames that is closed under taking subframes.*

Every subframe logic has the finite model property [4, Thm. 11.20]. Zakharyaschev showed that these logics are axiomatized by formulas having a very particular property:

DEFINITION 2.2. *A subframe formula $\beta(\mathfrak{F})$ of a finite frame $\mathfrak{F}$ is a formula satisfying for each descriptive frame $\mathfrak{G}$,*

$$\mathfrak{G} \not\models \beta(\mathfrak{F}) \Leftrightarrow \mathfrak{F} \text{ is a p-morphic image of a subframe of } \mathfrak{G}. \tag{1}$$

THEOREM 2.3 ([13]). *An intermediate logic is a subframe logic iff it is axiomatizable by subframe formulas of finite frames.*

Zakharyaschev's subframe formulas are special canonical formulas that are in $[\land, \rightarrow]$-form. In [2, 3] subframe formulas were constructed in NNIL-form. We will review this construction.

Let us first recall from [11] and [12] some facts about NNIL-formulas. NNIL-formulas are known to have the following normal form:

DEFINITION 2.4. NNIL-*formulas in* normal form *are defined as:*

$$\varphi := \bot \mid p \mid \varphi \land \varphi \mid \varphi \lor \varphi \mid p \rightarrow \varphi$$

Furthermore we have the following characterization theorem showing that NNIL-formulas are exactly the ones that are preserved under submodels.

THEOREM 2.5 ([11]). *1. If $\varphi \in$ NNIL, then φ is preserved under submodels, that is,*

$$\mathfrak{M}, w \models \varphi \Longrightarrow \mathfrak{N}, w \models \varphi \tag{2}$$

for all submodels $\mathfrak{N}$ of a model $\mathfrak{M}$ and all w in $\mathfrak{N}$.

2. *If φ is preserved under submodels, then there exists $\psi \in$ NNIL such that IPC $\vdash \psi \leftrightarrow \varphi$.*

This quickly leads to

COROLLARY 2.6. *NNIL-formulas are preserved under taking subframes of (Kripke and descriptive) frames, that is,*

$$\mathfrak{F} \models \varphi \implies \mathfrak{G} \models \varphi$$

for all subframes $\mathfrak{G}$ of (Kripke or descriptive) frames $\mathfrak{F}$.

We fix n propositional variables $p_1, \ldots, p_n$. We call the sequence $i_1 \ldots i_n$ associated with a point w in an n-model by taking i_k to be 1 if p_i is true in w, and 0 otherwise, the *color* of w, and denote it by $col(w)$. We write $i_1 \ldots i_n \leq j_1 \ldots j_n$ iff $i_k \leq j_k$ for each $k = 1, \ldots, n$, and $i_1 \ldots i_n < j_1 \ldots j_n$ if $i_1 \ldots i_n \leq j_1 \ldots j_n$ and $i_1 \ldots i_n \neq j_1 \ldots j_n$. A finite model $\mathfrak{M} = (W, R, V)$ is said to be *colorful* if $\mathfrak{M}$ is a finite $|W|$-model and for each $w \in W$, there is a propositional variable p_w such that $v \models p_w$ iff wRv.

Let $\mathfrak{F}$ be a finite rooted frame. For every point w in $\mathfrak{F}$ we introduce a propositional letter p_w and let V be such that $V(p_w) = R(w)$. We denote the colorful model $(\mathfrak{F}, V)$ by $\mathfrak{M}$. Next we inductively define the subframe formula $\beta(\mathfrak{F})$ in NNIL form [2, 3].

DEFINITION 2.7. *For every $v \in W$, let*

$$prop(v) := \{p_k : v \models p_k\} \text{ and } notprop(v) := \{p_k : v \not\models p_k\}.$$

We define $\beta(\mathfrak{F})$ by induction. If v is a maximal point of $\mathfrak{M}$, then let

$$\beta(v) := \bigwedge prop(v) \to \bigvee notprop(v)$$

Otherwise, if $w \in W$ is not maximal, let $w_1, \ldots, w_m$ be its immediate successors with $\beta(w_i)$ already defined for every w_i. Then let

$$\beta(w) := \bigwedge prop(w) \to \bigvee notprop(w) \vee \bigvee_{i=1}^{m} \beta(w_i).$$

Let r be the root of $\mathfrak{F}$. We define $\beta(\mathfrak{F}) = \beta(r)$.

THEOREM 2.8 ([2, 3]). *Let $\mathfrak{F}$ be a finite rooted frame. Then $\beta(\mathfrak{F})$ is a subframe formula of $\mathfrak{F}$.*

Thus, we are justified in calling $\beta(\mathfrak{F})$ the subframe formula of $\mathfrak{F}$ in NNIL form, and we can conclude

THEOREM 2.9. *An intermediate logic is a subframe logic iff it is axiomatizable by NNIL-formulas. In particular, every subframe logic is axiomatizable by NNIL-formulas of the form $\beta(\mathfrak{F})$.*

NNIL-formulas allow for a new approach to the characterization of the frame classes of subframe logics. The start of this approach is the following observation concerning *monotonic maps*, i.e., maps that preserve orders on frames and also colors on models.

LEMMA 2.10 ([3]). *NNIL-formulas φ are backwards preserved by monotonic maps,* that is, for any two models $\mathfrak{M} = (W, R, V)$ and $\mathfrak{N} = (W', R', V')$ and any monotonic map $f : W' \to W$ and $x \in W'$,
$$\mathfrak{M}, f(x) \models \varphi \Longrightarrow \mathfrak{N}, x \models \varphi. \tag{3}$$

Proof. Assume φ to be in normal form. We prove the lemma by induction on φ. Only the case $\varphi = p \to \psi$ is non-trivial. Suppose $\mathfrak{M}, f(x) \models p \to \psi$ and $\mathfrak{N}, y \models p$ for some y with $xR'y$. Since f is monotonic, $f(x)Rf(y)$ and $\mathfrak{M}, f(y) \models p$, thus $\mathfrak{M}, f(y) \models \psi$. By the induction hypothesis, we obtain $\mathfrak{N}, y \models \psi$, as required. ⊣

In Definition 2.7 we recalled subframe formulas in NNIL form. The NNIL-subframe formula $\beta(\mathfrak{F})$ of a finite rooted frame $\mathfrak{F}$ is constructed using a colorful model on $\mathfrak{F}$. Clearly, the same definition can be used to construct a formula $\beta(\mathfrak{M})$ of an arbitrary (not necessarily colorful) finite rooted model $\mathfrak{M} = (W, R)$. We will refer to $\beta(\mathfrak{M})$ as the *subframe formula of* $\mathfrak{M}$ and we usually write $\beta(w)$ instead of $\beta(\mathfrak{M}_w)$ for some $w \in W$. As a matter of fact, the refutation criterion of Definition 2.2 does not generally apply to subframe formulas of non-colorful models but it is the goal of this section to develop a refutation criterion that applies to these formulas as well. For the new approach we often need to unravel models to trees.

Recall that the *standard unraveling* $\mathfrak{M}_{\text{tree}} = (W_t, R_t, V_t)$ of a rooted model $\mathfrak{M} = (W, R, V)$ with root r is defined as

- $W_t = \{(r, w_1, \ldots, w_k) \mid w_1, \ldots, w_k \in W \text{ and } rR^+w_1 \ldots R^+w_k\}$;

- $\sigma R_t \tau$ iff σ is an initial segment of τ;

- $(r, w_1, \ldots, w_k) \in V_t(p)$ iff $w_k \in V(p)$.

We will write r for the root (r) of $\mathfrak{M}_{\text{tree}}$, and identify the root of $\mathfrak{M}_{\text{tree}}$ and that of $\mathfrak{M}$. If $\mathfrak{M}$ is finite it is more natural to take immediate successors instead of R^+, this gives smaller models. In this case, which will be the most usual one for us, we write $T_\mathfrak{M}$ instead of $\mathfrak{M}_{\text{tree}}$. Clearly, the map $\alpha : W_t \to W$ (called a *natural map*) defined as $\alpha(r, w_1, \ldots, w_k) = w_k$ is a surjective p-morphism between $\mathfrak{M}_{\text{tree}}$ and $\mathfrak{M}$. In this paper we will call a frame (W, R) *tree-like* if, for each $w \in W$, $R^{-1}(w)$ is a finite linear order.

LEMMA 2.11. *Let $\mathfrak{N} = (W', R', V')$ be a finite model with root r and suppose $\mathfrak{M} \not\models \beta(r)$ for some model $\mathfrak{M} = (W, R, V)$. Then there exists a monotonic map f from $T_\mathfrak{N}$ into $\mathfrak{M}$.*

Proof. It is not hard to see that $\beta(T_\mathfrak{N}) = \beta(r)$ since nodes in $\mathfrak{N}$ and $T_\mathfrak{N}$ have essentially the same immediate successors. The function f is defined stepwise upwards from the root r in such a way that for every x in $T_\mathfrak{N}$, $f(x)$ has the color of x and $\mathfrak{M}, f(x) \not\models \beta(x_i)$ for all immediate successors x_i of x. Suppose f has been defined already for some x and suppose y is an immediate successor of x. By the fact that $\mathfrak{M}, f(x) \not\models \beta(y)$ and the form of $\beta(y)$ it is clear that $f(x)$ has a successor w that has the same color as y and $\mathfrak{M}, w \not\models \beta(y_i)$ for all immediate successors y_i of y. Define $f(y) = w$. Note that the unraveling makes sure that we never assign different values to the same node since we can arrive there from the root in only one way. ⊣

THEOREM 2.12. *Let $\mathfrak{F}$ be a (descriptive or Kripke) frame and $\mathfrak{N}$ be a finite model with root r. Then, $\mathfrak{F} \not\models \beta(r)$ iff there is a monotonic map from $T_\mathfrak{N}$ into some model $\mathfrak{M}$ on $\mathfrak{F}$.*

Proof. If $\mathfrak{F} \not\models \beta(r)$, then $\mathfrak{M} \not\models \beta(r)$ for some model $\mathfrak{M}$ on $\mathfrak{F}$. By Lemma 2.11, there exists a monotonic map f from $T_\mathfrak{N}$ into $\mathfrak{M}$. Conversely, assume that there is a monotonic map f from $T_\mathfrak{N}$ into a model $\mathfrak{M}$ on $\mathfrak{F}$. By a straightforward induction on the depth of x in $T_\mathfrak{N}$ one shows that $T_\mathfrak{N}, x \not\models \beta(x)$, so in particular $T_\mathfrak{N}, r \not\models \beta(r)$. Since $\beta(r) \in$ NNIL we obtain $\mathfrak{M}, f(r) \not\models \beta(r)$ by Theorem 2.10. ⊣

We can get a still more useful result which no longer refers to models on the descriptive frames. For this we need a new concept.

DEFINITION 2.13. *A monotonic map f from a model $\mathfrak{N}$ to a frame $\mathfrak{F} = (W, R)$ is said to be color-consistent if $f(w) R f(u)$ implies $col(w) \leq col(u)$ for all w, u in $\mathfrak{N}$.*

THEOREM 2.14. *Let $\mathfrak{N}$ be a finite model with root r and let $\mathfrak{F}$ be a (descriptive or Kripke) frame. Then, $\mathfrak{F} \not\models \beta(r)$ iff there is a monotonic color-consistent map from $T_\mathfrak{N}$ into $\mathfrak{F}$.*

Proof. The left to right direction follows from Theorem 2.12, as a monotonic map into a model is color-consistent. Indeed, if $f(w)Rf(v)$ then $col(f(w)) \leq col(f(v))$ and so $col(w) \leq col(v)$ since f is color-consistent. For the other direction assume that f is a monotonic color-consistent map from $T_\mathfrak{N}$ into $\mathfrak{F} = (W, R, \mathcal{P})$. Let $w_1, \ldots, w_k$ be the elements of $T_\mathfrak{N}$ and $x_1 = f(w_1), \ldots, x_k = f(w_k)$.

CLAIM: There exist $U_1, \ldots, U_k \in \mathcal{P}$ such that for each i, j, we have that $x_i \in U_i$, and $x_j \in U_i$ iff $x_i R x_j$.

Proof of Claim 2. For each i, j such that $\neg x_i R x_j$ take U_{i_j} to be an up-set in $\mathcal{P}$ containing x_i but not x_j whose existence is guaranteed by refinement. Take for U_i the intersection of all those U_{i_j}. ⊣

Now, since f is color-consistent, we can define $V(p_j) = \bigcup \{U_i \mid 1 \leq i \leq k, T_\mathfrak{N}, w_i \models p_j\}$, where each U_i is as in the claim. The function f is now a monotonic map from $T_\mathfrak{N}$ into the model $\mathfrak{M}$ defined by V on $\mathfrak{F}$, and therefore $\mathfrak{F} \not\models \beta(r)$ by Theorem 2.12. ⊣

3 The Finite Model Property for NNIL-axioms

In this section we prove directly that logics axiomatized by NNIL-formulas (i.e. all subframe logics) have the finite model property. The proof is quite different from previous proofs like the one of Theorem 11.20 in [4], which uses canonical formulas. Our proof relies heavily on Theorem 2.14. Moreover, we use the fact that each IPC-formula φ can be brought into a frame-normal form of implication complexity ≤ 2. This results seems to be more or less folklore, although a closely related form is used in [9] and [8] where syntactic proofs are given. In any case, we will give a semantic proof of this fact here.

Given any formula φ, for each variable p and constant $\bot$ occurring in φ we let $s_p = p$ and $s_\bot = \bot$, and for each compound subformula ψ of φ we introduce a fresh variable s_ψ. Inductively define formulas φ'_+ and $\varphi' = \varphi'_+ \to s_\varphi$ as follows:

- If $\varphi = p$, then define $\varphi'_+ = \top$ and $\varphi' = \top \to p$.
- If $\varphi = \bot$, then define $\varphi'_+ = \top$ and $\varphi' = \top \to \bot$.
- If $\varphi = \psi * \chi$ for $* \in \{\wedge, \vee, \to\}$, then define

$$\varphi'_+ = \psi'_+ \wedge \chi'_+ \wedge ((s_\psi * s_\chi) \leftrightarrow s_\varphi) \text{ and } \varphi' = \varphi'_+ \to s_\varphi.$$

Observe that most conjuncts in φ'_+ are NNIL-formulas, except for subformulas of the form $(p_i \to q_i) \to r_i$. We will now show that the formula φ' can be viewed as a normal form for IPC-formulas over frames, as φ and φ' are frame-equivalent to each other.

LEMMA 3.1. *For any frame $\mathfrak{F}$, we have that $\mathfrak{F} \models \varphi \iff \mathfrak{F} \models \varphi'$.*

Proof. To prove the lemma, we first prove the following claim.

CLAIM: For any formula φ, any model $\mathfrak{M}$ and any node w in $\mathfrak{M}$, we have that $\mathfrak{M}, w \models \varphi'_+ \implies \mathfrak{M}, w \models \varphi \leftrightarrow s_\varphi$.

Proof of Claim 3. We prove the claim by induction on φ. If $\varphi = p$ or $\bot$, then $s_\varphi = \varphi$ by definition, thus the claim holds trivially.

Suppose $\varphi = \psi * \chi$ for $* \in \{\wedge, \vee, \to\}$. Assume that $\mathfrak{M}, w \models \varphi'_+$, i.e., $\mathfrak{M}, w \models \psi'_+ \wedge \chi'_+ \wedge ((s_\psi * s_\chi) \leftrightarrow s_\varphi)$. By the induction hypothesis, $\mathfrak{M}, w \models \psi \leftrightarrow s_\psi$ and $\mathfrak{M}, w \models \chi \leftrightarrow s_\chi$, implying $\mathfrak{M}, w \models (\psi * \chi) \leftrightarrow (s_\psi * s_\chi)$. Since $\mathfrak{M}, w \models (s_\psi * s_\chi) \leftrightarrow s_\varphi$, we obtain $\mathfrak{M}, w \models (\psi * \chi) \leftrightarrow s_\varphi$, as required. ⊣

Now, to prove the direction "$\implies$" of the lemma, it suffices to prove that $\mathfrak{M}, w \models \varphi \implies \mathfrak{M}, w \models \varphi'$ holds for any model $\mathfrak{M} = (W, R, V)$ and any node $w \in W$. Now, suppose $\mathfrak{M}, w \models \varphi$ and $\mathfrak{M}, u \models \varphi'_+$ for some $u \in W$ with wRu. By Claim 3, $\mathfrak{M}, u \models \varphi \leftrightarrow s_\varphi$, thus $\mathfrak{M}, u \models s_\varphi$, thereby $\mathfrak{M}, w \models \varphi'$.

For the direction "$\impliedby$", suppose $(\mathfrak{F}, V), w \not\models \varphi$ for some valuation V on $\mathfrak{F}$ and $w \in W$. Let V' be a valuation on $\mathfrak{F}$ be such that $V'(s_\psi) = V(\psi)$ for every subformula ψ of φ.

CLAIM: $(\mathfrak{F}, V') \models \varphi'_+$.

Proof of Claim 3. We prove the claim by induction on the subformulas ψ of φ. If $\psi = p$ or $\bot$, then $\psi'_+ = \top$ and the claim holds trivially. Suppose $\psi = \theta * \chi$. Then $\psi'_+ = \theta'_+ \wedge \chi'_+ \wedge ((s_\theta * s_\chi) \leftrightarrow s_\psi)$. By the induction hypothesis, we have that $(\mathfrak{F}, V') \models \theta'_+ \wedge \chi'_+$. Moreover, by the definition, $V'(s_\theta) = V(\theta)$, $V'(s_\chi) = V(\chi)$ and $V(\psi) = V'(s_\psi)$, which by a simple inductive argument imply that $V'(s_\theta * s_\chi) = V(\theta * \chi) = V'(s_\psi)$. Thus $(\mathfrak{F}, V') \models (s_\theta * s_\chi) \leftrightarrow s_\psi$. ⊣

To complete the proof, we will show that $(\mathfrak{F}, V') \not\models \varphi'$, which can be reduced to showing that $(\mathfrak{F}, V'), w \not\models \varphi'_+ \to s_\varphi$. By Claim 3, we have that $(\mathfrak{F}, V'), w \models \varphi'_+$. It then follows from Claim 3 that $(\mathfrak{F}, V'), w \models \varphi \leftrightarrow s_\varphi$. Since $V' \upharpoonright \mathsf{Prop}(\varphi) = V \upharpoonright \mathsf{Prop}(\varphi)$, the assumption implies that $(\mathfrak{F}, V'), w \not\models \varphi$, which gives $(\mathfrak{F}, V'), w \not\models s_\varphi$, as desired. ⊣

To reduce the length of the proof of the main theorem we now introduce a definition and prove a sequence of lemmas.

DEFINITION 3.2. *A submodel $\mathfrak{N} = (W', R, V)$ of a model $\mathfrak{M} = (W, R, V)$ is said to be color-preserving if, for any $w \in W'$, any $u \in W$, wRu implies that there exists $v \in W'$ such that wRv and $col(v) = col(u)$.*

LEMMA 3.3. *Let α be a p-morphism from a model $\mathfrak{M} = (W, R, V)$ onto a model $\mathfrak{M}' = (W', R', V')$. If $\mathfrak{N} = (W_0, R, V)$ is a color-preserving submodel of $\mathfrak{M}$, then the image $\mathfrak{N}'$ of $\mathfrak{N}$ under α is also a color-preserving submodel of $\mathfrak{M}'$.*

If $\mathfrak{N}$ is a color-preserving submodel of $\mathfrak{M}$ and $\mathfrak{N}'$ is a color-preserving submodel of $\mathfrak{N}$, then $\mathfrak{N}'$ is a color-preserving submodel of $\mathfrak{M}$.

Proof. Let $w \in W_0$ and $u' \in W'$ such that $\alpha(w)R'u'$. Since α is a p-morphism, there exists $u \in W$ such that $\alpha(u) = u'$ and wRu. By the assumption, $\mathfrak{N}$ is a color-preserving submodel of $\mathfrak{M}$, thus there exists $v \in W_0$ such that wRv and $col(v) = col(u)$. It follows that $\alpha(v)$ is in $\mathfrak{N}'$, $\alpha(w)R'\alpha(v)$ and $col(\alpha(v)) = col(v) = col(u) = col(\alpha(u)) = col(u')$, as required. The proof of the second property is similar. ⊣

LEMMA 3.4. *Every tree-like model $\mathfrak{M}$ has a tree-like color-preserving submodel $\mathfrak{N}$ of finite depth with the same root.*

Proof. Let $\mathfrak{M} = (W, R, V)$ and r is the root. Let $\mathfrak{N}$ be the submodel of $\mathfrak{M}$ on the set

$$W' = \{r\} \cup \{w \in W \mid col(v) < col(w) \text{ for the immediate predecessor } u \text{ of } w\}.$$

Since $\mathfrak{M}$ is a tree, W' is well-defined. The model $\mathfrak{N}$ has finite depth since all chains in $\mathfrak{N}$ are strictly increasing in color.

It remains to check that $\mathfrak{N}$ is a color-preserving submodel of $\mathfrak{M}$. For any $w \in W'$, any $u \in W$ such that wRu, clearly there exists a (unique) predecessor v of u in W' with $col(v) = col(u)$. Either $v = w$ or wR^+v, and we are done in both cases. ⊣

LEMMA 3.5. *Every tree-like model $\mathfrak{M}$ of finite depth has a finite tree-like color-preserving submodel $\mathfrak{N}$ with the same root.*

Proof. Assume that $\mathfrak{M}$ is of depth n. Putting $\mathfrak{N}_0 = \mathfrak{M}$, we inductively select a sequence of submodels $\mathfrak{N}_1, \ldots, \mathfrak{N}_n$ from $\mathfrak{M}$ as follows. Assume that $\mathfrak{N}_i$ has been defined. Consider each node w of depth $i+2$. There are only finitely many non-isomorphic subtrees generated by the immediate successors of w. Of each such isomorphism type we keep only one subtree above w and remove all the others. Let $\mathfrak{N}_{i+1}$ be the resulting submodel of $\mathfrak{N}_i$. It may be good to spell out the construction of $\mathfrak{N}_1$. The points of depth 2 have only maximal points as immediate successors. We delete all of these except for one of each color. Finally, $\mathfrak{N}_n$ is a finite model with the same root as $\mathfrak{M}$.

To prove that $\mathfrak{N}_n$ is a color-preserving submodel of $\mathfrak{M}$, we will prove that $\mathfrak{N}_{i+1} = (W_{i+1}, R, V)$ is a color-preserving submodel of $\mathfrak{N}_i = (W_i, R, V)$ for each i. Suppose $w \in W_{i+1}$, $u \in W_i$ and wRu. If $u \in W_{i+1}$, then we are done. Otherwise, u is in a subtree T of $\mathfrak{N}_i$ that is missing in $\mathfrak{N}_{i+1}$. By the construction there remains an isomorphic copy of T in $\mathfrak{N}_{i+1}$ above w and the node corresponding to u in this isomorphic copy will have the same color as u. So we are also done. ⊣

COROLLARY 3.6. *Every model $\mathfrak{M}$ has a finite color-preserving submodel $\mathfrak{N}$ with the same root.*

Proof. We construct $\mathfrak{N}$ in stages. First unravel $\mathfrak{M}$ to obtain a tree-like model $\mathfrak{M}_{\text{tree}}$ with the same root. Second, apply Lemma 3.4 to $\mathfrak{M}_{\text{tree}}$ to obtain a tree-like color-preserving submodel of finite depth with the same root. Lemma 3.5 then gives a finite tree-like color-preserving submodel $\mathfrak{N}_0$ with the same root. Then $\mathfrak{N}_0$ is a color-preserving submodel of $\mathfrak{M}_{\text{tree}}$. Let α be the natural p-morphism from $\mathfrak{M}_{\text{tree}}$ onto $\mathfrak{M}$, and $\mathfrak{N}$ the image of the finite model $\mathfrak{N}_0$ under α. α maps the root of $\mathfrak{M}_{\text{tree}}$ to the root of $\mathfrak{M}$, thus $\mathfrak{N}$ and $\mathfrak{M}$ have the same root. By Lemma 3.3, we obtain that $\mathfrak{N}$ is a finite color-preserving submodel of $\mathfrak{M}$. ⊣

THEOREM 3.7. *Each subframe logic L has the finite model property: if $L \not\vdash \varphi$, there is a finite Kripke model $\mathfrak{M}$ on an L-frame with $\mathfrak{M} \not\models \varphi$.*

Proof. The bulk of the proof will be devoted to showing that, if $\beta(r)$ is the subframe formula of a finite rooted model $\mathfrak{M}^r$, then the logic L_r axiomatized by $\beta(r)$ has the finite model property. By Theorem 2.14, L_r is the logic of the descriptive frames into which $T_{\mathfrak{M}^r}$ cannot be mapped by a monotonic color-consistent map. Assume that $L_r \nvdash \varphi$. Then φ is falsified on a model $\mathfrak{M}$ on a rooted descriptive L_r-frame $\mathfrak{F}$ into which $T_{\mathfrak{M}^r}$ cannot be mapped by a monotonic color-consistent map. By Lemma 3.6 we obtain a finite color-preserving submodel $\mathfrak{N}$ of $\mathfrak{M}$.

The underlying frame $\mathfrak{G}$ of $\mathfrak{N}$ is an L_r-frame. If not, then by Theorem 2.14 there is a monotonic color-consistent map f from $T_{\mathfrak{M}^r}$ into $\mathfrak{G}$. Since $\mathfrak{N}$ is a submodel of $\mathfrak{M}$, the map f can be regarded as a map with codomain $\mathfrak{F}$. It is easy to see that $f : T_{\mathfrak{M}^r} \to \mathfrak{F}$ is monotonic color-consistent and so we reached a contradiction to $\mathfrak{F}$ being an L_r-frame.

It remains to show that $\mathfrak{N}$ falsifies φ. By Lemma 3.1, we may assume without loss of generality that φ is in the frame-normal form $\varphi'_+ \to s$, and the root u of $\mathfrak{M}$ makes φ'_+ true and s false. By the construction, the root v is also the root of $\mathfrak{N}$, and $\mathfrak{N}, v \nvDash s$. It is then sufficient to prove that $\mathfrak{N}, v \vDash \varphi'_+$. As pointed out already, most conjuncts in φ'_+ are NNIL-formulas, and thus remain true in the submodel $\mathfrak{N}$. It is left to check that v makes the formulas of the form $(p \to q) \to r$ true in $\mathfrak{N}$. Let w be an element of $\mathfrak{N}$. Assuming $\mathfrak{N}, w \nvDash r$, we need to show that $\mathfrak{N}, w \nvDash p \to q$. Now, since $\mathfrak{M}, w \vDash (p \to q) \to r$ and $\mathfrak{M}, w \nvDash r$, we have $\mathfrak{M}, w \nvDash p \to q$, so there must exist a successor u of w in $\mathfrak{M}$ such that $\mathfrak{M}, u \vDash p$ and $\mathfrak{M}, u \nvDash q$. Since $\mathfrak{N}$ is a color-preserving submodel of $\mathfrak{M}$, there is a successor u_0 of w in $\mathfrak{N}$ such that $\mathfrak{N}, u_0 \vDash p$ and $\mathfrak{N}, u_0 \nvDash q$, which implies that $\mathfrak{N}, w \nvDash p \to q$.

We just obtained a finite model of L_r that falsifies φ. In general, a subframe logic L will, by Theorem 2.9, be axiomatized by a set $\{\beta(r_i) \mid i \in I\}$ of subframe formulas associated with a set $\{\mathfrak{N}_i \mid i \in I\}$ of finite models. It is the logic of the descriptive frames into which no $T_{\mathfrak{N}_i}$ associated with a $\beta(r_i)$ for $i \in I$ can be mapped by a monotonic color-consistent map. Application of the above construction will then lead to a finite model $\mathfrak{N}$ falsifying φ into which none of the $T_{\mathfrak{N}_i}$ for $i \in I$ can be mapped in the appropriate way. By Theorem 2.14, the frame of this model satisfies L. ⊣

We conclude by presenting to Albert the unsolved problem to characterize the formula class preserved under taking color-preserving submodels.

BIBLIOGRAPHY

[1] G. Bezhanishvili and S. Ghilardi. An algebraic approach to subframe logics. Intuitionistic case. *Ann. Pure Appl. Logic*, 147(1-2):84–100, 2007.
[2] N. Bezhanishvili. *Lattices of Intermediate and Cylindric Modal Logics*. PhD thesis, University of Amsterdam, 2006. Available at *http://www.illc.uva.nl/Research/Publications/Dissertations/DS-2006-02.text.pdf*.
[3] N. Bezhanishvili and D. de Jongh. Stable logics in intuitionistic logic. *Notre Dame Journal of Formal Logic*, 92, 2015. To appear, available at *http://www.illc.uva.nl/Research/Publications/Reports/PP-2014-19.text.pdf*.
[4] A. Chagrov and M. Zakharyaschev. *Modal logic*, volume 35 of *Oxford Logic Guides*. The Clarendon Press, New York, 1997.
[5] D. van Dalen. Intuitionistic Logic. In D. Gabbay and F. Guenthner, editors, *Handbook of Philosophical Logic*, volume 3, pages 225–339. Kluwer, Reidel, Dordrecht, 1986.
[6] D. de Jongh and A. Visser. Embeddings of Heyting Algebras. In Wilfrid Hodges, Martin Hyland, Charles Steinhorn, and John Truss, editors, *Logic: from foundations to applications, European logic coll. 1993*, Oxford Science Publications, pages 187–213. Clarendon Press, Oxford, 1996.
[7] K. Fine. Logics containing K4. II. *J. Symbolic Logic*, 50(3):619–651, 1985.
[8] E. Jerabek. A note on the substructural hierarchy. *Mathematical Logic Quarterly*, 62(1-2):102–110, 2016.
[9] R. Statman. Intuitionistic propositional logic is polynomial-space complete. *Theor. Comp. Sc.*, 9:67–72, 1979.
[10] A. Visser Evaluation, provably deductive equivalence in Heyting's Arithmetic. *Technical Report 4, Dept. of Philosophy, Utrecht University*, 1985.

[11] A. Visser, D. de Jongh, J. van Benthem, and G. Renardel de Lavalette. NNIL a study in intuitionistic logic. In A. Ponse, M. de Rijke, and Y. Venema, editors, *Modal logics and Process Algebra: a bisimulation perspective*, pages 239–326, 1995.
[12] F. Yang. Intuitionistic subframe formulas, NNIL-formulas and n-universal models. Master's Thesis, MoL-2008-12, ILLC, University of Amsterdam, 2008.
[13] M. Zakharyaschev. Syntax and semantics of superintuitionistic logics. *Algebra and Logic*, 28(4):262–282, 1989.

A Reflection on Collection

LESZEK A. KOŁODZIEJCZYK

Amicus Albert, amica veritas... sed amica etiam collectio.

Dedicated to Albert Visser on the occasion of his 65th birthday.

Abstract

In a recent paper, Visser showed that over the theory $\mathrm{EA} + \mathrm{Con}(\mathrm{EA})$, the collection scheme $\mathrm{B}\Sigma_1$ is provably equivalent to $\mathrm{RFN}_{\exists^{\leq}\Pi_1}(\mathrm{EA})$. We generalize this result to $\mathrm{B}\Sigma_n$ for $n \geq 1$ and prove some further properties of the resulting reflection principles.[1]

It is well-known that fragments of Peano arithmetic axiomatized by the induction scheme have natural characterizations in terms of reflection principles: over elementary arithmetic EA, the theory $\mathrm{I}\Sigma_n$ coincides with $\mathrm{RFN}_{\Pi_{n+2}}(\mathrm{EA})$ (that is, uniform Π_{n+2} reflection for EA) for each $n \geq 1$. Also Π_1 and Π_2 reflection correspond to natural fragments of arithmetic: $\mathrm{RFN}_{\Pi_2}(\mathrm{EA})$ is EA plus the totality of the function $\mathrm{supexp}(x) = $ "x times iterated exp", while for every provably Σ_1-complete theory T, $\mathrm{RFN}_{\Pi_1}(\mathrm{T})$ is simply $\mathrm{Con}(\mathrm{T})$.

Nearly as important as induction is the *collection scheme*:

$$\forall a\, [\forall x \leq a\, \exists y\, \varphi(x,y) \to \exists w\, \forall x \leq a\, \exists y \leq w\, \varphi(x,y)],$$

where the formula φ may contain parameters. For each $n \geq 1$, the theory $\mathrm{B}\Sigma_n$, axiomatized by $\mathrm{I}\Delta_0$ and collection for Σ_n formulas, is intermediate between $\mathrm{I}\Sigma_{n-1}$ and $\mathrm{I}\Sigma_n$ ([PK78], cf. e.g. [HP93, Chapter I]). It is entirely natural to wonder whether $\mathrm{B}\Sigma_n$ also has a characterization as a reflection principle of some sort. In fact, this very question was asked explicitly by Beklemishev [Bek].

In the recent paper [Vis15], Visser considered the question for $\mathrm{B}\Sigma_1$. Generally speaking, the answer turns out to be positive. Among a number of elegant results obtained in [Vis15], perhaps the cleanest statement is [Vis15, Theorem 4.10(i)][2]: over the theory $\mathrm{EA}+\mathrm{Con}(\mathrm{PA}^-)$, $\mathrm{B}\Sigma_1$ is equivalent to $\mathrm{RFN}_{\exists^{\leq}\Pi_1}(\mathrm{PA}^-)$. Here PA^- is the very weak fragment of EA consisting just of axioms for (nonnegative parts of) discretely ordered rings, and $\exists^{\leq}\Pi_1$ is the class of Π_1 formulas preceded by bounded existential quantifiers.

Obviously, this leads to the question whether Visser's characterization of $\mathrm{B}\Sigma_1$ can be generalized to the schemes $\mathrm{B}\Sigma_n$ for higher n. In this short paper, we prove that such a generalization is indeed possible, and can be proved similarly to Visser's result, albeit with one or two additional tricks. Our main theorem is that, provably in EA, for every extension T of EA, $\mathrm{RFN}_{\exists^{\leq}\Pi_n}(\mathrm{T})$ is

---

[1] Supported in part by Polish National Science Centre grant no. 2013/09/B/ST1/04390.
[2] As duly noted in [Vis15], this particular statement was suggested by a referee.

exactly equivalent to $\mathrm{RFN}_{\Pi_n}(\mathrm{T})$ plus $\mathrm{B}\Sigma_n$. Interestingly, for $n > 2$ our proof genuinely requires T to extend a theory with exponentiation rather than just, say, PA^-.

This paper consists of three sections. Section 1 contains a proof of our main theorem. In Section 2, we explain how to reformulate the theorem in terms of reflection for disjunctions of Π_n sentences. Finally, in Section 3 we take first steps towards understanding what happens if we nest $\exists^{\leq}\Pi_n$ reflection principles.

We assume that the reader has some familiarity with fragments of Peano Arithmetic as presented e.g. in [HP93]. Throughout the paper, we use the box notation $\Box_\tau(\psi)$ for "ψ is provable from τ", and we write $\psi(\dot{a})$ to indicate that a numeral denoting a has been substituted into the formula $\psi(x)$.

1 Reflection

For a sentence τ and for $n \geq 1$, the uniform reflection principle $\mathrm{RFN}_{\exists^{\leq}\Pi_n}(\tau)$ is:

$$\forall a \, (\Box_\tau(\psi(\dot{a})) \to \psi(a)),$$

where ψ ranges over $\exists^{\leq}\Pi_n$ formulas. As it stands, this is a scheme. However, we can use a universal Π_n formula to define satisfaction for $\exists^{\leq}\Pi_n$ by means of a formula which is itself "$\exists^{\leq}\Pi_n$ with a size parameter". This lets us express $\mathrm{RFN}_{\exists^{\leq}\Pi_n}(\tau)$ as a single sentence, and moreover, check that it is in fact equivalent to the *global*[3] $\exists^{\leq}\Pi_n$ reflection principle for τ:

$$\forall \psi \in (\exists^{\leq}\Pi_n \cap \mathrm{Sent}) \, (\Box_\tau(\psi) \to \mathrm{Tr}_{n+1}(\psi)),$$

where Tr_{n+1} is the usual truth definition for Σ_{n+1} sentences. This global form is the one we will actually use below. It is crucial that we apply Tr_{n+1} rather than a truth definition for Π_n sentences, in other words, that we do not try to put ψ in Π_n form before evaluating it.

Our main result is:

THEOREM 1.1. *For every $n \geq 1$, EA proves:*

$$\forall \tau \in \mathrm{Sent} \, [\Box_\tau(\mathrm{EA}) \to (\mathrm{RFN}_{\exists^{\leq}\Pi_n}(\tau) \equiv (\mathrm{RFN}_{\Pi_n}(\tau) \wedge \mathrm{B}\Sigma_n))].$$

REMARK. Since a proof always uses only finitely many axioms, the statement of the theorem can obviously be generalized to the case where τ is an infinitely axiomatized theory (given for instance by an elementary recursive enumeration of its axioms). The same applies to Theorem 2.1 below.

Proof. We reason in EA, assume $\Box_\tau(\mathrm{EA})$, and consider the two directions of the equivalence.

($\Rightarrow$) $\mathrm{RFN}_{\exists^{\leq}\Pi_n}(\tau)$ obviously implies $\mathrm{RFN}_{\Pi_n}(\tau)$, so the only thing to prove is that it also implies $\mathrm{B}\Sigma_n$. We assume that $n \geq 3$: the result for $n=1$ follows from [Vis15], and the one for $n=2$ is slightly simpler than the general case.

We consider Σ_n collection formulated as the scheme

$$\forall a \, \exists x \leq a \, \forall y \, (\pi(x,y) \to \forall x' \leq a \, \exists y' \leq y \, \pi(x',y')) \tag{1}$$

---

[3] As illustrated here, a "global" reflection principle differs from a "uniform" one in that it actually quantifies over sentences from a given class. In the cases considered in this paper, the strength of the global principles would not change if the quantifier ranged over pairs of a formula and an assignment.

where π is Π_{n-1} (possibly with parameters which we suppress for the sake of simplicity) This scheme implies the usual formulation of Σ_n collection in PA^-, and is equivalent to it in $\mathrm{I}\Sigma_{n-1}$ [AD88][4].

For any given a, formalizing a routine proof gives:

$$\Box_\tau \, [\exists x \leq \dot{a} \, \forall y \, (\pi(x,y) \to \forall x' \leq \dot{a} \, \exists y' \leq y \, \pi(x',y'))]. \tag{2}$$

We would like to apply $\mathrm{RFN}_{\exists \leq \Pi_n}(\tau)$ to the formula in square brackets, but we observe that the formula $\forall x' \leq \dot{a} \, \exists y' \leq y \, \pi(x',y'))$ is not Π_n. To deal with this issue, write $\pi(x',y')$ as $\forall z \, \sigma(x',y',z)$ with $\sigma \in \Sigma_{n-2}$. Observe that (2) implies

$$\Box_\tau \, [\exists x \leq \dot{a} \, \forall y \, (\pi(x,y) \to \forall x' \leq \dot{a} \, \forall z \, \exists y' \leq y \, \forall z' \leq z \, \sigma(x',y',z'))], \tag{3}$$

and this in turn implies

$$\Box_\tau \, [\exists x \leq \dot{a} \, \forall y \, (\pi(x,y) \to \forall x' \leq \dot{a} \, \forall z \, \forall s \leq (z+1)^{y+1} \, \exists y' \leq y \, \sigma(x',y',s_{y'}))]. \tag{4}$$

By moving the $\forall x' \, \forall z \, \forall s$ block in (4) outside the parentheses, we obtain an $\exists^{\leq} \Pi_n$ sentence. Thus, $\mathrm{RFN}_{\exists \leq \Pi_n}(\tau)$ gives:

$$\exists x \leq a \, \forall y \, (\pi(x,y) \to \forall x' \leq a \, \forall z \, \forall s \leq (z+1)^{y+1} \, \exists y' \leq y \, \sigma(x',y',s_{y'})). \tag{5}$$

To recover (1) from (5), apply $\mathrm{B}\Sigma_{n-1}$ twice (the first application takes the form of the Π_{n-2} finite axiom of choice and yields a formula with quantifiers ordered as inside the square brackets in (3)).

($\Leftarrow$) Assume $\mathrm{B}\Sigma_n + \mathrm{RFN}_{\Pi_n}(\tau)$. Let $\exists x \leq t \, \pi(x)$ be an $\exists^{\leq} \Pi_n$ sentence, where π is Π_n. Since t has to be a closed term and τ implies EA, we may assume without loss of generality that t is a numeral denoting some number a.

Assume $\Box_\tau [\exists x \leq \dot{a} \, \pi(x)]$. Already PA^- knows that $x \leq \dot{a}$ is equivalent to $\bigvee_{m=0}^{a}(x = \dot{n})$, so we have $\Box_\tau [\bigvee_{m=0}^{a} \pi(\dot{m})]$. The disjunction in square brackets has the form

$$\bigvee_{m=0}^{a} \forall z_1 \, \exists z_2 \ldots Q z_n \, \delta(\dot{m}, \bar{z}),$$

where δ is Δ_0 and $\forall z_1 \, \exists z_2 \ldots Q z_n \, \delta$ is π. By prenexification, this may be rewritten in a Π_n way as

$$\forall z_1^0 \ldots \forall z_1^a \, \exists z_2^0 \ldots \exists z_2^a \ldots Q z_n^0 \ldots Q z_n^a \, \bigvee_{m=0}^{a} \delta(\dot{m}, \bar{z}^m).$$

Denote this sentence by ξ. We have $\Box_\tau(\xi)$, and hence by $\mathrm{RFN}_{\Pi_n}(\tau)$ also $\mathrm{Tr}_n(\xi)$. To derive the formula $\exists x \leq \dot{a} \, \pi(x)$, we have to pull the $\exists x \leq a$ quantifier, which in $\mathrm{Tr}_n(\xi)$ lies hidden in the evaluation of the outermost disjunction of the Δ_0 part, out to the front of the sentence. This requires applying $\mathrm{B}\Sigma_n$ (in the guise of the finite axiom of choice) $\lceil n/2 \rceil$ times. ⊣

REMARK. The introduction of the bounded $\forall s$ quantifier in the ($\Rightarrow$) direction of the proof really makes use of the assumption that τ proves the totality of exponentiation rather than just e.g. $\Box_\tau(\mathrm{PA}^-)$. We do not know how to get rid of this assumption except for $n=1, 2$.

---

[4] A careful proof of the equivalence for $n=1$ can also be found in [Vis15].

COROLLARY 1.2. *The following equivalences hold over* EA:

(i) $\mathrm{RFN}_{\exists^{\leq}\Pi_1}(\mathrm{EA}) \equiv (\mathrm{Con}(\mathrm{EA}) + \mathrm{B}\Sigma_1)$,

(ii) $\mathrm{RFN}_{\exists^{\leq}\Pi_n}(\mathrm{EA}) \equiv \mathrm{B}\Sigma_n$, *for* $n \geq 2$.

Proof. Part (i) follows from Theorem 1.1 and the fact that $\mathrm{RFN}_{\Pi_1}(\mathrm{EA})$ is equivalent to $\mathrm{Con}(\mathrm{EA})$. Part (ii) follows from Theorem 1.1 and the fact that (for $n \geq 2$) $\mathrm{RFN}_{\Pi_n}(\mathrm{EA})$ follows from $\mathrm{B}\Sigma_n$ (see introduction). ⊣

2 Disjunctions

An $\exists^{\leq}\Pi_n$ sentence is more or less a disjunction of Π_n sentences, so it is to be expected that $\mathrm{RFN}_{\exists^{\leq}\Pi_n}$ should have an alternative formulation as reflection for disjunctions of Π_n sentences. Below, we show that this is indeed true for the principle $\mathrm{RFN}_{\bigvee \Pi_n}$ formulated as:

$$\forall \langle \pi_0, \ldots, \pi_a \rangle \left(\forall i \leq a\, (\pi_i \in \Pi_n) \wedge \Box_\tau [\bigvee_{i=0}^{a} \pi_i] \to \exists i \leq a\, \mathrm{Tr}_n(\pi_i) \right).$$

This time, we really need to consider a global rather than a uniform reflection principle, as the naïve formulation of uniform $\bigvee \Pi_n$ reflection is the same thing as Π_n reflection. As with the global variant of $\mathrm{RFN}_{\exists^{\leq}\Pi_n}$, it is crucial that we do not put the disjunction $\bigvee_{i=0}^{a} \pi_i$ in Π_n form before evaluating it.

THEOREM 2.1. *For every* $n \geq 1$, EA *proves*:

$$\forall \tau \in \mathrm{Sent}\, [\Box_\tau(\mathrm{EA}) \to (\mathrm{RFN}_{\exists^{\leq}\Pi_n}(\tau) \equiv \mathrm{RFN}_{\bigvee \Pi_n}(\tau))].$$

Proof. As in the proof of Theorem 1.1, we reason in EA and assume $\Box_\tau(\mathrm{EA})$.

($\Leftarrow$) Assume $\mathrm{RFN}_{\bigvee \Pi_n}(\tau)$ and $\Box_\tau[\exists x \leq \dot{a}\, \pi(x)]$ for π a Π_n formula. Then we also have $\Box_\tau[\bigvee_{m=0}^{a} \pi(\dot{m})]$, so we can apply $\mathrm{RFN}_{\bigvee \Pi_n}(\tau)$ to deduce that one of the disjuncts $\pi(\dot{m})$ is true and thus $\exists x \leq a\, \pi(x)$.

($\Rightarrow$) By Theorem 1.1, $\mathrm{RFN}_{\exists^{\leq}\Pi_n}(\tau)$ implies both $\mathrm{RFN}_{\Pi_n}(\tau)$ and $\mathrm{B}\Sigma_n$. Assuming $\Box_\tau[\bigvee_{i=0}^{a} \pi_i]$ for $\pi_i \in \Pi_n$, we can argue as in the ($\Leftarrow$) direction of Theorem 1.1: put $\bigvee_{i=0}^{a} \pi_i$ in Π_n form, use $\mathrm{RFN}_{\Pi_n}(\tau)$ to deduce that the Π_n form is true, and then use $\lceil n/2 \rceil$ applications of $\mathrm{B}\Sigma_n$ to conclude that one of $\pi_0, \ldots, \pi_a$ is true as well. ⊣

Thus, perhaps surprisingly, even though $\mathrm{RFN}_{\Pi_2}(\mathrm{EA})$ is just EA + supexp, $\mathrm{RFN}_{\bigvee \Pi_2}(\mathrm{EA})$ gives all of $\mathrm{B}\Sigma_2$ and in particular implies $\mathrm{I}\Sigma_1$. Similarly, for higher n, $\mathrm{RFN}_{\Pi_n}(\mathrm{EA})$ is just $\mathrm{I}\Sigma_{n-2}$, but $\mathrm{RFN}_{\bigvee \Pi_n}(\mathrm{EA})$ is $\mathrm{B}\Sigma_n$ and thus implies $\mathrm{I}\Sigma_{n-1}$. On the other hand, analysis of a typical proof showing $\mathrm{I}\Sigma_n \equiv \mathrm{RFN}_{\Pi_{n+2}}(\mathrm{EA})$ (see e.g. [Bek05]) reveals that reflection principle formulated like $\mathrm{RFN}_{\bigvee \Pi_n}(\mathrm{EA})$ but for $\bigvee(\Sigma_n \wedge \Pi_n)$ sentences rather than just $\bigvee \Pi_n$ sentences already gives $\mathrm{I}\Sigma_n$.

3 Iterations

Given a parametrized family of reflection principles, it is natural to ask how the principles interact, for instance what happens when they are nested or what provability logic they give rise to. For the $\mathrm{RFN}_{\exists^{\leq}\Pi_n}$ principles, these questions seem worthy of a separate study. Below, we only prove two preliminary results: we characterize the strength of finite iterations of $\mathrm{RFN}_{\exists^{\leq}\Pi_n}$ for fixed n

applied to EA, and we point out that the $\text{RFN}_{\exists^{\leq}\Pi_n}$ principles do not satisfy the same provability logic as the RFN_{Π_n} principles.

In this section, we assume that all theories contain EA, and we use the polymodal notation known from provability logic. Thus, for instance:

$$\langle n \rangle \tau := \text{RFN}_{\Pi_{n+1}}(\text{EA} + \tau),$$
$$[n]\tau := \neg \langle n \rangle \neg \tau,$$

so that $\langle n \rangle \top$ is $\text{I}\Sigma_{n-1}$ for $n \geq 2$ (where $\top$ is a constant for truth). Recall that the "+1" in the index of RFN appears because $\langle n \rangle \tau$ stands for "τ is consistent with Π_n truth", which is equivalent to "τ does not prove a false Π_{n+1} sentence".

To denote our reflection principles for the $\exists^{\leq}\Pi_n$ classes, we put a $\sim$ over the modal operators. In other words:

$$\widetilde{\langle n \rangle} \tau := \text{RFN}_{\exists^{\leq}\Pi_{n+1}}(\text{EA} + \tau),$$
$$\widetilde{[n]} \tau := \neg \widetilde{\langle n \rangle} \neg \tau.$$

We first describe what happens when $\widetilde{\langle n \rangle}$ for fixed n is applied to $\top$ more than once. Below, a superscript ^{k} indicates k-fold iteration of a given operator or group of operators.

THEOREM 3.1. *(i) For every $k \geq 1$, EA proves:*

$$\widetilde{\langle 0 \rangle}^k \top \equiv \left(\text{B}\Sigma_1 \wedge \langle 0 \rangle^k \top\right).$$

(ii) For $n > 0$ and for every $k \geq 1$, EA proves:

$$\widetilde{\langle n \rangle}^k \top \equiv \left(\text{B}\Sigma_{n+1} \wedge (\langle n \rangle \langle n+1 \rangle)^{k-1} \top\right).$$

Proof. The proofs of both (i) and (ii) are by induction on k.

The base case for (i), $k = 1$, follows from Corollary 1.2 part (i). In the inductive step for (i), note that by Theorem 1.1 and the inductive assumption, $\widetilde{\langle 0 \rangle}^{k+1} \top$ is equivalent to

$$\text{B}\Sigma_1 \wedge \text{Con}(\text{EA} + \text{B}\Sigma_1 + \langle 0 \rangle^k \top). \tag{6}$$

However, for any Π_1 sentence π, the theory $\text{EA} + \text{B}\Sigma_1 + \pi$ is interpretable in $\text{EA} + \pi$, and this fact is easily proved[5] in EA. Hence, (6) is in fact equivalent to

$$\text{B}\Sigma_1 \wedge \text{Con}(\text{EA} + \langle 0 \rangle^k \top),$$

which is what we had to show.

The proof for (ii) is essentially similar. The base case follows from Corollary 1.2 part (ii), so consider the inductive step. By Theorem 1.1 and the inductive assumption, $\widetilde{\langle n \rangle}^{k+1} \top$ is equivalent to

---

[5]One way to see this is to observe that, by a remark following Lemma 9 of [KY15], there is a formula $\varphi(x)$ which provably in EA defines the whole universe if $\text{I}\Sigma_1$ holds and a proper cut closed under exp if $\text{I}\Sigma_1$ fails; thus, $\text{EA} + \pi$ implies $(\text{EA} + \pi + \text{B}\Sigma_1)^\varphi$.

$$\mathrm{B}\Sigma_{n+1} \wedge \mathrm{RFN}_{\Pi_{n+1}} \left(\mathrm{B}\Sigma_{n+1} \wedge (\langle n \rangle \langle n+1 \rangle)^{k-1} \top \right). \tag{7}$$

Clearly, (7) implies
$$\mathrm{B}\Sigma_{n+1} \wedge \langle n \rangle \left(\langle n+1 \rangle \top \wedge (\langle n \rangle \langle n+1 \rangle)^{k-1} \top \right), \tag{8}$$

because $\langle n \rangle$ is $\mathrm{RFN}_{\Pi_{n+1}}$ and $\langle n+1 \rangle \top$ is $\mathrm{I}\Sigma_n$. By [Bek04, Lemma 4], it is always the case that $\langle n+1 \rangle \psi \wedge \langle n \rangle \tau$ is equivalent to $\langle n+1 \rangle (\psi \wedge \langle n \rangle \tau)$, so in particular the argument of the outermost $\langle n \rangle$ in (8) can be rewritten as $\langle n+1 \rangle (\langle n \rangle \langle n+1 \rangle)^{k-1} \top$. This proves that (8) is in fact equivalent to $\mathrm{B}\Sigma_{n+1} \wedge (\langle n \rangle \langle n+1 \rangle)^k \top$.

To see that (8) $\to$ (7) also holds, note that $(\langle n \rangle \langle n+1 \rangle)^{k-1} \top$ is a Π_{n+1} sentence, and $\mathrm{B}\Sigma_{n+1}$ is Π_{n+2}-conservative over $\mathrm{I}\Sigma_n$ provably in $\mathrm{I}\Sigma_1$ [HP93, Cor. IV.4.7]. ⊣

COROLLARY 3.2. *Over EA, the theory*
$$\left\{ \widetilde{\langle 0 \rangle}^k \top : k \in \mathbb{N} \right\}$$
has the same Π_1 consequences as $\langle 1 \rangle \top$, and for $n \geq 1$, the theory
$$\left\{ \widetilde{\langle n \rangle}^k \top : k \in \mathbb{N} \right\}$$
has the same Π_{n+1} consequences as $\langle n+1 \rangle \langle n+1 \rangle \top$.

Proof. The proofs of both statements are based on a similar idea, so let us consider the more subtle case of $n > 0$. We make use of a technical definition introduced in [Bek04]: for $n, k \in \mathbb{N}$, and for τ a sentence,
$$\mathcal{Q}_0^n(\tau) = \langle n \rangle \tau,$$
$$\mathcal{Q}_{k+1}^n(\tau) = \langle n \rangle (\tau \wedge \mathcal{Q}_k^n(\tau)).$$

By [Bek04], $\langle n+1 \rangle \tau$ is always Π_{n+1}-conservative over $\{ \mathcal{Q}_k^n(\tau) : k \in \mathbb{N} \}$. An easy induction on $k \geq 1$ shows that $\mathcal{Q}_{k-1}^n (\langle n+1 \rangle \top)$ is equivalent to $(\langle n \rangle \langle n+1 \rangle)^k \top$.

By Paris' proof of the Paris-Friedman theorem ([Par81], cf. [HP93, Thm IV.1.61]), any model $\mathcal{M}$ of $\langle n+1 \rangle \top$ has a cofinal Σ_{n+1}-elementary extension to a model $\mathcal{K}$ of $\mathrm{B}\Sigma_{n+1}$. It is not difficult to verify that if $\mathcal{M} \models \langle n+1 \rangle \tau$ for some τ, then $\mathcal{K} \models \langle n+1 \rangle \tau$ as well. Therefore, adding $\mathrm{B}\Sigma_{n+1}$ to $\langle n+1 \rangle \tau$ does not change even the Π_{n+2} consequences.

The following theories thus have the same Π_{n+1} consequences:

(i) $\mathrm{B}\Sigma_{n+1} + \langle n+1 \rangle \langle n+1 \rangle \top$,

(ii) $\langle n+1 \rangle \langle n+1 \rangle \top$,

(iii) $\{ (\langle n \rangle \langle n+1 \rangle)^k \top : k \in \mathbb{N} \}$,

(iv) $\mathrm{B}\Sigma_{n+1} + \{ (\langle n \rangle \langle n+1 \rangle)^k \top : k \in \mathbb{N} \}$.

By Theorem 3.1 this completes the proof. ⊣

Our second result is that the $\widetilde{\langle n \rangle}$ operators do not satisfy one of the axioms of the polymodal logic GLP, known to axiomatize the provability logic of the $\langle n \rangle$ operators over PA ([Jap86, Ign93]), and to be sound for $\langle n \rangle$ already over EA. GLP consists of the usual Gödel-Löb logic GL for each individual n and axioms of two additional types:

$$\langle m \rangle \tau \to \langle n \rangle \tau, \qquad \text{for } n \leq m, \tag{9}$$

$$\langle n \rangle \tau \to [m] \langle n \rangle \tau, \qquad \text{for } n < m. \tag{10}$$

It is straightforward to verify that each $\widetilde{\langle n \rangle}$ satisfies GL and also that the $\sim$-analogue of (9) is satisfied. However:

THEOREM 3.3. *The statements* $\widetilde{\langle 0 \rangle} \top \to \widetilde{[1]} \widetilde{\langle 0 \rangle} \top$ *and* $\widetilde{\langle 0 \rangle} \top \to \widetilde{[2]} \widetilde{\langle 0 \rangle} \top$ *are false in* $\mathbb{N}$.

Proof. By Corollary 1.2 part (i), $\widetilde{\langle 0 \rangle} \top$ is simply $\mathrm{B}\Sigma_1 + \mathrm{EA} + \mathrm{Con}(\mathrm{EA})$, which is clearly true in $\mathbb{N}$. On the other hand, $\widetilde{[n]} \widetilde{\langle 0 \rangle} \top$, or $\neg \widetilde{\langle n \rangle} \neg \widetilde{\langle 0 \rangle} \top$, is by Theorem 1.1 equivalent to:

$$\mathrm{B}\Sigma_{n+1} \to \neg \mathrm{RFN}_{\Pi_{n+1}}(\mathrm{EA} \wedge (\neg \mathrm{B}\Sigma_1 \vee \neg \mathrm{Con}(\mathrm{EA}))).$$

The antecedent of this implication is true in $\mathrm{B}\Sigma_n$ is true in $\mathbb{N}$ for any $n \in \mathbb{N}$. However, the consequent is false for $n \leq 2$: by [Par70], even $\mathrm{EA} + \neg \mathrm{B}\Sigma_1$, and thus *a fortiori* $\mathrm{EA} + (\neg \mathrm{B}\Sigma_1 \vee \neg \mathrm{Con}(\mathrm{EA}))$, does not prove any false Π_3 sentence. ⊣

Acknowledgment. I am very grateful to Lev Beklemishev and Joost Joosten for reading preliminary drafts of this paper and suggesting quite a few improvements.

BIBLIOGRAPHY

[AD88] Z. Adamowicz and R. Kossak, *A note on $\mathrm{B}\Sigma_n$ and an intermediate induction scheme*, Zeitschrift für Mathematische Logik und Grundlagen der Mathematik **34** (1988), 261–264.

[Bek] L. D. Beklemishev, list of open problems maintained at http://www.mi.ras.ru/~bekl

[Bek04] L. D. Beklemishev, *Provability algebras and proof-theoretic ordinals, I*, Annals of Pure and Applied Logic **128** (2004), 103–124.

[Bek05] L. D. Beklemishev, *Reflection principles and provability algebras in formal arithmetic*, Russian Mathematical Surveys **60** (2005), 197–268.

[HP93] P. Hájek and P. Pudlák, *Metamathematics of First-Order Arithmetic*, Springer-Verlag, 1993.

[Ign93] K. N. Ignatiev, *On strong provability predicates and the associated modal logics*, Journal of Symbolic Logic **58** (1993), 249–290.

[Jap86] G. K. Japaridze, *Modal-logical means of studying provability*, Ph.D. thesis, Moscow State University, 1986, in Russian.

[PK78] J. B. Paris and L. A. S. Kirby, *Σ_n collection schemas in arithmetic*, Logic Colloquium '77, Studies in Logic and the Foundations of Mathematics, vol. 96, North-Holland, 1978, pp. 199–209.

[KY15] L. A. Kołodziejczyk and K. Yokoyama, *Categorical characterizations of the natural numbers require primitive recursion*, Annals of Pure and Applied Logic **166** (2015), 219–231.

[Par70] C. Parsons, *On a number-theoretic choice schema and its relation to induction*, Intuitionism and Proof Theory. Proceedings of the Summer Conference at Buffalo, N.Y., 1968, Studies in Logic and the Foundations of Mathematics, vol. 57, North-Holland, 1970, pp. 459–473.

[Par81] J. B. Paris, *Some conservation results for fragments of arithmetic*, Model Theory and Arithmetic, Lecture Notes in Mathematics, no. 890, Springer, 1981, pp. 251–262.

[Vis15] A. Visser, *Oracle bites theory*, The Facts Matter. Essays on Logic and Cognition in Honour of Rineke Verbrugge (S. Ghosh and J. Szymanik, eds.), College Publications, 2015, pp. 133–147.

Expansions of Pseudofinite Structures and Circuit and Proof Complexity

JAN KRAJÍČEK

I am honored to have a chance to contribute to this volume celebrating a personal anniversary of Albert Visser.

Abstract

I shall describe a general model-theoretic task to construct expansions of pseudofinite structures and discuss several examples of particular relevance to computational complexity. Then I will present one specific situation where finding a suitable expansion would imply that, assuming a one-way permutation exists, the computational class $\mathcal{NP}$ is not closed under complementation.

1 Task

Consider the following situation: $\mathbf{M}$ is a nonstandard model of true arithmetic (in the usual language of arithmetic $0, 1, +, \cdot, \leq$), n is a nonstandard element of $\mathbf{M}$, L is a finite language and $W \in \mathbf{M}$ is its interpretation on the universe $[n] = \{1, \ldots, n\}$; W can be identified with a subset of $[n^k]$ for some $k \in \mathbf{N}$. We shall denote the resulting structure $\mathbf{A}_W$; it is coded by an element of $\mathbf{M}$ that is $\leq 2^{n^k}$. Without a loss of generality we shall assume that L contains constants $1, n$, the ordering relation $\leq$ interpreted as in $\mathbf{M}$, and ternary relation symbols $\oplus$ and $\odot$ for the graphs of addition and multiplication inherited from $\mathbf{M}$. Because $\mathbf{M}$ is a model of true arithmetic $\mathbf{A}_W$ is pseudofinite: it satisfies the theory of all finite L-structures.

Paris and Dimitracopoulos [20] studied the problem of for how large $m > n$ does the theory of the arithmetic structure on $[n]$ determine the theory of the arithmetic structure on $[m]$ and proved that it does not for $m = 2^n$. They also pointed out various links between questions of this type and complexity theory problems around the collapse of the polynomial time hierarchy. Ajtai [1] showed (among other similar results) that if M is a countable nonstandard model of PA and L is finite then for any L-structure $\mathbf{A}_W$ there are two sets $U, U' \subseteq [n]$, both elements of $\mathbf{M}$, such that $\mathbf{M}$ thinks that $|U|$ is odd and $|U'|$ is even while, as structures, $(\mathbf{A}_W, U) \cong (\mathbf{A}_W, U')$ (the isomorphism is not in $\mathbf{M}$, of course)[1]. Krajíček and Pudlák [18] showed (improving upon earlier results of Hájek [10] and Solovay [22]) that for any nonstandard $t \leq n \in \mathbf{M}$ one can construct $\mathbf{M}' \supseteq \mathbf{M}$ containing a proof of contradiction in PA of length bounded by n^t without adding any new elements to interval $[0, n]$. Máté [19] considered the full second order structure on $[n]$ (coded in $\mathbf{M}$) and reformulated the statement that $\mathcal{NP} \neq co\mathcal{NP}$ as a statement about non-preservation of the theory of the structure in an expansion coming from a model $\mathbf{M}' \supseteq \mathbf{M}$.

The most interesting results of this kind (to this author) were obtained initially by Ajtai [1, 2]. In the first paper he established that parity of n bits cannot be computed by AC^0 circuits (proved inde-

---
[1] We shall discuss another example with parity later.

pendently by Furst, Saxe and Sipser [8]) and he reports there that his first proof of the lower bound was by model theory of arithmetic although he eventually chose to present the result combinatorially. In the second paper he proved that propositional formulas PHP_m formalizing the pigeonhole principle do not have polynomial size constant depth Frege proofs. That proof is by constructing a suitable model of arithmetic[2]. We shall discuss these two examples in the next section.

In this, mostly expository, note we are interested in the general question of how to construct expansions of $\mathbf{A}_W$ with particular properties. Before formulating this more specifically we will consider in the next section three examples. The examples go back to Ajtai [1, 2] and (essentially) Máté [19] but two of them are not presented in the literature with enough details and are formulated with unnecessarily strong hypotheses. In the subsequent section we shall discuss a specific open problem whose solution would have interesting implications for computational complexity.

The note is self-contained modulo a basic knowledge of logic and complexity theory. Notions not explained here can be likely found in [13].

2 Examples

Parity example. It is well-known that the parity of a string of bits cannot be computed by AC^0 circuits, [1, 8]. That is,

(1) For any $d \geq 1$ and large enough m, any depth d, size $\leq m^d$ circuit with m inputs must compute erroneously the parity of some m-bit string.

Computability by AC^0 circuits is equivalent to first-order definability in the presence of an extra structure (of a fixed signature, the extra structure depending just on the size of the universe). In particular, (1) is equivalent to

(2) For any finite language L and any formula $\Phi(X)$ in the language $L(X)$, L augmented by a unary predicate $X(x)$, for m large enough and any L-structure $\mathbf{B}$ with universe $[m]$ it holds:

- There is $U \subseteq [m]$ such that the equivalence:

$$(\mathbf{B}, U) \models \Phi(U) \text{ iff } |U| \text{ is odd}$$

fails.

Using overspill in $\mathbf{M}$, (2) is equivalent to the same statement for any L-structure on any nonstandard $[n]$ in $\mathbf{M}$ (with $U \in \mathbf{M}$ and its parity defined in $\mathbf{M}$). And that can be further formulated as follows. For $u \in [n]$ denote $U^{<u} := \{v \in U \mid v < u\}$ and $U^{\leq u} := \{v \in U \mid v \leq u\}$.

(3) For any nonstandard $n \in \mathbf{M}$, any finite language L, any $L(X)$ formula $\Phi(X)$ and any L-structure $\mathbf{A}_W \in \mathbf{M}$ with universe $[n]$ it holds:

There is $U \subseteq [n]$, $U \in \mathbf{M}$, and $u \in U$ such that the following holds:

---

[2]The original manuscript was only about models of bounded arithmetic. After Ajtai learned that Paris and Wilkie [21] linked provability of PHP in bounded arithmetic with a conjecture of Cook and Reckhow [6] that formulas PHP_m are hard for Frege systems, shown eventually false by Buss [5], he added a few hand-written pages showing how his result implies a lower bound for constant depth Frege systems.

(a)
$$(\mathbf{A}_W, U^{<u}) \models \Phi(U^{<u}) \text{ iff } (\mathbf{A}_W, U^{\leq u}) \models \Phi(U^{\leq u}).$$

(b) For $t \in U^{<u}$,
$$(\mathbf{A}_W, U^{<t}) \models \Phi(U^{<t}) \text{ iff } (\mathbf{A}_W, U^{\leq t}) \models \neg\Phi(U^{\leq t}).$$

Let $Y(x)$ be a new unary predicate and consider first-order theory T_1^Φ in the language $L(X, Y)$, $L(X)$ augmented by Y, with axioms:

1. The least number principle axioms:
$$\exists x\, \alpha(x) \to \exists x \forall y < x\, (\alpha(x) \wedge \neg \alpha(y))$$
   for all formulas α in the language $L(X, Y)$ (as we will evaluate formulas over a structure with universe $[n]$ they are de facto bounded),

2. $\exists x \in X\ (x \in Y) \not\equiv \Phi(X^{\leq x})$,

3. axiom $\Psi(Y)$, where Ψ is the formula:
$$Y \subseteq X \wedge \min X \in Y \wedge (\forall y \in X\ suc_X(\min X) = y \to y \notin Y) \wedge$$
$$\forall x, y \in X\ suc_X(suc_X(x)) = y \to x \in Y \equiv y \in Y$$
   where for $x \in X$, $suc_X(x)$ is $\min\{z \in X \mid x < z\}$, if it exists.

Claim 1: *Statement (3) for a given language L, formula Φ, set $U \in \mathbf{M}$ and some $u \in U$ is equivalent to the existence of $V \subseteq [n]$ (V not necessarily in $\mathbf{M}$) such that the expanded structure $(\mathbf{A}_W, U, V)$ satisfies T_1^Φ (U interprets X and V interprets Y).*

For the only if direction note that for U and $u \in U$ satisfying (3) a suitable V can be defined in $\mathbf{M}$ already: take for V the subset of U consisting of its elements on odd positions. Axiom 1. holds because $\mathbf{M}$ satisfies it for all formulas, axiom 3. holds by the definition of V and axiom 2. holds as it is witnessed by $x := u$.

For the if direction assume that $(\mathbf{A}_W, U, V) \models T_1^\Phi$. Take for u the minimal x witnessing axiom 2.; it exists by the least number principle. Utilizing axiom 3. we see that the pair U, u satisfies statement (3).

PHP example. In this example we aim at a proof complexity lower bound. Given $m \geq 2$, consider a propositional formula PHP_m formed using atoms p_{ij}, $i \in [m]$ and $j \in [m-1]$ that is the disjunction of the following formulas:

- $\bigvee_i \bigwedge_j \neg p_{ij}$,
- $\bigvee_{i_1 \neq i_2} \bigvee_j (p_{i_1 j} \wedge p_{i_2 j})$,
- $\bigvee_i \bigvee_{j_1 \neq j_2} (p_{ij_1} \wedge p_{ij_2})$.

Having a falsifying assignment $p_{ij} := a_{ij} \in \{0,1\}$ we could define the graph of an injective map from $[m]$ to $[m-1]$:
$$\{(i,j) \mid a_{ij} = 1\}$$
which is impossible. So PHP_m is a tautology.

The depth of a formula (in DeMorgan language) is the maximal number of connected blocks of alike connectives on a path of subformulas of the formula (the depth of PHP_m is thus 3). A Frege system F is a sound and implicationally complete finite collection of inference rules and axiom schemes. Its depth d subsystem F_d is allowed to use only formulas of depth $\leq d$.

A path in a depth d formula of size (i.e. the number of symbols) s can be naturally determined by a d-tuple of numbers $\leq s$, giving d pointers to which subformula the path moves when alternating from one connective to another. This implies that such a formula can be coded by a d-ary function on $[s]$ giving information about atoms (or constants) in which individual paths end. Hence an F_d proof of PHP_n of size polynomial in n can be coded by a relation S on $[n]$. The qualification *naturally* used above means that, given d, one can define in $([n], R, S)$ the satisfaction relation between formulas in the proof S and a truth assignment[3] R.

Using this, and overspill as before, the statement

(4) For all $d \geq 3$ and all large enough $m \geq 2$ there is no F_d proof of PHP_m of size $\leq m^d$,

is equivalent to the following statement about **M** and any nonstandard $n \in \mathbf{M}$:

(5) for all $d \geq 3$ and any relation S on $[n]$, $S \in \mathbf{M}$, S does not encode an F_d proof of PHP_n.

Let L be a finite language and let Z be a new symbol for a binary relation. Consider the theory T_2:

- Induction axioms for all $L(Z)$-formulas as in T_1^Φ,
- $\neg PHP(Z)$ axiom:
$$[\forall x \exists y < n\ Z(x,y)] \land [\forall x < x' \forall y < n\ \neg Z(x,y) \lor \neg Z(x',y)] \land$$
$$[\forall x \forall y < y' < n\ \neg Z(x,y) \lor \neg Z(x,y')].$$

Claim 2: *Assume that for any finite L and any L-structure $\mathbf{A}_W$ with universe $[n]$ (any nonstandard $n \in \mathbf{M}$) there exists an expansion $(\mathbf{A}_W, R)$ satisfying T_2 with $Z := R$. Then statement (5) is true.*

To see this assume that (5) fails, i.e. some relation S on $[n]$ encodes an F_d-proof of PHP_n. Let $(\mathbf{A}_W, R)$ be a model of T_2, with W containing S. Using the truth definition for depth $\leq d$ formulas show that under the assignment
$$p_{ij} := 1 \text{ if } R(i,j) \text{ and } p_{ij} := 0 \text{ otherwise}$$
all propositional axioms and the formula $\neg PHP_n$ are true. Hence, by induction, there has to be an inference whose all hypotheses are true while its conclusion is not. But that is impossible.

Note that Claim 2 is formulated as an implication and not equivalence as Claim 1; we shall return to this issue in the next section.

---
[3] Details left out in this example can be found in [13].

TAUT example. Let $TAUT \subseteq \{0,1\}^*$ be the set of propositional tautologies in the DeMorgan language. A propositional proof system (abbreviated to PPS) in the sense of Cook and Reckhow [6] is a polynomial time binary relation P on $\{0,1\}^*$ such that

- $\forall x, y \ (P(x,y) \to x \in TAUT)$ (soundness),

- $\forall x \in TAUT \ \exists y \ P(x,y)$ (completeness).

A PPS P is p-bounded if there exists $k \geq 1$ such that the $\exists y$ in the completeness can be bounded by $|y| \leq |x|^k$. A p-bounded PPS exists iff $\mathcal{NP} = co\mathcal{NP}$, see [6]. The main task is therefore to establish for all PPSs a super-polynomial lengths-of-proofs lower bound. This may be far away at present but lower bounds for specific PPSs have interesting consequences as well (e.g. for independence results or for SAT algorithms).

As before we shall consider strings of length polynomial in n coded by relations on $[n]$. For a given PPS P and $k \geq 1$ there exists by Fagin's theorem[4] a first order formula $\Theta_k(X, Y, Y')$ such that $\exists Y' \Theta_k(X, Y, Y')$ defines on $[n]$ the relation $P(x, y) \wedge |y| \leq |x|^k$ for formulas $X \subseteq [n]$; Y and Y' are new relation names of arity depending on P and k.

Denote by $SAT(Z, X)$ a first order formula defining the satisfaction relation between an assignment $Z \subseteq [n]$ and a DNF formula $X \subseteq [n]$ (we want to avoid coding of evaluations of general formulas in this discussion). Let L be an arbitrary finite language not containing predicate symbols X, Z and define a theory T_3^k with axioms:

- $\neg SAT(Z, X)$,

- axiom scheme $\Theta_k(\alpha, \beta, \beta') \to SAT(\gamma, \alpha)$, where $\alpha, \beta, \beta', \gamma$ range over all $L(X, Z)$ formulas.

Claim 3: *Assume that for any finite L, any L-structure $\mathbf{A}_W$ with universe $[n]$ and any standard $k \geq 1$ there is $F \subseteq [n]$, $F \in \mathbf{M}$ a DNF formula that is a tautology in $\mathbf{M}$, such that $(\mathbf{A}_W, F)$ has an expansion $(\mathbf{A}_W, F, R)$ satisfying T_3^k with $X := F$ and $Z := R$. Then P is not p-bounded.*

If P were p-bounded with exponent $|x|^k$ then by overspill the formula F would have a P-proof in $\mathbf{M}$ of size polynomial in n, and $\Theta_k(F, S, S')$ would hold for some relations $S, S' \in \mathbf{M}$ on $[n]$ that we can put into W. Hence an expansion that is a model of T_3^k for $X := F$ is impossible.

3 Discussion

Claims 2 and 3 can be established as equivalence statements using the theory of propositional translations of Π_1^1 theories (cf. [13, Chpt.9]). For that argument to work one does not need that $\mathbf{M}$ is a model of true arithmetic but only that

(a) 2^{n^t} exists in $\mathbf{M}$ for some nonstandard $t \leq n$,

(b) $\mathbf{M}$ satisfies bounded arithmetic theory R_2^1 (which yields forms of collection and comprehension schemes needed for the construction in [13, Sec.9.4]),

---
[4] Invoking just Fagin's theorem here is not enough for various properties one needs often from Θ_k. The formalization needs to be "natural" again. The claim below holds for arbitrary Θ_k defining the relation, though.

(c) **M** is countable.

Ajtai [3, 4] formulated a general existence theorem for theories T going well beyond first order or Π_1^1 theories. Such T can be not only second order or third order, etc., but it can be a finite set theory over Ur-elements $[n]$ (and even more), and he allows not only expansions of $\mathbf{A}_W$ but expansions of end-extensions of $\mathbf{A}_W$. The existence of such a model of T is characterized by the non-existence of a proof of contradiction in T that is - in a specific, rather technical, sense - definable over $\mathbf{A}_W$. We refer the reader to Garlík [9] who found a simpler and more conceptual proof of Ajtai's theorem. The construction needs **M** satisfying (a)-(c) above and also

(d) L is finite.

In Claims 1 - 3 we stipulated that L is finite in order to avoid the discussion how it is coded. In these claims L can be, in fact, infinite as long as $\mathbf{A}_W$ is coded in **M**. But in [3, 4, 9] the hypothesis (d) is needed.

The intended goal of Claims 1 - 3 is to offer a strategy how to prove a lower bound in complexity theory by constructing a suitable expansion of $\mathbf{A}_W$. This has been done by Ajtai [1, 2] for the lower bounds explained in Claims 1 and 2. Ajtai [1, 2] works in a model of PA but the construction needs only assumptions (a) and (c) above and a variant of (b):

(b') **M** is a model of the theory PV and of the weak pigeonhole principle for p-time functions (as represented by PV-terms), denoted WPHP(PV).

See [13, Sec.15.2] for how the WPHP(PV) is used.

It is a challenge to construct a suitable model of T_3^k in the situation of Claim 3 for strong PPS P (or even for all PPSs). In an attempt to meet the challenge particular models **M'** (extending a cut in **M** of elements with length subexponential in n) were constructed in [15, 17] and [14] (two different constructions). They were constructed under the assumptions that a one-way permutation exists. These models **M'** satisfy the following conditions:

1. $2^{2^{n^\epsilon}}$ exists in **M'** for some standard $\epsilon > 0$ (this is stronger that (a)),

2. **M'** is a model of the true $\forall \Pi_1^b$ theory in the language of PV (this does not imply either (b) or (b')). In particular, all PPSs are sound in **M'**.

3. L is infinite and coded in **M'** but $\mathbf{A}_W$ is not coded in **M'**,

4. there is a DNF formula $F \subseteq [n]$, $F \in \mathbf{M}$ (the original model of true arithmetic) that has the form:
$$\bigvee_{i \in [n]} \varphi_i(x, y^i)$$
with each $\varphi_i(x, y^i)$ having the form
$$\psi_i(x, y^i) \to \eta_i(y^i)$$
and such that:

(i) F is a tautology in both $\mathbf{M}$ and $\mathbf{M}'$,

(ii) $x, y^1, \ldots, y^n$ are mutually disjoint tuples of variables,

(iii) there exists assignments $A \in W$ and $B^i \in W$ for all $i \in [n]$ such that $\varphi_i(A, B^i)$ fails, i.e. $SAT((A, B^i), \neg \varphi_i)$, and thus $SAT((A, B^i), \psi_i) \wedge SAT(B^i, \neg \eta_i)$, hold in $\mathbf{M}'$.

(The reason why (i)-(iii) are not a priori contradictory is that without a collections scheme we have apparently no way to combine assignments A and all B^i together to get an assignment falsifying F.)

We now derive more properties of $\mathbf{M}'$ using the assumption that a PPS P is p-bounded and sufficiently strong[5]. We shall use the specific form of the formula F used in [14, 17]; we will formulate its properties explicitly but the interested reader is assumed to learn the definition of the formula in [14, 17] and check that it has the stated properties.

Claim 4: *Assume P is p-bounded and sufficiently strong. Then, in $\mathbf{M}'$, the following facts hold:*

(iv) There is a P-proof of $\bigvee_{i \in [n]} \varphi_i(A, y^i)$ in $\mathbf{M}'$ coded by a relation on $[n]$,

(v) for each $i \in [n]$ there are P-proofs of $(\psi_i(x, z) \wedge \psi_i(x, z')) \to z \equiv z'$ and of $\neg \varphi_i(A, y^i)$ in $\mathbf{M}'$ coded by relations on $[n]$,

(vi) formula $\bigwedge_{i \in [n]} \psi_i(A, y^i)$ is not P-refutable in $\mathbf{M}'$.

Statement (iv) follows from the p-boundedness of P and 2. and (i) above: F has a F-proof π in $\mathbf{M} \cap \mathbf{M}'$ and there is a p-time function $f(x, y)$ such that in $\mathbf{M}$ $f(a, p)$ is a P-proof of $F(a, y^1, \ldots, y^n)$ for all a and all P-proofs p of F, and hence by 2. $f(A, \pi)$ is the wanted proof. The first part of statement (v) follows again from the p-boundedness of P and the fact that formulas $(\psi_i(x, z) \wedge \psi_i(x, z')) \to z \equiv z'$ are tautologies in $\mathbf{M}$. It follows that $\psi_i(A, B^i) \to \psi_i(A, y^i)$ and thus $\psi_i(A, y^i)$ (using that the true sentences $\psi_i(A, B^i)$ have P-proofs) and $\neg \eta_i(B^i)$ have P-proofs in $\mathbf{M}'$ too. Statement (vi) is valid because of the specific formulas ψ_i have the property that it holds in $\mathbf{M}$ that

$$\forall x \exists y = ((y)_1, \ldots, (y)_n) \bigwedge_{i \in [n]} \psi(x, (y)_i) .$$

This implies that in $\mathbf{M}$ for no a and $k \in \mathbf{N}$ there is a P-proof of $\neg \bigwedge_{i \in [n]} \psi_i(a, y^i)$ of size $\leq n^k$, and by p-boundedness of P these facts have P-proofs in $\mathbf{M}$ (and hence in $\mathbf{M}'$) and P is sound in $\mathbf{M}'$.

Ideally we would like to bring the existence of $\mathbf{M}'$ with properties 1.-4.(with (iv)-(vi) added) to a contradiction. That would imply that the hypothesis of the existence of a one-way permutation used in [14, 17] contradicts the hypothesis of p-boundedness of P used in Claim 4, entailing a conditional lower bound for (possibly all) P. In light of Claim 4 it seems natural to try to extend model $\mathbf{M}'$ by adding a satisfying assignment for $\bigwedge_i \psi(A, y^i)$.

The existence of a satisfying assignment for $\bigwedge_i \psi(A, y^i)$ would follow if $\mathbf{M}'$ would satisfy a collection scheme for Σ_1^1 formulas on $\mathbf{A}_W$. But that is unlikely as the argument of Cook and

---

[5] We need that P contains resolution, P-proofs of true sentences can be constructed by a p-time algorithm (hence its soundness is true in $\mathbf{M}'$), and that P simulates modus ponens and substitutions of constants with only a polynomial increase in the proof length.

Thapen [7] implies that the true $\forall \Pi_1^b$ theory does not prove the scheme (unless factoring is not hard), and the same argument implies that one cannot argue just using the $\forall \Pi_1^b$ theory of $\mathbf{M}'$ that it has an extension where the collection holds (we would also need that it adds no new elements into $[n]$). The construction of $\mathbf{M}'$ in [14] depends just on the $\forall \Pi_1^b$ theory of $\mathbf{M}$ and hence it is unlikely that it can be used again with the ground model being $\mathbf{M}'$.

The use of the property that we add no new elements into $[n]$ in Claims 1 - 3 was solely to preserve the first order theory of $\mathbf{A}_W$ (in the preceding paragraph it would also imply that (iii) above remains true in an extension of $\mathbf{M}'$). There is an alternative for arranging that. In the forcing from [15, 17] one naturally adds many new elements into $[n]$ (and, in fact, does not include all of $[n]$ from $\mathbf{M}$ into $\mathbf{M}'$) but the first order theory of $\mathbf{A}_W$ is nevertheless often preserved. The set-up of the method presupposes some approximate counting available in the ground model (here $\mathbf{M}'$) and the model is assumed to be $\aleph_1$-saturated. The former can be arranged as one can modify the constructions so that the so called dual WPHP for p-time functions, dWPHP(PV), is true in $\mathbf{M}'$ too and that yields some approximate counting by Jeřábek [11, 12]. As for the latter condition: some saturation property of $\mathbf{M}'$ could be arranged (cf. [16]) if one could modify the forcing construction of $\mathbf{M}'$ so that it is defined by a compact family (in the sense of [17]) of random variables. That would be possible if one could establish a hard-core lemma for the computation model underlying the construction. That, together with the fact (see 3. above) that $\mathbf{A}_W$ is not coded in $\mathbf{M}'$, seem to be the main technical obstacles to apply Claim 3 to a general PPS.

Acknowledgements: I thank Leszek Kolodziejczyk (Warsaw) and Neil Thapen (Prague) for comments on a draft of this paper.

BIBLIOGRAPHY

[1] M. Ajtai, Σ_1^1 - formulae on finite structures, *Annals of Pure and Applied Logic*, **24**, (1983), pp.1-48.

[2] M. Ajtai, The complexity of the pigeonhole principle, in: *Proc. IEEE 29th Annual Symp. on Foundation of Computer Science*, (1988), pp. 346-355.

[3] M. Ajtai, Generalizations of the Compactness Theorem and Godel's Completeness Theorem for Nonstandard Finite Structures, in: Proc. of the 4th international conference on Theory and applications of models of computation, (2007), pp.13-33.

[4] M. Ajtai, A Generalization of Godel's Completeness Theorem for Nonstandard Finite Structures, preprint (2011).

[5] S. R. Buss, The propositional pigeonhole principle has polynomial size Frege proofs, *J. Symbolic Logic*, **52**, (1987), pp.916-927.

[6] S. A. Cook, and Reckhow, The relative efficiency of propositional proof systems, *J. Symbolic Logic*, **44(1)**, (1979), pp.36-50.

[7] S. A. Cook and N. Thapen, The strength of replacement in weak arithmetic, *ACM Transactions on Computational Logic*, **Vol 7:4**, (2006), pp.749-764.

[8] M. Furst, J. B. Saxe, and M. Sipser, Parity, circuits and the polynomial-time hierarchy, *Math. Systems Theory*, **17**, (1984), pp.13-27.

[9] M. Garlík, A New Proof of Ajtai's Completeness Theorem for Nonstandard Finite Structures, *Archive for Mathematical Logic*, **54(3-4)**, (2015), pp. 413-424.

[10] P. Hájek, On a new notion of partial conservativity, in: E. Boerger et.al. (eds.), Logic Colloquium 83, vol.2, pp.217-232. Berlin, Heidelberg, New York. Springer. (1983).

[11] E. Jeřábek, Dual weak pigeonhole principle, Boolean complexity, and derandomization, *Annals of Pure and Applied Logic*, **129**, (2004), pp. 1 – 37.

[12] E. Jeřábek, Approximate counting in bounded arithmetic, *J. of Symbolic Logic*, **72(3)**, (2007), pp. 959-993.

[13] J. Krajíček, *Bounded arithmetic, propositional logic, and complexity theory*, Encyclopedia of Mathematics and Its Applications, Vol. **60**, Cambridge University Press, (1995).

[14] J. Krajíček, On the proof complexity of the Nisan-Wigderson generator based on a hard $NP \cap coNP$ function, *J. of Mathematical Logic*, **11(1)**, (2011), pp.11-27.

[15] J. Krajíček, Pseudo-finite hard instances for a student-teacher game with a Nisan-Wigderson generator, *Logical methods in Computer Science*, **8 (3:09)**, (2012), pp.1-8.
[16] J. Krajíček, A saturation property of structures obtained by forcing with a compact family of random variables, *Archive for Mathematical Logic*, **52(1)**, (2013), pp.19-28.
[17] J. Krajíček, *Forcing with random variables and proof complexity*, London Mathematical Society Lecture Note Series, No.382, Cambridge University Press, (2011).
[18] J. Krajíček and P. Pudlák, On the Structure of Initial Segments of Models of Arithmetic, *Archive for Mathematical Logic*, **28(2)**, (1989), pp, 91-98.
[19] A. Máté, Nondeterministic Polynomial-Time Computations and Models of Arithmetic, *J. of ACM*, **37(1)**, (1990), pp.175-193.
[20] J. Paris and C. Dimitracopoulos, Truth definitions for Δ_0 formulae, in : *Logic and Algorithmic, l'Enseignement Mathematique*, **30**, (1982), pp.318-329, Genéve.
[21] J. Paris and A. J. Wilkie, Counting problems in bounded arithmetic, in: *Methods in Mathematical Logic*, LNM 1130, (1985), pp.317-340.Springer.
[22] R. Solovay, Injecting inconsistencies into models of PA, *Annals of Pure and Applied Logic*, **44**, (1989), pp.101-132.

Zero or Hero?

JOOP LEO

Hoe mooi om wakker te worden met één van de grote vragen van onze tijd.[1] Albert wrote this on 21 March 2015 in response to an email that I had sent him the night before. In this short note I will discuss the issue.

In his paper 'Why the theory R is special', Albert introduced *piecewise interpretations* [7]. Its definition contains expressions like

$$(R(u_{j_0}, \ldots, u_{j_{k-1}}))^{\tau,g} := F(R, f)(\vec{v}_{j_0}^{gj_0}, \ldots, \vec{v}_{j_{k-1}}^{gj_{k-1}})$$

When I had to use piecewise interpretations in a paper of my own, I thought of simplifying the definition a bit by letting the indices run from 1 to k instead of from 0 to $k-1$ [6]. But I wondered whether there might perhaps be a good reason not to do so. I sent Albert an email with the question what he considered as the advantage of starting counting indices from 0.

Albert replied that he agreed that starting with 1 (in its basic graphical representation) is indeed simpler. It saves one symbol and in LaTeX source code even three. Moreover, the last index is in this case also the length of the sequence. But despite this, Albert is a zero-counter. His reason is the von Neumann representation of ordinals in which every number corresponds with the sequence of its predecessors. For example, 0 corresponds with the empty sequence and 2 corresponds with the sequence 0, 1. In set theory you can conceive of a sequence of length n as a function with n as domain.

I was not immediately convinced by Albert's argument. Why would we start counting indices from 0, but not the people waiting in front of me to be served? And we had a William I, William II, William III, but no William 0.

Also in the internationally accepted civil (Gregorian) calendar, there is no year 0. One should be aware of that when determining someone's age who lived 2000 years ago. Seneca, for example, (presumably) was born in 4 BC. He became 65 in the year 62 AD and not in 61 AD.

There is a story about the Polish mathematician Sierpiński being worried that he had lost a piece of luggage. His wife said "No dear, all six pieces are there". Sierpiński replied: "That can't be true, I've counted them several times: zero, one, two, three, four, five." [2, p. 265]

Albert, however, would not get in such a situation as Sierpiński because he counts cardinals starting from 1, and ordinals from 0. But being second in a line, doesn't that simply mean for everyone that there is just one person in front of you? On the other hand, when an American makes an emergency call in Paris saying that there is a fire on the second floor, this might lead to a disastrous misunderstanding because what an American calls the second floor a European would call the first floor. Obviously, more needs to be said about counting.

Vincent van Oostrom told me about his ideas, which shed new light on the problem. First, he noted that for teachers it is handy to start counting from 1. On a blackboard a sequence $x_1, \ldots, x_n$

---

[1] *How nice to wake up with one of the big questions of our time.*

can be changed into the element x_i simply by erasing the last part and adding a dot. This was not a joke; he and some others really do this.

Second, Vincent suggested that the difficulty in choosing where to start might be caused by mixing up two different noties, namely *measuring* and *indexing*.

Measuring is usually continuous and starts with 0. This is necessary for counting; two halves give a whole.

On the other hand, for indices all kinds of things can be chosen. There is freedom of definition and naming. However, if you want to keep a correspondence with measuring, then the use of numbers is a natural choice. The number $\frac{1}{2}$ might be a good representative of the interval $[0, 1]$. Therefore, it is also appropriate as index for an object depicted within that interval.

There are of course pragmatic reasons for rejecting $\frac{1}{2}$ as index. It is not a whole number, but fractional, and it is perhaps a somewhat arbitrary number in the interval (although it is right in the middle). Moreover, 0 and 1 could also both serve as representatives of the interval $[0, 1]$. So, let us restrict ourselves to choosing between 0 and 1.

I also asked Wilfrid Hodges about his ideas about the role of zero in counting indices. The reason for asking him was that he uses both conventions in his publications.

Wilfrid said that in writing model theory for professionals, he counts from 0 because that's the normal custom. However, when writing for the general public he counts from 1 to n, because people reckon to count from the 1-st to the n-th, not from the 0-th to the $(n-1)$-th.

For intermediate cases, he tends to use 1 to n in the first draft, and then he changes to 0 to $n-1$ in later drafts if 1 to n turns out to be inconvenient. For example, suppose it is necessary to quantify over all subsets of the n-tuple.Then the easiest way to do that is to quantify over all subsets of the indexing set. If the indexing set is not $\{1,\ldots,n\}$ but $\{0,\ldots,n-1\}$, then using von Neumann ordinals, the index set is the ordinal n, and the set of all its subsets is the power set of n. This makes for a much neater account than quantifying over subsets of $\{1,\ldots,n\}$. He thinks this example is fairly typical of the reasons why model theorists and others count from 0: "Von Neumann ordinals just are incredibly neat."

I objected that if one would denote $\{1,\ldots,n\}$ as $\bar{n}$, then the set of all subsets of $\{1,\ldots,n\}$ is the power set of $\bar{n}$. For me this sounds (almost) as neat as that the set of all subsets of $\{0,\ldots,n-1\}$ is the power set of n. Wilfrid responded that a lot was hidden in my 'almost', and why opt for second best?

In computer science, there is also a tendency to count from zero. In languages like C, Java, and Haskell indices begin at 0. In other languages, like good old COBOL, counting begins at 1. On the other hand, FORTRAN and Algol 68 provide a flexibility that serves 0-counters, 1- counters, as well as other counters. If you declare in FORTRAN an array as `integer v(-1:3)`, then `v(2) = 0` sets the *fourth* element of v to zero. FORTRAN as well as Algol 68, however, serve the 1-counters slightly better because the default lower bound is one. If you declare in FORTRAN an array as `integer v(3)`, then `v(2) = 0` assigns the *second* element of v to zero.

Edsger W. Dijkstra wrote in 1982 a short manuscript titled 'Why numbering should start at zero' [3]. Dijkstra starts with comparing the following conventions to denote the sequence $2, 3, \ldots, 12$:

a $2 \leq i < 13$

b $1 < i \leq 12$

c $2 \leq i \leq 12$

d $1 < i < 13$

He argues that the best way to denote the sequence is (a). The denotations (a) and (b) have the advantage above the denotations (c) and (d) that the difference between the bounds equals the length of the sequence, and the advantage that for adjacent subsequences the lower bound of the one is the upperbound of the other. Furthermore, exclusion of the lower bound as in (b) and (d) would force to use -1 as the lower bound for a sequence starting with 0, which is ugly.

Subsequently, Dijkstra argues that adhering to the preferred denotation of sequences yields for the range of n indices two possibilities: $0 \leq i < n$ and $1 \leq i < n+1$, of which he regards the first as nicer. Therefore, he proposes to start at zero: the index equals the number of elements preceding it in the sequence. He concludes: "the moral of the story is that we had better regard —after all those centuries!—zero as a most natural number."

Is Dijkstra's argument convincing? I have my doubts. Is it not ugly to use for the range two different operators instead of only one? If so, then denoting the range as $1 \leq i \leq n$ might still be the best choice.

Some programmers have another reason for counting from 0. In lower level languages and in languages like C, the first element in an array can be found at the starting position of the array. If we start counting with index 0, then x[i] is found at offset i from the starting position of the array. This makes compilation a bit easier.

In browsing through Albert's papers I noticed a salient detail. In some co-authored papers the one counting convention is used everywhere and in some others a mix of counting conventions. For example, in 'The interpretability logic of all reasonable arithmetical theories', which he wrote with Joost Joosten, counting starts from zero [5]. In the paper 'ω-models of finite set theory', written with Ali Enayat and James Schmerl, we see a formula starting with $\exists x_0, x_1, \ldots, x_{n+2}$, as well as a formula starting with $\exists x_1, x_1, \ldots, x_n$ and a sequence $x_1 E x_2 \cdots E x_{n-1} E x_n$ [1]. Similary, in the paper 'On the termination of Russell's description elimination algorithm', which Albert wrote together with Clemens Grabmayer, Vincent van Oostrom, and myself, we see sequences like $x_1, \ldots, x_n$, and rewrite sequences like $a_0 \longrightarrow_R a_1 \longrightarrow_R a_2 \longrightarrow_R a_3 \longrightarrow \ldots$ [4].

In a future century, some overzealous historian investigating the great contributions of Albert to logic might interpret this as fingerprints of the respective authors of these papers.

What can we conclude from this short analysis? Honestly, I don't know. Despite the strong arguments of the zero-counters, I personally still find it more natural to start *always* from one. As long as we count people in a check-out line as $1, 2, 3, \ldots$, it simply feels strange to count indices as

$0, 1, 2, \ldots$. If this paper was not part of Albert's liber amicorum, I would probably have asked again for his opinion before submitting.

Acknowledgments Many thanks to Wilfrid Hodges and Vincent van Oostrom for their input.

BIBLIOGRAPHY

[1] Ali Enayat, James Schmerl, and Albert Visser. ω-models of finite set theory. In Juliette Kennedy and Roman Kossak, editors, *Set Theory, Arithmetic, and Foundations of Mathematics: Theorems, Philosophies*, pages 43 – 65. Cambridge University Press, New York, 2011.

[2] John Conway and Richard Guy. *The Book of Numbers*. Springer-Verlag New York, 1996.

[3] Edsgar W. Dijkstra. Why numbering should start at zero, 1982.

[4] Clemens Grabmayer, Joop Leo, Vincent van Oostrom, and Albert Visser. On the termination of Russell's description elimination algorithm. *The Review of Symbolic Logic*, 4(3):367 – 393, 2011.

[5] Joost Joosten and Albert Visser. The interpretability logic of all reasonable arithmetical theories. *Erkenntnis*, 53(1/2):3–26, 2000.

[6] Joop Leo. Coordinate-free logic. Technical Report 331, Utrecht University Logic Group Preprint Series, 2015.

[7] Albert Visser. Why the theory R is special. In Neil Tennant, editor, *Foundational Adventures, Essays in Honor of Harvey M. Friedman*, volume 22, pages 1–17, London, 2009. College Publications.

Dense Chains of Σ_n Sentences with Strong Conservativity Properties

V. YU. SHAVRUKOV

Abstract

Per Lindström constructed an effective dense implication chain $(\xi_r)_{r \in \mathbb{Q}}$ of Σ_n sentences Π_n-conservative over PA with the property that $\neg \xi_q$ is Σ_n-conservative over PA $+ \xi_p$ when $p < q$ We discuss further properties of such sequences and review some related results. In particular, we describe how sentences from the sequence partition the poset of prime filters on the lattice of Σ_n sentences modulo PA. We point out the lack of Σ_n vs Π_n symmetry for these sequences An alternative construction of ξ_r is also included.

to Albert at 65

1 Introduction

A sentence φ is Γ-*conservative over* a theory T if $T \vdash \gamma$ for each Γ sentence γ such that $T + \varphi \vdash \gamma$. Arana [1] and D'Aquino & Knight [3] prefer an equivalent formulation:

PROPOSITION 1.1. *Suppose the class Γ of sentences is closed under conjunction modulo T-provability, and let $\check{\Gamma}$ be the class of negations of sentences in Γ.*
Then φ is Γ-conservative over T if and only if the consistency of $T + \Psi$ implies the consistency of $T + \Psi + \varphi$ for each $\Psi \subseteq \check{\Gamma}$. ⊣

In this paper, the only contenders for the role of Γ are the familiar 1st order arithmetical classes Σ_n and Π_n with $n > 0$. Guaspari [4] appears to be the first systematic study of Σ_n and Π_n conservativity, while Lindström [8, Chapter 5] presents the recent state of the art.

This paper is devoted to a discussion of two ostensibly symmetric theorems. The first one of these was established by Lindström [9, Theorem, case $\Gamma = \Pi_n$]. An important precursor is Theorem 1.7 in Arana [1].

THEOREM Π (Lindström). *Let $n > 0$ and let T be an r.e. consistent extension of PA. There is a Π_n formula $\eta(r)$ ($r \in \mathbb{Q}$) such that*

(i) $T \vdash \forall p, q \in \mathbb{Q}\, (p \leq q \to (\eta(q) \to \eta(p)))$;

(ii) *for each p, $\eta(p)$ is Σ_n-conservative over T;*

(iii) *if $p < q$ then $\neg \eta(q)$ is Π_n-conservative over $T + \eta(p)$.*

The $n = 1$ case of the following theorem is due to Solovay [13, 4.2], and the general case appears in Lindström [9, Theorem, case $\Gamma = \Sigma_n$].

THEOREM Σ (Solovay and Lindström). *Let $n > 0$ and let T be an r.e. consistent extension of PA. There is a Σ_n formula $\xi(r)$ ($r \in \mathbb{Q}$) such that*

(i) $T \vdash \forall p, q \in \mathbb{Q}\, (p \leq q \to (\xi(q) \to \xi(p)))$;
(ii) *for each p, $\xi(p)$ is Π_n-conservative over T*;
(iii) *if $p < q$ then $\neg\xi(q)$ is Σ_n-conservative over $T + \xi(p)$.*

Observe that the statements of the two theorems are obtained from one another by swapping Σ_n and Π_n.[1] Both theorems generalize the existence of *doubly conservative* Γ sentences γ: Such γ are $\check{\Gamma}$-conservative over T while $\neg\gamma$ is Γ-conservative over T (Lindström [8, Theorem 5.3(a)]). The formulas $\eta(r)$ and $\xi(r)$ are best thought of as effective implication chains of sentences $(\eta(r))_{r \in \mathbb{Q}}$ and $(\xi(r))_{r \in \mathbb{Q}}$. The requirements placed by the theorems on elements of the chains may not be immediately transparent. Both the motivation and the only uses of either theorem thus far consist in constructions of large chains of prime filters in Σ_n/T, the lattice of Σ_n sentences modulo T-provability (Solovay [13], Arana [1], D'Aquino & Knight [3], Lindström [9]). The existence of such chains however has corollaries of its own — see D'Aquino & Knight [3, section 5].

Section 2 reviews the properties of the two chains of sentences in terms of their behaviour in the poset of prime filters of Σ_n/T and explains how said properties relate to chains of prime filters. Next we explain why Theorem Σ is "better" than Theorem Π: In section 3, any sequence $(\xi(r))_{r \in \mathbb{Q}}$ as in Theorem Σ gives rise to a sequence $(\eta(r))_{r \in \mathbb{Q}}$ as in Theorem Π in a straightforward fashion: $\eta(r) \equiv \neg\xi(-r)$. Section 4 provides a counterexample illustrating why going in the opposite Π-to-Σ direction is not generally viable. Finally, in section 5 we share an alternative recipe for putting together the formula $\xi(r)$.

I am grateful to the late Per Lindström for many thoughtful discussions and extensive correspondence, and to Julia Knight for her patient explanations. The referees supplied two extensive reports highlighting a series of ticklish issues together with an embarrassing collection of inaccuracies, and suggested numerous improvements.

2 Properties of $\eta(r)$ and $\xi(r)$

CONVENTION 2.1. Let us agree for the rest of the paper that $n > 0$ and that T is an r.e. consistent theory extending PA in the language of PA: while Theorems Π and Σ are clearly unaffected by language extensions, some of the steps in our arguments, such as Fact 2.3 below, do depend on language purity.

One should first observe that all elements of the chains $(\eta(r))_{r \in \mathbb{Q}}$ and $(\xi(r))_{r \in \mathbb{Q}}$ satisfying Theorems Π and Σ respectively are independent of T: Each element is irrefutable in T by clauses (ii) and unprovable in T by clauses (iii) of respective theorems.

Furthermore, it is easily seen that each $\neg\xi(q)$ from Theorem Σ is Σ_n-conservative over T: Fix some $p < q$. If $T + \neg\xi(q) \vdash \sigma$ where σ is Σ_n, then $T + \xi(p) \vdash \sigma$ by clause (iii), while $T + \neg\xi(p) \vdash T + \neg\xi(q) \vdash \sigma$ by clause (i), so $T \vdash \sigma$. A symmetric argument shows that in Theorem Π, each $\neg\eta(q)$ is Π_n-conservative over T.

---
[1] The reader looking for mnemonic hints may observe that η looks a little like Π while ξ sounds a little like Σ.

Lindström [9, §3] defines $E_{n,T}$, the n-E-tree of T, as the poset of (proper) prime filters of Σ_n/T — which we shall be referring to as *primes* or *points* — ordered by inclusion. (Recall that a proper filter f is *prime* if $\sigma \vee \tau \in$ f always implies $\sigma \in$ f or $\tau \in$ f.) When $n = 1$, this is E_T, the E-tree of T (see Jensen & Ehrenfeucht [6, Theorem 6] or Simmons [11, Definition 1.22]). Any prime filter contains some minimal prime, and is contained in some maximal one. All n-E-trees are ↑-forestlike, which is to say that the set of predecessors of each point is linearly ordered (Jensen & Ehrenfeucht [6, Theorem 6(i)] or Simmons [11, Lemma 1.23]). When T is Σ_1-sound, $E_T = \vec{E}_{1,T}$ is a rooted tree.

The *picture* $\sigma^\star$ of a Σ_n sentence σ in $E_{n,T}$ is the collection of those primes f that contain σ. Thus f $\in \sigma^\star \Leftrightarrow$ f $\ni \sigma$. Since the ordering on $E_{n,T}$ is inclusion of primes, the picture of any Σ_n sentence is closed upwards. Symmetrically, the *picture* $\pi^\star$ of a Π_n sentence π in $E_{n,T}$ is the collection of those f $\not\ni \neg\pi$. The picture $\pi^\star$ is then the complement of $(\neg\pi)^\star$. Pictures of Π_n sentences in $E_{n,T}$ are closed downwards.

LEMMA 2.2. *Let σ be a Σ_n and π a Π_n sentence. Then*

(a) *σ is Π_n-conservative over $T + \pi$ if and only if for each point f $\in \pi^\star$ there is a point g $\geq$ f with g $\in \pi^\star \cap \sigma^\star$.*

(b) *π is Σ_n-conservative over $T + \sigma$ if and only if for each f $\in \sigma^\star$ there is a point g $\leq$ f with g $\in \sigma^\star \cap \pi^\star$.*

In particular, σ is Π_n-conservative over T iff $\sigma^\star$ covers all maximal points of $E_{n,T}$, and π is Σ_n-conservative over T iff $\pi^\star$ contains all minimal points.

Comment. This is easily seen with the help of Proposition 1.1 (put $\Psi =$ f in clause (a) and $\Psi = \{\neg\varsigma \mid \varsigma \notin$ f $\}$ in (b)). Alternatively, employ the following fact: ⊣

FACT 2.3 (Guaspari [4, Note (3) to Theorems 6.5 and 6.6]). *Let $\preceq^e_k$ stand for the relation of Σ_k-elementary end-extension. Let φ be any sentence.*

(a) *φ is Π_n-conservative over T iff for each $\mathcal{M} \models T$ there is $\mathcal{N} \succeq^e_{n-1} \mathcal{M}$ such that $\mathcal{N} \models T + \varphi$.*

(b) *φ is Σ_n-conservative over T iff for each $\mathcal{M} \models T$ there is $\mathcal{I} \preceq^e_{n-1} \mathcal{M}$ such that $\mathcal{I} \models T + \varphi$.* ⊣

It follows from Lemma 2.2(a) that the implication chain $(\eta(r))_{r\in\mathbb{Q}}$ from Theorem II is characterized by min $E_{n,T} \subseteq \eta(r)^\star$ for each rational r, together with the requirement that each point f $\in \eta(p)^\star$ such that $p < q$ sees above itself a point g $\in \eta(p)^\star - \eta(q)^\star$:

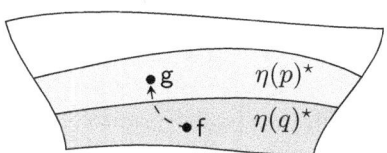

Similarly, in view of Lemma 2.2(b), the picture of each $\xi(r)$ from $(\xi(r))_{r\in\mathbb{Q}}$ covers all maximal points of $E_{n,T}$, and below each point $\mathsf{f} \in \xi(p)^\star$ there is a point $\mathsf{g} \in \xi(p)^\star - \xi(q)^\star$ whenever $p < q$:

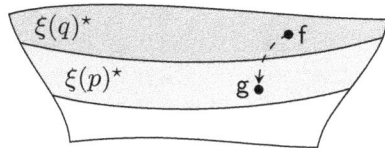

Now suppose we are given points $\mathsf{e} < \mathsf{f}$ in $E_{n,T}$ such that $\mathsf{e} \notin \xi(p)^\star$ and $\mathsf{f} \in \xi(q)^\star$, where $p < q$. Then there is a point $\mathsf{g} < \mathsf{f}$ with $\mathsf{g} \in \xi(p)^\star - \xi(q)^\star$. Since $E_{n,T}$ is a $\uparrow$-forest, g and e must be comparable. As $\mathsf{e} \notin \xi(p)^\star \ni \mathsf{g}$, one must have $\mathsf{e} < \mathsf{g}$. Thus for any pair of comparable points, one above the $\xi(p)^\star - \xi(q)^\star$ strip, the other below it, there is a point in between that belongs to that strip:

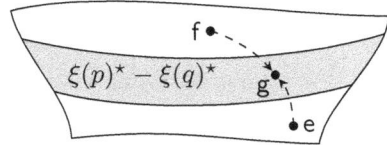

We refer to this as *the border property* (of the $\xi(r)$ family). The reader will note that the lower part of the last figure looks like the characteristic property of the $\eta(r)$ family. More on that in the next section.

Arana [1] and D'Aquino & Knight [3] utilize a version of Theorem Π to construct an infinite chain within $E_{n,T}$: Using effective transfinite recursion and Barwise–Kreisel compactness, Arana [1, Theorem 1.7] strings together an effective family $(\eta_a)_{a\in H}$ of Π_n sentences that obeys (analogues of) clauses (ii) and (iii) of Theorem Π, where H is an appropriate recursive ordering of type $\omega_1^{\mathrm{CK}} \cdot (1+\eta)$, the operation $\cdot$ is lexicographic product, and $\eta = \mathrm{otp}\,\mathbb{Q}$ is the countable dense ordertype without endpoints. The sequence $(\eta_a)_{a\in H}$ then becomes a skeleton which D'Aquino & Knight [3, Lemma 5.2] flesh out with a descending chain $(\mathsf{f}_a)_{a\in H}$ of primes such that $\neg\eta(b) \in \mathsf{f}_a$ iff $a \leq_H b$ — in fact, these authors construct models of arithmetic rather than primes. That is done with the help of ordinary 1st order compactness. Since H embeds η, the existence of a densely ordered chain of primes follows.

Theorem Σ enables a different approach: Select a point $\mathsf{f} \in E_{n,T}$ such that $\xi(r) \in \mathsf{f}$ for some $r \in \mathbb{Q}$. It follows from clause (iii) of Theorem Σ that for any rational $p \leq r$, the smallest prime among those predecessors of f that contain $\xi(p)$ fails to contain any of $\xi(q)$ with $q > p$ — recall that the set of predecessors is always linearly ordered by inclusion. In this way Solovay [13, 4.3] and Lindström [9, §3] detect a densely ordered subset within every maximal chain (*branch*) of $E_{n,T}$. Relativization of the base theory T places a densely ordered subset into every non-singleton endsegment of every branch.

In both cases, one obtains $2^{\aleph_0}$-sized chains of primes by filling in the Dedekind cuts.

3 From Σ to Π

The next proposition shows that if, for a partially ordered D, an indexed family $(\xi_r)_{r\in D}$ enjoys the properties of Theorem Σ, then the family $(\neg\xi_r)_{r\in D}$ automatically satisfies the properties required by Theorem Π for the order-antipode of D — with the possible exception of minimal points of D if those are present.

PROPOSITION 3.1. *Let $(D, \leq)$ be a partial ordering. Suppose $(\xi_r)_{r \in D}$ is a family of Σ_n sentences with the properties of Theorem Σ:*

(i) *if $p \leq q$ then $T \vdash \xi_q \to \xi_p$;*
(ii) *for each p, ξ_p is Π_n-conservative over T;*
(iii) *if $p < q$ then $\neg \xi_q$ is Σ_n-conservative over $T + \xi_p$.*

Then $(\xi_r)_{r \in D}$ also satisfies

(iv) *$\neg \xi_p$ is Σ_n-conservative over T unless p is a minimal element of D,*

and

(v) *if $p < q$ then ξ_p is Π_n-conservative over $T + \neg \xi_q$.*

Proof. (iv) To show that $\neg \xi_p$ is Σ_n-conservative over T, consider an arbitrary model $\mathscr{M} \models T$. In view of Fact 2.3(b) it suffices to find $\mathscr{I} \preceq^e_{n-1} \mathscr{M}$ with $\mathscr{I} \models T + \neg \xi_p$. We may assume $\mathscr{M} \models \xi_p$. Select $q < p$ in D. By clause (i) we then also have $\mathscr{M} \models \xi_q$. By (iii), $\neg \xi_p$ is Σ_n-conservative over $T + \xi_q$, so Fact 2.3(b) implies that there is $\mathscr{I} \preceq^e_{n-1} \mathscr{M}$ such that $\mathscr{I} \models T + \xi_q + \neg \xi_p$. In particular, $\mathscr{I} \models T + \neg \xi_p$ as required.

(v) Take any $\mathscr{M} \models T + \neg \xi_q$. Suppose for the sake of interest that $\mathscr{M} \models \neg \xi_p$. Since by clause (ii) ξ_p is Π_n-conservative over T there is by Fact 2.3(a) a $\mathscr{K} \succeq^e_{n-1} \mathscr{M}$ with $\mathscr{K} \models T + \xi_p$. By (iii) and Fact 2.3(b) there is an $\mathscr{N} \preceq^e_{n-1} \mathscr{K}$ such that $\mathscr{N} \models T + \neg \xi_q + \xi_p$. Observe that one of $\mathscr{N}$ and $\mathscr{M}$ must be a Σ_{n-1}-elementary end-extension of the other. Since $\mathscr{N} \models \xi_p$, $\mathscr{M} \models \neg \xi_p$, and ξ_p is Σ_n, one has $\mathscr{M} \preceq^e_{n-1} \mathscr{N}$. We have shown that $\mathscr{M}$ has a Σ_{n-1}-elementary end-extension $\mathscr{N} \models T + \neg \xi_q + \xi_p$ (if $\mathscr{M} \models \xi_p$, just put $\mathscr{N} = \mathscr{M}$). By Fact 2.3(a) this shows that ξ_p is Π_n-conservative over $T + \neg \xi_q$. ⊣

The proof of Proposition 3.1 essentially repeats the pictorial considerations of section 2 and presents yet more evidence that, for some purposes, primes can play the role of wireframes for models of arithmetic.

Since there are no minimal elements in $\mathbb{Q}$, Theorem Σ combines with Proposition 3.1 to show that standing the sequence $(\xi_r)_{r \in \mathbb{Q}}$ on its head and applying negation, we obtain an instantiation of Theorem Π — this is already pretty much clear from the border property of section 2:

COROLLARY 3.2. *Suppose a chain $(\xi_r)_{r \in \mathbb{Q}}$ of Σ_n sentences satisfies clauses (i)–(iii) of Theorem Σ. Then the chain of Π_n sentences $(\eta_r)_{r \in \mathbb{Q}}$ satisfies clauses (i)–(iii) of Theorem Π, where $\eta_r \equiv \neg \xi_{-r}$.* ⊣

4 An Example

The success of the proof of clause (v) of Proposition 3.1 can be attributed to the forest-likeness of $E_{n,T}$. In a forest, the net result of going up and then backtracking a little way down is still going up rather than sideways. Theorem 1 in Bennet [2] shows the existence of two Σ_n sentences ς_0 and ς_1 such that $T \nvdash \varsigma_0 \vee \varsigma_1$ while ς_i is Π_n-conservative over $T + \neg \varsigma_{1-i}$ for both i. Thus in $E_{n,T}$ there is a point $e \notin \varsigma_0^\star \cup \varsigma_1^\star$ with, by Lemma 2.2(a), two points $f_i \geq e$ such that $f_i \in \varsigma_i^\star - \varsigma_{1-i}^\star$ for both i. The points f_0 and f_1 cannot be comparable by the upwards persistence of $\varsigma_i^\star$. This shows that $E_{n,T}$ is a properly branching forest, in other words, it is not a forest "in the other direction". Therefore one should not expect a construction similar to that of Proposition 3.1 to reverse the implication of Corollary 3.2 and obtain a sequence satisfying Theorem Σ from an arbitrary sequence saisfying Theorem Π.

This section provides a counterexample showing the failure of Π-to-Σ reversal for the two-element chain. While this is, strictly speaking, an illustration to rather than an example of non-reversibility of Corollary 3.2, there is little doubt that the technique extends to larger chains.

EXAMPLE 4.1. *There are Π_n sentences θ_0 and θ_1 such that*

(i) $T \vdash \theta_1 \to \theta_0$;

(ii) *for both i, θ_i is Σ_n-conservative over T;*

(iii) *$\neg \theta_1$ is Π_n-conservative over $T + \theta_0$;*

(iv) *for both i, $\neg \theta_i$ is Π_n-conservative over T,*

and yet

(v) *θ_0 is not Σ_n-conservative over $T + \neg \theta_1$.*

Note that clauses (i)–(iii) say that the pair (θ_0, θ_1) fully complies with all clauses of Theorem Π w.r.t. the ordering $0 < 1$, clauses (i) and (iv) show that clauses (i) and (ii) of Theorem Σ do hold for $(\neg\theta_1, \neg\theta_0)$ and $1 < 0$, while clause (iii) of Theorem Σ fails for the latter chain in view of the present clause (v).

Details. For arithmetical sentences φ and ψ, write $\varphi \blacktriangleleft_n \psi$ if $T + \psi$ proves every Π_n consequence of $T + \varphi$ while $T + \varphi$ proves every Σ_n consequence of $T + \psi$. According to Lindström [8, Theorem 6.14], the relation $\varphi \blacktriangleleft_1 \psi$ coincides with faithful interpretability of $T + \varphi$ in $T + \psi$ when T is a pure r.e. extension of PA, which it presently is (Convention 2.1).

Consider the partially ordered set $D = (\{0, 1, 2, 3\}, \leq_D)$ with $\leq_D$ given by

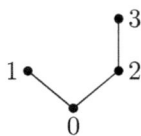

Suppose we succeed in procuring a collection of Σ_n sentences $(\sigma_a)_{a \in D}$ with the properties

(vi) $T \vdash \sigma_0$;

(vii) $T \vdash \neg(\sigma_1 \,\&\, \sigma_2)$;

(viii) $T \vdash \sigma_b \to \sigma_a$ for all $a \leq_D b$ in D; and

(ix) $\lambda_a \blacktriangleright_n \lambda_b$ for all $a \leq_D b$ in D, where

$$\lambda_c \equiv \sigma_c \,\&\, \bigwedge\nolimits_{D \ni d \not\leq_D c} \neg \sigma_d.$$

Note that $T \vdash \neg(\lambda_a \mathbin{\&} \lambda_b)$ for all $a \neq b$ in D. It is also straightforward to verify that $T \vdash \sigma_a \leftrightarrow \bigvee_{D \ni b \geq_D a} \lambda_b$ for all $a \in D$: for example,

$$T \vdash \sigma_2 \leftrightarrow \sigma_2 \vee \sigma_3 \qquad \text{(by property (viii))}$$
$$\leftrightarrow (\sigma_2 \mathbin{\&} \neg \sigma_3) \vee \sigma_3$$
$$\leftrightarrow (\sigma_2 \mathbin{\&} \neg\sigma_3 \mathbin{\&} \neg\sigma_1) \vee (\sigma_3 \mathbin{\&} \neg\sigma_1) \qquad \text{(by properties (vii) and (viii))}$$
$$\leftrightarrow \lambda_2 \vee \lambda_3.$$

In particular, in view of property (vi), we have $T \vdash \bigvee_{a \in D} \lambda_a$. Furthermore, all $T + \lambda_a$ are consistent — $T + \lambda_a \vdash \bot$ for any of $a \in D$ would imply the inconsistency of each $T + \lambda_b$ because $\lambda_a \blacktriangleleft_n \lambda_0 \blacktriangleright_n \lambda_b$ by property (ix).

Put

$$\theta_0 \equiv \neg\sigma_1 \mathbin{\&} \neg\sigma_3 \equiv \lambda_0 \vee \lambda_2 \quad \text{and} \quad \theta_1 \equiv \neg\sigma_1 \mathbin{\&} \neg\sigma_2 \equiv \lambda_0:$$

We claim that θ_0 and θ_1 are as required.

Clause (i) is clear.

(ii) Suppose $T + \theta_1 \equiv T + \lambda_0 \vdash \varsigma$ for a Σ_n sentence ς. Then $T + \lambda_a \vdash \varsigma$ for all $a \in D$ by property (ix), for $0 \leq_D a$ and $\lambda_0 \blacktriangleright_n \lambda_a$. Therefore $T \vdash \varsigma$, so that θ_1 and hence θ_0 are Σ_n-conservative over T.

(iii) Suppose π is Π_n and $T + \theta_0 + \neg\theta_1 \equiv T + \lambda_2 \vdash \pi$. Then $T + \lambda_0 \vdash \pi$ as $\lambda_0 \blacktriangleright_n \lambda_2$, so $T + \theta_0 \equiv T + \lambda_0 \vee \lambda_2 \vdash \pi$, which shows that $\neg\theta_1$ is Π_n-conservative over $T + \theta_0$.

(iv) To show that $T + \neg\theta_0$ is Π_n-conservative over T, suppose $T + \neg\theta_0 \equiv T + \lambda_1 \vee \lambda_3 \vdash \pi$ where π is Π_n. Then $T + \lambda_2 \vdash \pi$ and $T + \lambda_0 \vdash \pi$ because $\lambda_2 \blacktriangleright_n \lambda_3$ and $\lambda_0 \blacktriangleright_n \lambda_1$. Hence $T \equiv T + \lambda_0 \vee \cdots \vee \lambda_3 \vdash \pi$.

(v) Recall that σ_2 is Σ_n and observe that $T + \neg\theta_1 + \theta_0 \equiv T + \lambda_2 \vdash \sigma_2$. We also have $T + \lambda_1 \vdash \neg\sigma_2$ by the construction of λ_1. Since $T + \lambda_1$ is consistent, we cannot have $T + \lambda_1 \vdash \sigma_2$. Hence $T + \neg\theta_1 \equiv T + \lambda_1 \vee \lambda_3$ also fails to prove σ_2. Thus θ_0 is not Σ_n-conservative over $T - \neg\theta_1$.

Now all that remains is to produce the collection $(\sigma_a)_{a \in D}$ satisfying properties (vi)–(ix). This can be done by pre-existent means: The following construction is essentially the same as the one featured in the alternative proof of Theorem 5.3(a) in Lindström [8, p. 89] and Exercise 5.17 ibidem. We borrow it in the form which is closely identical to that of Construction 4.1 in Lindström & Shavrukov [10] — the only differences are that we drop that paper's restriction to $n = 1$ and the orientation of the partial order on D is opposite to the one in [10]. Another caution to the reader is that [10], unlike the present paper, applies the expression 'Γ-conservative over' to theories that are not necessarily extensions of one another.

CONSTRUCTION 4.2. With the help of the Fixed Point Lemma we shall define a Solovay function[2]

---
[2] By *Solovay functions* we mean a format for self-referential constructions rather than computable functions approximating prefix-free Kolmogorov complexity of the argument.

$H : \omega \to D$. The sentences $(\sigma_a)_{a \in D}$ and $(\lambda_a)_{a \in D}$ then read as follows:

$$\sigma_a \equiv \exists x\, H(x) \geq_D a \quad \text{and} \quad \lambda_a \equiv \sigma_a \,\&\, \bigwedge\nolimits_{D \ni b \not\leq_D a} \neg \sigma_b.$$

With this definition, the T-provable monotonicity property (viii) of $(\sigma_a)_{a \in D}$ w.r.t. $\leq_D$ is immediate.

Let us briefly recall some details of Solovay function manufacture. Suppose there is a formula $H'(e; x) = z$ that we like to think of as defining a function. Suppose further that we have a formula $\lambda'(e, y)$ with just the free variables shown that has one or more distinguished occurrences of subformulas of the form $H'(e; t_x) = t_z$, where the displayed occurrences of e in these subformulas account for all occurrences of e in $\lambda'(e, y)$, and t_x and t_z are some terms — these may vary per occurrence. One then obtains the formula $\lambda(y)$ as a fixed point

$$T \vdash \lambda(y) \leftrightarrow \lambda'(\ulcorner \lambda(y) \urcorner, y),$$

where $\ulcorner \lambda(y) \urcorner$ is the gödelnumber of the formula $\lambda(y)$ with y free. (In our present instance, we have $\lambda_a \equiv \lambda(a)$ for $a \in D$.) The traditional shortcut consists in defining the single-argument function $H(x) = H'(\ulcorner \lambda(y) \urcorner; x)$ while pretending that we have access to the gödelnumber of $\lambda(y)$. The two-argument function $H'(e; x)$ remains behind the scenes of the narration together with the formula $\lambda'(e, y)$ on this and one other similar occasion in Construction 5.5.

Where ς is a Σ_n formula, we write $x : \varsigma$ for the formula saying that x upper-bounds witnesses to all leading existential quantifiers of ς, so that $T \vdash \varsigma \leftrightarrow \exists x\, x : \varsigma$. For $n = 1$, the formula $x : \varsigma$ is Δ_0, and it is Π_{n-1} in the presence of $\mathrm{B}\Sigma_{n-1}$ when $n > 1$. With the help of an appropriate truthdefinition (Hájek & Pudlák [5, I.1(d)]), the ς in $x : \varsigma$ can be treated as a variable over Σ_n sentences. The usual Σ_1 provability predicate for T is denoted by $\Box$. The formula $x : \Box \varphi$, which we take to imply that x exceeds the gödelnumber of φ, is then Δ_0.

Let the notation $a \lessdot_D b$ express that b is an immediate succesor of a in D. For $a \lessdot_D b$ in D, consider the Δ_n-in-$\mathrm{I}\Sigma_n$ condition

$$\uparrow_n(a, b, x) \quad \equiv \quad \text{there is a } \Pi_n \text{ sentence } \pi \text{ and } m \leq x \text{ such that } m : \Box(\lambda_b \to \pi) \text{ and } x : \neg\pi,$$
$$\text{and for all } \Sigma_n \text{ sentences } \varsigma \text{ such that } m : \Box(\lambda_a \to \varsigma) \text{ one has } x : \varsigma.$$

Put $H(0) = 0$ — this immediately validates property (vi). Given $H(x)$, we put $H(x+1) = b$ if $b \gtrdot_D a$ and $\uparrow_n(H(x), b, x)$ (if there is more than one such $b \in D$, select any one of those), and let $H(x+1) = H(x)$ if no such $b \in D$ exists.

Note that the T-provably total function H is Δ_n in T. Since H is T-provably $\leq_D$-monotone, it is clear that $T \vdash \lambda_a \leftrightarrow a = \lim H$ for all $a \in D$. Provable monotonicity of H also implies that T refutes $\exists x\, H(x) \geq_D 1 \,\&\, \exists x\, H(x) \geq_D 2$, so that property (vii) holds.

As for property (ix), observe that to show $a \leq_D b \Rightarrow \lambda_a \blacktriangleright_n \lambda_b$ for all $a, b \in D$, it suffices to show this for pairs with $a \lessdot_D b$ because the relation $\blacktriangleleft_n$ is reflexive and transitive, and D is finite.

The next lemma together with its proof reproduces Lemma 4.3 from Lindström & Shavrukov [10] with Σ_n/Π_n substituted for Σ_1/Π_1. To facilitate the proof, we recall a well-known fact:

FACT 4.3 (Small Reflection — see Hájek & Pudlák [5, Lemma III.4.40]). *$T \vdash m : \Box \varphi \to \varphi$ for all $m \in \omega$ and all sentences φ.* ⊣

LEMMA 4.4. *If* $a <_D b$ *then* $\lambda_a \blacktriangleright_n \lambda_b$.

Proof. First we show that $T \vdash \lambda_a \to \varsigma$ implies $T \vdash \lambda_b \to \varsigma$ for Σ_n sentences ς: Assume $T \vdash \lambda_a \to \varsigma$. Then $m' : \Box(\lambda_a \to \varsigma)$ for some $m' \in \omega$. The following argument takes place in T:

Towards contradiction, suppose λ_b and $\neg\varsigma$. Since D is forest-like, the only way for H to get to b is directly from a, so $\uparrow_n(a, b, x)$ holds for some x. In particular, there is $m \leq x$ with $m : \Box(\lambda_b \to \pi)$ and $x : \neg\pi$ where π is Π_n. As $m' : \Box(\lambda_a \to \varsigma)$ and $\neg\varsigma$, we must have $m < m'$. Thus $m' : \Box(\lambda_b \to \pi)$, λ_b and $\neg\pi$. But this contradicts Small Reflection.

Thus $T \vdash \lambda_b \to \varsigma$.

Up next, $T \vdash \lambda_b \to \pi \Rightarrow T \vdash \lambda_c \to \pi$ for Π_n sentences π: If $T \vdash \lambda_b \to \pi$ then there is some m such that $m : \Box(\lambda_b \to \pi)$. Argue in T:

Assume λ_a and $\neg\pi$. Choose x' so that $x' : \neg\pi$ and $H(x) = a$ for all $x \geq x'$. Suppose a Σ_n sentence ς is such that $m : \Box(\lambda_a \to \varsigma)$. Then ς is true by Small Reflection. Hence there is an $x \geq x'$ such that $x : \varsigma$ for all Σ_n sentences ς with $m : \Box(\lambda_a \to \varsigma)$, for there are at most finitely many such ς. Thus condition $\uparrow_n(a, b, x)$ holds and H will have to forever leave a which contradicts λ_a.

Therefore $T \vdash \lambda_a \to \pi$ which completes the proof of the Lemma, as well as that of property (ix) and Example 4.1. ⊣

It should be clear that Construction 4.2 works together with Lemma 4.4 for any finite tree-like rooted poset D, not just the one we found useful.

5 Constructing $\xi(r)$

Solovay [13, section 5] produces a formula $\xi(r)$ instantiating Theorem Σ for $n = 1$ and $T = $ PA in a fixed-point-free manner. His proof involves induction within a model of PA, of which each step calls on the Arithmetized Completeness Theorem for 1st order logic. The arguments appear to generalize without difficulty to larger n and T.

Lindström [9, Theorem, case $\Gamma = \Sigma_n$] employs the Fixed Point Lemma in the form of a Solovay function to conduct an otherwise methodologically minimalist construction of the formula $\xi(r)$ with the properties from Theorem Σ. That construction is closer in form and spirit to our Construction 4.2 and the underlying alternative proof of Theorem 3.5(a) and Exercise 5.17 from Lindström [8], and offers good prospects for relaxing strength requirements on the base theory T.

The present section presents yet another, somewhat eclectic manner of obtaining $\xi(r)$. It features a Solovay function that in all likelihood does not converge towards any point within its official range. With Solovay's construction it shares the goal of arranging a 'well spread apart' sequence of witnesses ([13, 5.3]), which is achieved by explicit intervention of a Σ_n indicator for Σ_{n-1}-elementary initial segments satisfying T.

Ingredients and utensils

Where $\Box$ is a provability predicate, the Σ_n formula $\Box^{\Sigma_n}\varphi$ says that there is a true Σ_n sentence ς such that $\Box(\varsigma \to \varphi)$. The following facts may present a good reason to recall Convention 2.1.

FACT 5.1. *Let F be a finite subtheory of T, and let $\Box_F$ be the provability predicate that represents F as a finite list of axioms. Then*

$$T \vdash \forall x \left(\Box_F^{\Sigma_n} \varphi(x) \to \varphi(x) \right)$$

for each formula $\varphi(x)$ and each $n \in \omega$.

Comment. This is a form of uniform reflection for finite subtheories, and follows from e.g. Theorem I.4.33 in Hájek & Pudlák [5]. ⊣

FACT 5.2. *A sentence φ is Π_n-conservative over T if and only if for each finite subtheory F of T one has $T \vdash \Diamond_F^{\Sigma_n} \varphi$ with $\Box_F$ as in Fact 5.1.*

Comment. This is a hierarchical generalization of the well-known Orey–Hájek Lemma (see e.g. Hájek & Pudlák [5, Theorems III.3.45 and III.3.46]) with the connection to provably Π_n-faithful interpretability (as in Guaspari [4, Theorem 3.4]) omitted. It essentially tells us that the set

$$\{ \Diamond_F^{\Sigma_n} \varphi \mid F \text{ a finite subtheory of } T \}$$

is cofinal among the Π_n consequences of φ modulo T. The adaptation of the proof from Π_1 to Π_n conservativity presents few challenges. ⊣

FACT 5.3 (A Σ_n indicator for Σ_{n-1}-elementary initial T-segments). *There exists a Σ_n formula $Y_n(x, y) = z$ such that*

(i) $I\Sigma_n \vdash \forall x, y\, \exists! z\, Y_n(x, y) = z$;

(ii) $I\Sigma_n \vdash \forall x, y, x', y'\, (x' \leq x \leq y \leq y' \to Y_n(x, y) \leq Y_n(x', y'))$;

(iii) *for any $\mathscr{M} \models T$ and any $a, b \in \mathscr{M}$ we have $\mathscr{M} \models Y_n(a, b) > \mathbb{N}$ iff there is $\mathscr{I} \preceq_{n-1}^e \mathscr{M}$ with $\mathscr{I} \models T$ and $a \in \mathscr{I} < b$;*

(iv) *for each $m \in \omega$ we have $T \vdash \forall x\, \exists y\, Y_n(x, y) > m$.*

Observe that since Y_n is $I\Sigma_n$-provably a function, the formula $Y_n(x, y) = z$, as well as $Y_n(x, y) > z$ and $Y_n(x, y) < z$, is in fact Δ_n in $I\Sigma_n$.

Reference. Clauses (i)–(iii) are Exercise 14.8 in Kaye [7]. Hint:

$$Y_n(a, b) = \min\Big(\{b\}$$
$$\cup \{x \mid \text{ for some } \Pi_{n-1} \text{ formula } \theta(\cdot, \cdot),\ x : \Box_T \forall u \exists v\, \theta(u, v) \text{ and } \forall w < b\, \neg\theta(a, w) \}\Big).$$

The proofs of clauses (i) and (ii) only require minimization applied to $\Delta_0(\Sigma_{n-1})$ formulas, so $I\Sigma_n$ is, if anything, too generous — see Hájek & Pudlák [5, Lemma I.2.14]. ($\Delta_0(\Sigma_k)$ is the class you get by wrapping Σ_k formulas into Boolean connectives and bounded quantifiers.)

Clause (iv) follows from the others as in the proof of Proposition 14.1 in Kaye [7]. ⊣

Our approach will require stratification of the theory T into a chain of finite subtheories.

LEMMA 5.4. *There is an ascending chain of finite theories $(T_m)_{m \in \omega}$ together with an $I\Sigma_1$-provably computable function of m producing the code for a finite axiom set for T_m (with the matching provability predicates $\Box_m$ for T_m using these codes to represent the axiom set) so that:*

(i) $T = \bigcup_{m \in \omega} T_m$;

(ii) $T_0 \vdash I\Sigma_n$;

(iii) *for each $m \in \omega$ and each Π_n formula θ, $T_{m+1} \vdash \forall x \, (\Box_m^{\Sigma_n} \theta(x) \to \theta(x))$;*

(iv) *for each $m \in \omega$, $T_m \vdash \forall x \exists y \, Y_n(x, y) > m$*

Comment. The required sequence $(T_m)_{m \in \omega}$ is easily arranged with the help of Facts 5.1 and 5.3(iv).

⊣

Cooking Directions

We shall find it technically convenient to restrict the rational variable r in the target formula $\xi(r)$ to the interval $[0, 1]$ of $\mathbb{Q}$ which we denote $[0, 1]_\mathbb{Q}$. Thus the variables p, q, r in what follows are understood to range over that interval. In order to instantiate Theorem Σ, we shall remove the endpoints 0 and 1 leaving us with a family indexed by elements of the interval $(0, 1)$ of $\mathbb{Q}$. Since $\mathbb{Q}$ is effectively $I\Sigma_1$-provably order-isomorphic to $(0, 1)_\mathbb{Q}$, our deviation is of no consequence to Theorem Σ.

CONSTRUCTION 5.5. We construct a Δ_n Solovay function $F : \omega \to [0, 1]_\mathbb{Q}$ together with an auxiliary function $J : \omega \to \omega$ via the Fixed Point Lemma. The role of $J(x)$ is to remember the last time before x that F changed its value. The formula $\xi(r)$, where r is understood to range over $[0, 1]_\mathbb{Q}$, is defined by

$$\xi(r) \equiv \exists x \, F(x) \geq r.$$

First, we introduce a useful abbreviation:

$$\Uparrow_n(x, k, p) \equiv x : \Box_k^{\Sigma_n} \neg \xi(p) \, \& \, F(x) < p \leq F(x) + \tfrac{1}{k+1} \, \& \, Y_n(J(x), x) > k.$$

Note that because of the condition $x : \Box_k^{\Sigma_n} \neg \xi(p)$, the integer code of any $p \in [0, 1]_\mathbb{Q}$ satisfying $\exists k \, \Uparrow_n(x, k, p)$ is bounded by (some very tame recursive function of) x, as is the minimal k, if any, satisfying $\exists p \, \Uparrow_n(x, k, p)$.

Put $F(0) = 0$, $J(0) = 0$, and

$$\bigl(F(x+1), J(x+1)\bigr) = \begin{cases} (p, x+1) & \text{where } p \text{ is any element of } [0,1]_\mathbb{Q} \text{ such that } \Uparrow_n(x, k, p) \\ & \text{with } k \text{ being minimal satisfying } \exists q \, \Uparrow_n(x, k, q), \\ (F(x), J(x)) & \text{if no pair } (k, p) \text{ satisfies } \Uparrow_n(x, k, p). \end{cases}$$

The formula $\Uparrow_n(x, k, p)$ is Δ_n in $I\Sigma_n$ in the values $F(x)$ and $J(x)$ by Fact 5.3. Since $I\Sigma_n \vdash B\Sigma_n$, the class of formulas that are Δ_n in $I\Sigma_n$ is closed under Boolean connectives and bounded quantification (see Hájek & Pudlák [5, Theorem I.2.5(3)]). The condition connecting $F(x+1)$ and $J(x+1)$ to $F(x)$ and $J(x)$ being Δ_n in $I\Sigma_n$, the functions F and J must also be Δ_n in $I\Sigma_n$, as does the formula $\Uparrow_n(x, k, p)$. The formula $\xi(r)$ is then Σ_n.

The following Lemma is straightforward:

LEMMA 5.6. T_0 *proves that*

(a) $\forall p, q \, (p \leq q \to (\xi(q) \to \xi(p)))$, *and*

(b) $\xi(0)$.

⊣

Next we collect together some simple properties of F and J:

LEMMA 5.7. T_0 proves that for all x, y

(a) $y \leq x \to (F(y) \leq F(x) \, \& \, J(y) \leq J(x))$;
(b) $F(J(x)) = F(x) \, \& \, J(J(x)) = J(x) \leq x$;
(c) $0 < J(x) \to F(J(x) - 1) < F(J(x))$;
(d) The following are equivalent:
   (i) $F(x) < F(y)$;
   (ii) $x < J(y)$;
   (iii) $J(x) < J(y)$.

Proof. (a)–(c) follow by straightforward Σ_n induction on x. We turn to (d):
(i) $\to$ (ii) If $x \geq J(y)$, then $F(x) \geq F(J(y)) = F(y)$ by clauses (a) and (b).
(ii) $\to$ (iii) follows at once from (b).
(iii) $\to$ (i) Under the assumption $J(x) < J(y)$ one has

$$F(x) = F(J(x)) \leq F(J(y) - 1) < F(J(y)) = F(y)$$

by clauses (b), (a), (c), and (b) again. ⊣

LEMMA 5.8. *For each $k \in \omega$,*

$$T_k \vdash \forall x, m, q \, (\Uparrow_n(x, m, q) \to m \geq k).$$

Proof. Reason in T_k:
By the definition of $F(x+1)$, $\Uparrow_n(x, m, q)$ entails $F(x) < F(x+1) = r \in (0, 1]_\mathbb{Q}$ with $\Uparrow_n(x, \ell, r)$ holding true for some $\ell \leq m$. In particular, we have $x : \Box_\ell^{\Sigma_n} \neg \xi(r)$, hence $\Box_m^{\Sigma_n} \neg \xi(r)$. Seeing as $F(x+1) = r$ implies $\xi(r)$ in T_0, we conclude from Lemma 5.4(iii) that $m \geq k$. ⊣

LEMMA 5.9. *Let $k \in \omega$ and suppose $0 \leq p < q \leq 1$ with $q - p \leq \frac{1}{k+1}$. Then*

$$T_{k+1} \vdash \xi(p) \to \Diamond_k^{\Sigma_n} \xi(q).$$

Proof. Argue in T_{k+1}:
Suppose towards contradiction that $\xi(p)$ and $\Box_k^{\Sigma_n} \neg \xi(q)$, with $\neg \xi(q)$ following by Lemma 5.4(iii). Construct two numbers

$$a = \min\{\, x \mid x : \Box_k^{\Sigma_n} \neg \xi(q) \, \& \, F(x) \geq p \,\} \quad \text{and} \quad b = \min\{\, y \mid Y_n(J(a), y) > k \,\}.$$

with the help of Lemma 5.4(ii). Note that b exists by Lemma 5.4(iv). We distinguish two cases:

Case 1: $a \geq b$.
In this case we have $Y_n(J(a), a) \geq Y_n(J(a), b) > k$ by Fact 5.3(ii) and the construction of b, hence $\Uparrow_n(a, k, q)$ follows from $p \leq F(a) < q$ contradicting Lemma 5.8.

Case 2: $a < b$.

Suppose we had $J(b) > J(a)$. Then $J(b) > a$ by Lemma 5.7(d), and $F(J(b)) > F(J(b) - 1)$ by Lemma 5.7(c). This entails $\Uparrow_n(J(b) - 1, \ell, r)$ for some ℓ and r. Hence

$$\ell < Y_n\big(J(J(b) - 1), J(b) - 1\big) \leq Y_n(J(a), b - 1) \leq k$$

by Lemma 5.7(a) and (b), Fact 5.3(ii), and the minimality property of b. But $\ell < k$ together with $\Uparrow_n(J(b) - 1, \ell, r)$ contradict Lemma 5.8.

Thus $J(a) = J(b)$. Now $Y_n(J(b), b) = Y_n(J(a), b) > k$. Taking into account $p \leq F(a) \leq F(b) < q$, we deduce $\Uparrow_n(b, k, q)$ which once again contradicts Lemma 5.8. ⊣

LEMMA 5.10. $\xi(p)$ *is* Π_n-*conservative over* T *for each* $p \in [0,1]_\mathbb{Q}$.

Proof. In view of Lemma 5.6(a), it suffices to show that $\xi(1)$ is Π_n-conservative over T.

Fix an arbitrary integer $m > 0$. Since the harmonic series $\sum_{k=0}^{\infty} \frac{1}{k+1}$ diverges, we can select a finite sequence $0 = p_N < \cdots < p_0 = 1$ of rationals such that $p_k - p_{k+1} < \frac{1}{m+k+1}$ for all $k < N$. For this sequence one has

$$T_{m+k+1} \vdash \xi(p_{k+1}) \to \Diamond^{\Sigma_n}_{m+k}\xi(p_k)$$

by Lemma 5.9. It therefore follows by induction on k from 0 to $N - 1$ that

$$T_{m+N} \vdash \xi(p_N) \to \Diamond^{\Sigma_n}_{m+N-1} \cdots \Diamond^{\Sigma_n}_{m+1}\Diamond^{\Sigma_n}_{m}\xi(p_0).$$

Hence $T_{m+N} \vdash \xi(0) \to (\Diamond^{\Sigma_n}_m)^N \xi(1)$, which implies $T \vdash \xi(0) \to \Diamond^{\Sigma_n}_m\xi(1)$, for $\Box^{\Sigma_n}_k$ clearly obeys K4 — see e.g. Smoryński [12, Lemma 3.3.7].

Now recall that $m > 0$ is arbitrary. By Fact 5.2 and Lemma 5.6(b), the present Lemma follows. ⊣

LEMMA 5.11. $\neg\xi(q)$ *is* Σ_n-*conservative over* $T + \xi(p)$ *whenever* $p < q$ *in* $[0,1]_\mathbb{Q}$.

Proof. By Fact 2.3(b), it will suffice to show that each $\mathscr{M} \models T + \xi(p)$ has a Σ_{n-1}-elementary initial segment $\mathscr{I} \models T + \xi(p) + \neg\xi(q)$. We may as well assume $\mathscr{M} \models \xi(q)$.

In $\mathscr{M}$, define two numbers

$$c = \min\{\, x \mid F(x) > p \,\} \quad \text{and} \quad d = \max\{\, y \mid F(y) < q \,\}.$$

We have $F(d+1) \geq q > F(d)$. Hence $\Uparrow_n(d, \ell, F(d+1))$, where $\ell > \mathbb{N}$ by Lemma 5.8, which implies $F(d+1) - F(d) \leq \frac{1}{\ell+1} < \frac{1}{\mathbb{N}}$. Similarly, $F(c) - F(c-1) < \frac{1}{\mathbb{N}}$. Since $F(c-1) \leq p < q \leq F(d+1)$ and both p and q are standard, it follows that $F(c) < F(d)$. Therefore $J(d) > c$ by Lemma 5.7(d). Fact 5.3(ii) then allows to conclude

$$Y_n(c, d) \geq Y_n(J(d), d) > \ell > \mathbb{N}.$$

By Fact 5.3(iii), there is $\mathscr{I} \preceq^e_{n-1} \mathscr{M}$ with $c \in \mathscr{I} < d$ and $\mathscr{I} \models T$. Since formulas that are Δ_n in T are absolute between $\mathscr{M}$ and $\mathscr{I}$, we have $\mathscr{I} \models F(c) > p$ and $\mathscr{I} \models F(e) < q$ for all $e \in \mathscr{I}$. Thus $\mathscr{I} \models \xi(p)\ \&\ \neg\xi(q)$ as required. ⊣

Lemmas 5.6(a), 5.10, and 5.11 constitute a proof of Theorem Σ. ⊣

BIBLIOGRAPHY

[1] A. Arana. Arithmetical independence results using higher recursion theory. *Journal of Symbolic Logic* 69 (2004) 1–8.
[2] C. Bennet. Lindenbaum algebras and partial conservativity. *Proceedings of the American Mathematical Society* 97 (1986) 323–327.
[3] P. D'Aquino, J. F. Knight. Coding in $I\Delta_0$. *Nonstandard Models of Arithmetic and Set Theory.* (A. Enayat, R. Kossak, eds.) American Mathematical Society, (2004) 23–36.
[4] D. Guaspari. Partially conservative extensions of arithmetic. *Transactions of the American Mathematical Society* 254 (1979) 47–68.
[5] P. Hájek, P. Pudlák. *Metamathematics of First-Order Arithmetic.* Springer-Verlag (1993).
[6] D. Jensen, A. Ehrenfeucht. Some problem in elementary arithmetics. *Fundamenta mathematicae* 92 (1976) 223–245.
[7] R. Kaye. *Models of Peano Arithmetic.* Clarendon Press (1991).
[8] P. Lindström. *Aspects of Incompleteness*, 2nd ed. Association for Symbolic Logic / A K Peters (2003).
[9] P. Lindström. A theorem on partial conservativity in arithmetic. *Journal of Symbolic Logic* 76 (2011) 341–347.
[10] P. Lindström, V. Yu. Shavrukov. The $\forall\exists$ theory of Peano Σ_1 sentences. *Journal of Mathematical Logic* 8 (2008) 251–280.
[11] H. Simmons. Existentially closed models of basic number theory. *Logic Colloquium 76.* (R. O. Gandy, J. M. E. Hyland, eds.) North-Holland (1977) 325–369.
[12] C. Smoryński. *Self-Reference and Modal Logic.* Springer-Verlag (1985).
[13] R. M. Solovay. Branches of the E-tree. Manuscript (1987).

Checking Your History

A.S. TROELSTRA

This offering on the occasion of Albert Visser's retirement is rather far removed from Albert's professional interests. But I know Albert as someone with an interest in history, a man of erudition and a scholar, and I trust he can appreciate this piece from that point of view.

The point I should like to make in this paper is that one should distrust other people's 'history' of a topic in principle. Yes, of course, it is practically impossible to check everything which is offered as the historical truth, but if you feel that it matters to your subject, make an attempt to get at the bottom of the matter.

When writing papers and books on mathematical subjects — for me a thing of the past now — I usually tried to add some historical notes. The formal systems known as Peano Arithmetic (PA) and Heyting Arithmetic (HA) often figured in my writings, and I soon discovered that Peano arithmetic did not occur in Peano's writings and was not invented by him. It took me somewhat longer to discover that Heyting Arithmetic as well did not occur as such in Heyting's writings. I found this out in 1978, when I was asked to write a historical introduction to Heyting's papers on the formalization of intuitionistic mathematics.

Turning after my retirement from mathematics to the history of biology, it did not come as a surprise that there too it is advisable to accept other people's historical 'facts' with some reserve, especially if the facts reported sound sensational or a bit too good to be true — my favourite term for this is 'highly coloured information'. I shall offer two illustrations of this.

1 The mysterious Quasdanovich

When I tried to compile a bibliography of natural history travel narratives, I went through R.J. Howgego's voluminous encyclopaedia of travellers to see whether there were any names which I had missed so far.

To my delight I discovered a certain Sigismund Mathias Quasdanovich, a Hungarian botanist who was professor of botany in Vienna in the last quarter of the eighteenth century. He had, apparently, travelled in South America and had also written on his travels. Howgego listed several of his works. Discovering a new author for inclusion is always interesting, and I had never heard of Quasdanovich before. Of course, the first thing was to see where I could find his writings. Checking digitalized library catalogues worldwide failed to uncover any location where his books were held — not even in Vienna. Jan-Frits Veldkamp, botanist at the herbarium of Naturalis in Leiden also got interested. He too did some checking and could find no trace of Quasdanovich in the literature. Even taking the fact into account that not all libraries have already succeeded in including the older literature in their digital catalogue, drawing such a complete blank seemed suspicious, and I came to the conclusion that Quasdanovich was probably a fictitious personage.

But how did this entry get into a serious work of reference such as Howgego's encyclopaedia?

A practical joke by Howgego seemed to me to be extremely unlikely.

A search on the internet soon revealed that the information on Quasdanovich almost literally derived from a lemma in *Appleton's Cyclopaedia of American Biography* (1887–1888). The lemma on Quasdanovich occurs in volume 3. Once again, how did this spurious lemma get into a serious encyclopaedia? A booby-trap for plagiarists, or a practical joke of the editor? Having got so far with my researches, I wrote to another friend, Michael Ohl of the Berliner Naturkunde Museum. He was as much intrigued as I was, and told the story to a colleague of the History Department of the Museum. Soon afterwards I received a copy of a paper by J.B. Dobson on spurious lemmas in *Appleton's Cyclopaedia*.

From this paper I learnt that from volume 3 onwards, *Appleton's Cyclopaedia* contains many spurious lemmas. Since the contributors to Appleton's Cyclopaedia did not sign their lemmas, one can only guess at the name of the person responsible for this. To guess at the motive is not so difficult: contributors were free to propose subjects for inclusion, and were paid by the amount of text they produced. Dobson produced a list of lemma's having been shown to be spurious up to the appearance of his article; the article on Quasdanovich is not listed there.

As this example shows, even if someone has proven the untrustworthiness of a source and has published this fact, the publication may be overlooked by later authors and so the misinformation is continued...

2   The penguins of Pierre Sonnerat

Let us now turn to my second example. When I came across the book by Pierre Sonnerat, *Voyage la Nouvelle-Guinée*, I consulted the internet for further information and came across a blog by the Australian zoologist Karl Shuker on 'Sonnerat's non-existent penguins (and Kookaburra) of New Guinea'. After Shuker had published on a spurious fact he had found in Sonnerat's writings — concerning the existence of a secretary bird in the Philippines — a colleague pointed out to him that there are more spurious facts in Sonnerat's book: Sonnerat claimed to have collected no less than three species of penguin in New Guinea. Shuker decided to investigate the matter further and came to the conclusions outlined below.

(1) Sonnerat claimed to have observed in New Guinea three species of penguin, as well as the Australian giant kingfisher *Dacelo novaeguineae* (synonym: *Dacelo gigas*), also known as the kookaburra or laughing jackass. Sonnerat gave pictures of all four, and the three penguin species can according to Shuker readily be identified from the pictures. Shuker identified the penguins as: (a) Sonnerat's *Manchot de la Nouvelle Guinée* = king penguin = *Aptenodytes patagonica*; (b) Sonnerat's *Manchot à collier de la Nouvelle Guinée* = emperor penguin = *Aptenodytes forsteri*; (c) Sonnerat's *Manchot Papua* = Gentoo penguin = *Pygoscelis papua*. Sonnerat's information is obviously false, since penguins do not occur in New Guinea, and the kookaburra occurs in Queensland, but not in New Guinea. Note that two of the scientific names mentioned, *Pygoscelis papua* and *Dacelo novaeguineae* seem to derive from Sonnerat's false information.

(2) According to Shuker, Sonnerat never actually visited New Guinea. His expedition was a complete fiction and was publicly exposed during his life.

Le Manchot papou

(3) Where did Sonnerat get his pictures of three penguin species? Shuker offers the following explanation, which he derived from a book by the Australian ornithologist Peggy Olsen: when Sonnerat visited Cape Town in 1770 he met with Sir Joseph Banks, who had accompanied Captain Cook on his first circumnavigation as a naturalist, and who gave him some bird specimens with instructions to deliver them to Philibert Commerson, a French naturalist working on Mauritius. Sonnerat duly transmitted the specimens to Commerson, who had his draughtsman Jossigny make drawings of them. But when Commerson died in 1773, Sonnerat appropriated the drawings and published them afterwards under his own name.

For his conclusions Shuker gives only two sources: the book by Peggy Olsen and Sonnerat's own book. Olsen, dealing with Australian ornithology, is primarily interested in the case of the kookaburra and mentions only one penguin; this is corrected by Shuker and changed to three penguins.

Initially I took Shuker's story as 'the truth about Sonnerat'. But my friend Michael Ohl was intrigued by the whole story and wanted to see more references. We both began to search for more sources discussing Sonnerat's claims. It soon transpired that Shuker's story was incomplete and not completely correct.

Le Manchot à Collier de la Nouvelle Guinée.

The first doubtful point is Shuker's identification of the collared penguin, the *Manchot collier de la Nouvelle Guinée*. On the basis of Sonnerat's picture and description J.R. Forster described the species in 1781 under the name *Aptenodytes torquata*. No one, apart from Sonnerat, has ever claimed to have seen this species. The truth is that we do not really know what species this picture represents; but certainly not the emperor penguin, which has a curved beak.

In his paper from 1900 the French ornithologist Émile Oustalet states that he is unable to allocate this picture to any of the known species of penguin.

The second point which needs correction is Sonnerat's voyage. Pierre Poivre, governor of Isle de France (Mauritius) and related to Sonnerat, had a pet project: breaking the Dutch monopoly in the spice trade by introducing clove and nutmeg trees in the French possessions, notably on Isle de France. A first expedition from 1769 to 1770 had been only partially successful. A second expedition with two ships, 'le Nécessaire' under M. Cordé and 'l'Isle de France' under the Chevalier de Coëtivy, was sent out; the expedition as a whole was directed by Mathieu Provost and lasted from June 29, 1771 to June 4, 1772. Sonnerat participated in this expedition. The ships sailed until they had reached two small islands, Pulau Gebe and Pulau Fau somewhat to the east of the Bird's head Peninsula (Dutch: Vogelkop, Indonesian: Kepala Burung) of New Guinea. More precisely, Pulau Fau lies about halfway between Halmahera and Waigeo. Provost knew from the previous expedition that in this area, not under Dutch control, he could acquire the seedlings of clove and nutmeg and this time he was completely successful. Sonnerat stayed on Pulau Fau while provost hunted for seedlings. At Pulau Fau Sonnerat's narrative ends, and the book closes with a description of plants and animals said to occur on New Guinea. So the title of the book, claiming a voyage to New Guinea, stretched the truth only a little...

Le Manchot de la Nouvelle Guinée.

A minor correction to Shuker's blog is that Sonnerat did indeed meet Joseph Banks in the Cape Colony, but in 1771, not in 1770. Banks certainly handed Sonnerat some skins of birds, with the request to transmit them to Commerson on Mauritius, and we may assume that the skin of a kookaburra was one of them. But it is not plausible that Sonnerat also got the skins of penguins from Banks. Cook on his first circumnavigation did not touch places where the species depicted could have been acquired. The more likely explanation of the origin of the penguin specimens is offered by Oustalet and Lysaght. Commerson had accompanied Bougainville on his circumnavigation, and left Bougainville at Isle de France when the expedition was nearly over. One of Bougainville's ships had visited the Falklands, where the king penguin and the gentoo penguin breed, and it is quite likely that material of these two species had been acquired there.

Already Oustalet suggested that the penguins depicted in Sonnerat's book derive from materials collected on Bougainville's voyage, most probably on the Falklands, and Averil Lysaght motivated this in more detail in her paper from 1952. Georges Cuvier, in a study of Commerson and Sonnerat (1845), observed that many of the drawings in Sonnerat's two books, the *Voyage à la Nouvelle-Guinée* already mentioned and his *Voyage aux Indes Orientales et à la Chine* (1782) derive from drawings in the papers of Commerson at the Mus/'eum d'Histoire Naturelle in Paris. As to the drawings of penguins, Oustalet also consulted the papers of Commerson and noted that two of the three drawings of penguins there are signed by Jossigny; the third is in another hand, far less competently drawn and bears a legend in handwriting: 'Pingouin du Cap', that is to say *Spheniscus demersus*, in English known as the jackass penguin or black-footed penguin. Oustalet observes it is difficult to accept that name as correct. He omits saying that the third drawing was the original for the engraving of the dubious *Aptenodytes torquata*, though I suspect that is the case. However, I have not been able to inspect Commerson's papers.

So far, there has been a lot of talk about Sonnerat in this paper; now it is time to say something about his life.

Pierre Sonnerat was the godson of the botanist and colonial administrator Pierre Poivre (1719–1786), who was a nephew of Sonnerat's mother. He was born in Lyon on August 18, 1748, and died in Paris on March 31, 1814.

Sonnerat was trained as a designer in the Lyonese textile industry. He travelled to Mauritius (then called Isle de France) in 1769, in the service of the French marine as 'écrivain de vaisseau' (writer on a ship), where his relative Pierre Poivre was 'intendant' (governor). At Mauritius he got interested in natural history and had the opportunity to acquire some skills by contact with Philibert Commerson (1727–1773), the gifted naturalist who had accompanied Bougainville on the greater part of his circumnavigation. On the last leg of the voyage he had left Bougainville in order to investigate the natural history of the Mascarenes. This was a project of Poivre, and Poivre had provided Commerson with a skilled draughtsman, Paul Philippe Sauguin de Jossigny. On his arrival Sonnerat was set to work as an assistant to Jossigny.

From March to April 1771 Sonnerat visits the Cape, where he meets with Joseph Banks. Then follows the successful expedition to the Spice Islands already mentioned above. Sonnerat leaves Mauritius in October 1772, in the company of Pierre Poivre and his family; he is at the Cape from November 24, 1772 till February 24, 1773. While at the Cape he collects in the company of C.P. Thunberg. Commerson stays at Mauritius, very ill, and dies in March 1773. After his return to France Sonnerat is elected an associated member of the Academy of Sciences, Letters and Arts of Lyons, and in January 1774 he is elected correspondent (of Michel Adanson) of the Académie Royale des Sciences in Paris. Thanks to powerful protection he makes a remarkable promotion to

'Sous-Commissaire de la Marine' (deputy commissioner of the navy). July 1775 he returns to Mauritius, where the new intendant is inimical. Sonnerat obtains leave to proceed to Pondicherry and from there to Canton, in order to make observations and collect for the Cabinet du Roi (the King's cabinet of natural curiosities). In China he only sees a corner of Canton, a big disappointment. The journey lasts from 1775 to 1777. He returns to Pondicherry to work as Sous-Commissaire de la Marine. While in Pondicherry he collects material on Indian society, customs and religion.

From July to October 1778 Pondicherry is besieged by the British; after capitulation Sonnerat returns to Isle de France. In the period 1779–1780 he had the opportunity to explore the south-eastern coast of Madagascar. In 1781 he returns to France and is promoted to 'Commissaire des Colonies'. He marries in 1782 or 1783 and in 1782 appears his second book *Voyage aux Indes Orientales*. In July 1786 he returns to Pondicherry as Commissaire des Colonies. 1788–1789 he visits Sri Lanka twice.

From 1790 to 1793 he is 'Commandant' (head) of the French trading post at Yanam, at the mouth of the Godavari River, in the present Indian state Puducherry. In 1793 Yanam is captured by the British. Sonnerat is a prisoner of war from 1793 to 1813. Thanks to the good offices of Joseph Banks, he is repatriated in 1813 and December of the same year he meets Banks in London. He dies the next year in Paris.

While held prisoner at Yanam, he worked on a sequel to his *Voyage aux Indes Orientales*, which was never published; the manuscript seem to have been lost. There has been a lot of confusion as to Sonnerat's biography, due to fragmentary records and the unreliable information concerning his career which Sonnerat himself provided. Madeleine Ly-Tio-Fane, a historian from Mauritius, wrote a detailed and carefully researched biography of Sonnerat, in which she corrected many false statements made concerning Sonnerat. This biography is difficult to get hold of, since it was published on Mauritius by the author (Michael found one copy in a library in Frankfurt; the only copy in the Netherlands which I could locate was in Leiden). Some important facts about Sonnerat not yet known to Ly-Tio-Fane were published on the internet by Morel in 2013.

It has been suggested that Sonnerat 'stole' material for his own book from Commerson while at Mauritius. But Sonnerat left Mauritius before Commerson died; hence it is more likely that he gained access to Commerson's papers in Paris, after they had been sent to Paris on Commerson's death. Ly-Tio-Fane emphasizes that in the eighteenth century natural history research was seen as a joint undertaking. That may be true, but Sonnerat does not show many scruples in using other people's material without informing us about his sources.

Sonnerat liked to stretch the truth: after having described in his *Voyage la Nouvelle Guinée* at some length the famous sea coconut of the Seychelles, *Lodoicea maldivica,* he states: 'the tree which I just described, seems to be a female individual. I have not encountered any other [than female individuals], nor have others who like me have travelled in these islands, where I was in July, which undoubtedly is the time of perfect maturity of the fruit...' (p.9). There is a strong suggestion in this passage that he has actually visited the islands of the Seychelles and seen many sea coconuts in the wild; but as we know from the ship's log, he only passed by!

References

Appleton's Cyclopaedia of American Biography. 6 volumes. J.G. Crandall and J. Fiske (editors) 1887–1888. D. Appleton & Co., New York. The first three volumes are dated 1887, volumes 4–6 are dated 1888.

Cuvier, G. 1845. *Histoire des Sciences Naturelles.* Fortin, Masson & Co, Paris. Vol.5, p.93–97.

Dobson, J.B. 1993. The spurious articles in Appleton's Cyclopaedia of American Biography — Some new discoveries and considerations, *Biography* 16 (1993), p.388–408.

Forster, J.R. 1781. Historia Aptenodytae generis avium orbi australi proprii, *Commentationes Societatis Regiae Scientiarum Gottingensis* 3, p.121–148.

Howgego R.J. 2003. *Encyclopaedia of exploration to 1800: a comprehensive reference guide to the history of and literature of exploration, travel and colonization from the earliest times to the year 1800.* Hordern House Rare Book, Potts Point, Australia, p.868–869.

Lysaght, A. 1952. Manchots de l'Antarctique en Nouvelle-Guinée, *L'Oiseau et la Revue Française d'Ornithologie* 22, p.120–124.

Lysaght, A. 1956. Why did Sonnerat record the Kookaburra, *Dacelo gigas* (Boddaert) from New Guinea? *The Emu* 56, p.224–225.

Ly-Tio-Fane, M. 1976, *Pierre Sonnerat 1748–1814. An Account of his Life and Work.* Printed for the author, Cassis, Mauritius.

Morel, J.P. 2013. Eléments biographiques sur Pierre Sonnerat. Prémières années sous le regard de Pierre Poivre, file dating from October 2013 on http://www.pierre-poivre.fr/Bio-Sonnerat.pdf, accessed 11-02-2015. This text contains some important corrections to the facts stated in Ly-Tio-Fane 1976.

Nouvelle Biographie Générale, lemma 'Sonnerat, (Pierre)' in vol. 44 (1865), col.179–180.

Olsen, P. 2001. *Feather and Brush: Three Centuries of Australian Bird Art.* CSIRO Publishing, Clayton (Victoria), Australia.

Oustalet, É. 1900. Note sur une particularité de conformation de la patte chez les manchots, *Bulletin du Muséum d'Histoire naturelle* 6, p.218–222.

Shuker, K. 2013. Sonnerat's non-existent penguins (and Kookaburra) of New Guinea. Blog *ShukerNature*, http://www.karlshuker.blogspot.com[1], accessed 02-12-2014.

Sonnerat, P. 1776. *Voyage à la nouvelle Guinée: dans lequel on trouve la description des Lieux, des observations physiques et morales, & des détails relatifs à l'histoire naturelle dans le regne animal & le regne végétal.* Ruault, Paris. There exist translations into German (Leipzig 1777) and English (London 1780, 1781).

Sonnerat, P. 1782. *Voyage aux Indes Orientales et à la Chine: fait par ordre du roi, depuis 1774 jusqu'en 1781: dans lequel on traite des moeurs, de la religion, des sciences & des arts des Indiens, des Chinois, des Pégouins & des Madégasses; suivi d'observations sur le cap de Bonne-Espérance, les isles de France & de Bourbon, les Maldives, Ceylan, Malacca, les Philippines & les Moluques, & recherches sur l'histoire naturelle de ces pays.* L'auteur, Froulé, Nyon and Barrois, Paris. There exist translations into German (1783, Zürich) and English (1788–1789, Calcutta).

---

[1] http://www.karlshuker.blogspot.co.uk/2013/05/sonnerats-non-existent-penguins-and.html

On the Diameter of Lascar Strong Types after Ludomir Newelski

DOMENICO ZAMBELLA

Abstract

We present a theorem of mathematical logic which only assumes the notions of structure, elementary equivalence, and compactness (saturation).

In [New03] Newelski proved that type-definable Lascar strong types have finite diameter. This exposition is based on the proof in [Pel08] up to a minor difference: the notion of *weak c-free* of [NP06] is replaced with the notion of *non-drifting* that is introduced here.

1 Introduction

Few recent results in mathematical logic have a statement that is accessible to logicians outside a specific area. One of them is the theorem on the diameter of Lascar strong types. The theorem concerns a graph that can naturally be defined in any infinite structure (yes, even in models of fragments of arithmetic!).

The problem can be presented in different ways that are equivalent. We choose the one that requires the fewer prerequisites. For a Galois-theoretical perspective, close to Lascar's original approach [Las82], we refer the reader to e.g. [Pel08]. We assume the reader knows what a saturated model is and we fix one. This is denoted by $\mathcal{U}$ and will be our universe for the rest of the paper. We denote its cardinality by κ which we assume to be uncountable and larger than the cardinality of the language. We also fix a set $A \subseteq \mathcal{U}$ of small cardinality, where **small** means $< \kappa$. There would be no loss of generality in assuming $A = \varnothing$. Indeed, A is fixed throughout the following so it could be absorbed in the language and forgotten about. However, we display it all along. We denote by $L(A)$ the set of formulas with parameters in A. By $|L(A)|$ we denote the cardinality of the set of sentences in $L(A)$. This cardinality does not play a role in the proof and assuming $|L(A)| = \omega$ may help on the first reading.

Let z be a tuple of variables of ordinal **length** $|z| < \kappa$. Though the theorem is also interesting for infinite tuples, the length of z does not play any role in the proof. Again, for a first reading one can assume z is a single variable. If $a, b \in \mathcal{U}^{|z|}$ we write $a \equiv_A b$ if $\varphi(a) \leftrightarrow \varphi(b)$ holds (in $\mathcal{U}$) for every $\varphi(z) \in L(A)$. In words we say that a and b have the same type over A.

A **definable set** is a set of the form $\varphi(\mathcal{U}) = \{a \in \mathcal{U}^{|z|} : \varphi(a)\}$ for some formula $\varphi(z) \in L(\mathcal{U})$. A **type** is a set of formulas $p(z) \subseteq L(B)$ for some $B \subseteq \mathcal{U}$ of small cardinality. A **type-definable set** is a set of the form $p(\mathcal{U})$, that is, the intersection of $\varphi(\mathcal{U})$ for $\varphi(z) \in p(z)$.

It may be useful (though not essential) to interpret this in topological terms. The sets $\varphi(\mathcal{U}) = \{a \in \mathcal{U}^{|z|} : \varphi(a)\}$ for $\varphi(z) \in L(A)$ form a base for a topology. This topology is zero-dimensional and it is compact because $\mathcal{U}$ is saturated. It is never T_0 as any pair of tuples $a \equiv_A b$ have exactly the same neighborhoods, such a pair exists for cardinality reasons. However it is immediate that the topology induced on the quotient $\mathcal{U}^{|z|}/\equiv_A$ is Hausdorff (this is the so-called *Kolmogorov quotient*). In this topology the closed sets are those of the form $p(\mathcal{U})$ where $p(z) \subseteq L(A)$ is any type.

In what follows, by **model** we understand an elementary substructure of $\mathcal{U}$ of small cardinality. The **Lascar graph over** A has $\mathcal{U}^{|z|}$ as the set of vertices and an edge between all pairs of vertices $a, b \in \mathcal{U}^{|z|}$ such that $a \equiv_M b$ for some model M containing A. We write $d_A(a, b)$ for the distance between a and b in the Lascar graph over A. Let us spell this out: $d_A(a, b) \leq n$ if there is a sequence $a_0, \ldots, a_n$ such that $a = a_0$, $b = a_n$, and $a_i \equiv_{M_i} a_{i+1}$ for some models M_i containing A. We write $d_A(a, b) < \infty$ if a and b are in the same connected component of the Lascar graph over A.

DEFINITION 1.1. *For every* $a \in \mathcal{U}^{|z|}$

1. $a \stackrel{L}{\equiv}_A b$ *if* $d_A(a, b) < \infty$;
2. $\mathscr{L}(a/A) = \{b : a \stackrel{L}{\equiv}_A b\}$.

We call $\mathscr{L}(a/A)$ *the Lascar strong type of a over A. If $a \stackrel{L}{\equiv}_A b$ we say that a and b have the same Lascar strong type over A.*

We are ready to state Newelski's theorem which we prove in the next section.

THEOREM 1.2. *For every* $a \in \mathcal{U}^{|z|}$ *the following are equivalent*

1. $\mathscr{L}(a/A)$ *is type-definable;*
2. $\mathscr{L}(a/A) = \{c : d_A(a, c) < n\}$ *for some* $n < \omega$.

Newelski's original proof has been simplified over the years. Most proofs have a definite topological dynamics flavor (the liaison with topological dynamics was clarified in [New09]). Below we give a streamlined version of the proof in [Pel08] (see also [Cas11, Theorem 9.22]).

More recent contributions to the subject have investigated the descriptive set theoretic complexity of the relation of having the same Lascar strong types. This is beyond the scope of this short note so we refer the interested reader to [KPS13], [KMS14] and [KM14].

It is interesting to note that if $\mathscr{L}(a/A)$ is type-definable for every $a \in \mathcal{U}^{|z|}$ then the equivalence relation $\stackrel{L}{\equiv}_A$ is also type-definable. This might be surprising at first, so we sketch a proof below (not required for the main theorem).

The equivalence relation $\stackrel{L}{\equiv}_A$ is **invariant** over A, that is, invariant over automorphisms that fix A. Its equivalence classes, are **Lascar invariant** over A, that is, invariant over automorphisms that fix some model containing A. There are at most $2^{2^{|L(A)|}}$ sets that are Lascar invariant over A. Then $\stackrel{L}{\equiv}_A$ is a **bounded** equivalence relation, that is, it has $< \kappa$ equivalence classes. It is not difficult to verify that $\stackrel{L}{\equiv}_A$ is the finest bounded equivalence relation invariant over A.

Let $e_A(x, z) \subseteq L(A)$ be the union of all types over A that define a bounded equivalence relation. It suffices to prove that if $\mathscr{L}(a/A)$ is type-definable then $\mathscr{L}(a/A) = e_A(\mathcal{U}, a)$. In fact, if this holds for every $a \in \mathcal{U}^{|z|}$, then $e_A(x, z)$ defines $\stackrel{L}{\equiv}_A$.

If $\mathscr{L}(a/A)$ is type-definable, by Theorem 1.2 and Proposition 2.1 it is defined by the type $d_A(a, z) < n$ for some n. The same is true for every $b \equiv_A a$. Hence $d_A(x, z) < n$ defines an equivalence relation whose restriction to $\mathscr{O}(a/A) = \{b : b \equiv_A a\}$ has boundedly many classes. By a well-known fact (see for example [Sim15, Proposition 5.11]), there is a bounded equivalence relation type-definable over A that coincides with $d_A(x, z) < n$ on $\mathcal{O}(a/A)$. Then $e_A(\mathcal{U}, a) \subseteq \mathscr{L}(a/A)$ follows. The converse inclusion is trivial.

In his seminal paper [Las82] Lascar asked for (not literally but in an equivalent way) examples where the relation of having the same Lascar strong type is not type-definable. By the theorem above this is equivalent to asking for structures where the diameter of a connected component of the Lascar graph is infinite. In tame structures, like stable and simple ones, the diameter is always finite. The first example with infinite diameter was constructed by Ziegler [CLPZ01] and later more natural examples were found [CP12].

One may ask the same question for models of first-order arithmetic. Models of full induction have definable Skolem functions, hence the connected components of the Lascar graph have diameter ≤ 1 which means that the Lascar strong types are trivial, i.e. they are just types. However, for weaker fragments of arithmetic the question is worth asking and may have non trivial answers.

2 Lascar strong automorphisms

It may not be immediately obvious that the relation $d_A(z, y) \leq n$ is type-definable. From this the easy direction of the main theorem follows.

PROPOSITION 2.1. *For every $n < \omega$ there is a type $p_n(z, y) \subseteq L(A)$ equivalent to $d_A(z, y) \leq n$.*

Proof. In a saturated structure types are closed under existential quantification, therefore it suffices to prove the proposition with $n = 1$. Let $\lambda = |L(A)|$ and let $w = \langle w_i : i < \lambda \rangle$ be a tuple of distinct variables. Then $p_1(z, y) = \exists w\, p(w, z, y)$ where

$$p(w, z, y) = q(w) \cup \left\{ \varphi(z, w) \leftrightarrow \varphi(y, w) : \varphi(z, w) \in L(A) \right\}$$

and $q(w) \subseteq L(A)$ is a consistent type with the property that all its realizations enumerate a model containing A.

Now we only need to prove that such a type exists. Let $\langle \psi_i(x, w_{\restriction i}) : i < \lambda \rangle$ be an enumeration of the formulas in $L_{x,w}(A)$, where x is a single variable. Let

$$q(w) = \{ \exists x\, \psi_i(x, w_{\restriction i}) \to \psi_i(w_i, w_{\restriction i}) : i < \lambda \}.$$

Any realization of $q(w)$ satisfy the Tarski-Vaught test therefore it enumerates a model containing A. Vice versa it is clear that we can realize $q(w)$ in any model containing A. ⊣

We write **Aut**$(\mathcal{U}/A)$ for the set of automorphisms of $\mathcal{U}$ that fix A. We write **Autf**$(\mathcal{U}/A)$ for the subgroup of Aut$(\mathcal{U}/A)$ generated by the automorphisms that fix some model M containing A. The "f" in the symbol stands for *fort*, the French for *strong*. It is immediate to verify that Autf$(\mathcal{U}/A)$ is a normal subgroup of Aut$(\mathcal{U}/A)$.

Recall that saturated models are homogeneous, hence any $a \equiv_B b$ are conjugated over B, that is, there is an $f \in \text{Aut}(\mathcal{U}/B)$ such that $fa = b$. Then it is easy to verify that $a \equiv^L_A b$ if and only if $fa = b$ for some $f \in \text{Autf}(\mathcal{U}/A)$.

The following notions apply generally to any group G acting on some set X and and to any set $\mathscr{D} \subseteq X$. Below we always have $G = \text{Autf}(\mathcal{U}/A)$ and $X = \mathcal{U}^{|z|}$. We say that $\mathscr{D}$ **is drifting** if for every finitely many $f_1, \ldots, f_n \in G$ there is a $g \in G$ such that $g[\mathscr{D}]$ is disjoint from all the $f_i[\mathscr{D}]$. We say that $\mathscr{D}$ **is quasi-invariant** if for every finitely many $f_1, \ldots, f_n \in G$ the sets $f_i[\mathscr{D}]$ have non-empty intersection. Note parenthetically that $\mathscr{D}$ is quasi-invariant if and only if $\neg \mathscr{D} = X \smallsetminus \mathscr{D}$ is not c-free in the sense of [Pel08] or not generic in the sense of [KMS14]. We say that a formula or a type is drifting or quasi-invariant if the set it defines is.

The union of drifting sets need not be drifting. However, the following lemma says it cannot be quasi-invariant.

LEMMA 2.2. *The union of finitely many drifting sets in not quasi-invariant.*

Proof. It is convenient to prove an apparently more general claim. If $\mathscr{D}_1, \ldots, \mathscr{D}_n$ are all drifting and $\mathscr{L}$ is such that for some finite $F \subseteq G$

♯ $$\mathscr{L} \subseteq \bigcup_{f \in F} f[\mathscr{D}_1 \cup \cdots \cup \mathscr{D}_n],$$

then $\mathscr{L}$ is not quasi-invariant. (The statement is slightly awkward since a superset of a quasi-invariant set must be quasi-invariant.)

The claim is vacuously true for $n = 0$. Let n be positive, let $\mathscr{C} = \mathscr{D}_1 \cup \cdots \cup \mathscr{D}_{n-1}$, and assume the claim holds for $n - 1$. Since $\mathscr{D}_n$ is drifting there is a $g \in G$ such that $g[\mathscr{D}_n]$ is disjoint from $f[\mathscr{D}_n]$ for every $f \in F$, which implies that

$$\mathscr{L} \cap g[\mathscr{D}_n] \subseteq \bigcup_{f \in F} f[\mathscr{C}].$$

Hence for every $h \in G$ there holds

$$hg^{-1}[\mathscr{L}] \cap h[\mathscr{D}_n] \subseteq \bigcup_{f \in F} hg^{-1} f[\mathscr{C}].$$

Rewriting ♯ as

$$\mathscr{L} \subseteq \bigcup_{f \in F} f[\mathscr{C}] \cup \bigcup_{h \in F} f[\mathscr{D}_n],$$

we observe that

$$\mathscr{L} \cap \bigcap_{h \in F} hg^{-1}[\mathscr{L}] \subseteq \bigcup_{f \in F} f[\mathscr{C}] \cup \bigcup_{f \in F} hg^{-1} f[\mathscr{C}].$$

By the induction hypothesis, the r.h.s. cannot be quasi invariant. Hence neither is $\mathscr{L}$, proving the claim and with it the lemma. ⊣

The following is a consequence of Baire's category theorem. We sketch a proof for the convenience of the reader.

LEMMA 2.3. *Let $p(x) \subseteq L(B)$ and $p_n(x) \subseteq L(A)$, for $n < \omega$, be consistent types such that*

1. $$p(x) \to \bigvee_{n < \omega} p_n(x)$$

Then there is an $n < \omega$ and a formula $\varphi(x) \in L(A)$ consistent with $p(x)$ such that

2. $$p(x) \wedge \varphi(x) \to p_n(x)$$

Proof. Negate 2 and choose inductively for every $n < \omega$ a formula $\psi_n(x) \in p_n(x)$ such that $p(x) \wedge \neg\psi_0(x) \wedge \cdots \wedge \neg\psi_n(x)$ is consistent. By compactness, we contradict 1. ⊣

Finally we can prove the Theorem 1.2 which we restate for convenience.

THEOREM 2.2. *For every $a \in \mathscr{U}^{|z|}$ the following are equivalent*

1. $\mathscr{L}(a/A)$ *is type-definable;*

2. $\mathscr{L}(a/A) = \{c : d_A(a,c) < n\}$ for some $n < \omega$.

Proof. Implications 2⇒1 holds by Proposition 2.1. We prove 1⇒2. Suppose $\mathscr{L}(a/A)$ is type-definable, say by the type $l(z)$. Let $p(z,y)$ be some consistent type (to be defined below) such that and $p(z,y) \to l(z) \wedge l(y)$. Then, in particular

$$p(z,y) \to \bigvee_{n<\omega} d_A(z,y) < n.$$

By Proposition 2.1 and Lemma 2.3, there is some $n < \omega$ and some $\varphi(z,y) \in L(A)$ consistent with $p(z,y)$ such that

♯1 $\qquad p(z,y) \wedge \varphi(z,y) \to d_A(z,y) < n.$

Below we define $p(z,y)$ so that for every $\psi(z,y) \in L(A)$

♯2 $\qquad p(z,a) \wedge \psi(z,a) \qquad$ is non-drifting whenever it is consistent.

Drifting and quasi-invariance are relative to the action of $\mathrm{Aut}\,\mathrm{f}(\mathscr{U}/A)$ on $\mathscr{U}^{|z|}$. Then, in particular, $p(z,a) \wedge \varphi(z,a)$ is non-drifting and the theorem follows. In fact, let $a_0, \ldots, a_k \in \mathscr{L}(a/A)$ be such that every set $p(\mathscr{U},c) \cap \varphi(\mathscr{U},c)$ for $c \in \mathscr{L}(a/A)$ intersects some $p(\mathscr{U},a_i) \cap \varphi(\mathscr{U},a_i)$. Let m be such that $d_A(a_i,a_j) \leq m$ for every $i,j \leq k$. From ♯1 we obtain that $d_A(a,c) \leq m + 2n$. As $c \in \mathscr{L}(a/A)$ is arbitrary, the theorem follows.

The required type $p(z,y)$ is union of a chain of types $p_\alpha(z,y)$ defined as follows

$$p_0(z,y) = l(z) \cup l(y);$$

♯3 $\qquad p_{\alpha+1}(z,y) = p_\alpha(z,y) \cup \{\neg\psi(z,y) \in L(A) : p_\alpha(z,a) \wedge \psi(z,a) \text{ is drifting}\};$

$$p_\alpha(z,y) = \bigcup_{n<\alpha} p_n(z,y) \quad \text{for limit } \alpha.$$

Clearly, the chain stabilizes at some stage $\leq |L(A)|$ yielding a type which satisfies ♯2. So we only need to prove consistency. We prove that $p_\alpha(z,a)$ is quasi-invariant (so, in particular, consistent). Suppose that $p_n(z,a)$ is quasi-invariant for every $n < \alpha$ but, for a contradiction, $p_\alpha(z,a)$ is not. Then for some $f_1, \ldots, f_k \in \mathrm{Aut}\,\mathrm{f}(\mathscr{U}/A)$

$$p_\alpha(z,a) \cup \bigcup_{i=1}^k p_\alpha(z,f_i a)$$

is inconsistent. By compactness there is some $n < \alpha$ and some $\psi_i(z,y)$ as in ♯3 such that

$$p_n(z,a) \to \neg \bigwedge_{j=1}^{\neg n} \bigwedge_{i=1}^k \neg\psi_j(z,f_i a)$$

As $p_n(z,a)$ is quasi-invariant, from Lemma 2.2 we obtain that $p_n(z,f_i a) \wedge \psi_j(z,f_i a)$ is non-drifting for some i,j. Clearly we can replace $f_i a$ with a, then this contradicts the construction of $p_\alpha(z,y)$ and proves the theorem. ⊣

Acknowledgement

We are indebted to the anonymous referee for many useful comments and for a neat proof of Lemma 2.2.

BIBLIOGRAPHY

[Cas11]   Enrique Casanovas. *Simple theories and hyperimaginaries*, volume 39 of *Lecture Notes in Logic*. Association for Symbolic Logic; Cambridge University Press, 2011.

[CLPZ01]   E. Casanovas, D. Lascar, A. Pillay, and M. Ziegler. Galois groups of first order theories. *J. Math. Log.*, 1(2):305–319, 2001.

[CP12]   Annalisa Conversano and Anand Pillay. Connected components of definable groups and o-minimality i. *Adv. Math.*, 231(2):605–623, 2012.

[KM14]   Itay Kaplan and Benjamin D. Miller. An embedding theorem of $\mathbb{E}_0$ with model theoretic applications. *J. Math. Log.*, 14(2):14–22, 2014.

[KMS14]   Itay Kaplan, Benjamin D. Miller, and Pierre Simon. The borel cardinality of lascar strong types. *J. Lond. Math. Soc. (2)*, 90(2):609–630, 2014.

[KPS13]   Krzysztof Krupiński, Anand Pillay, and Sławomir Solecki. Borel equivalence relations and lascar strong types. *J. Math. Log.*, 13(2), 2013.

[Las82]   Daniel Lascar. On the category of models of a complete theory. *J. Symbolic Logic*, 47(2):249–266, 1982.

[New03]   Ludomir Newelski. The diameter of a lascar strong type. *Fund. Math.*, 176(2):157–170, 2003.

[New09]   Ludomir Newelski. Topological dynamics of definable group actions. *J. Symbolic Logic*, 74(1):50–72, 2009.

[NP06]   Ludomir Newelski and Marcin Petrykowski. Weak generic types and coverings of groups. i. *Fund. Math.*, 191(3):201–225, 2006.

[Pel08]   Rodrigo Peláez. *About the Lascar group*. PhD thesis, Universitat de Barcelona, Departament de Lógica, História i Filosofia de la Ciéncia, 2008.

[Sim15]   Pierre Simon. *A Guide to NIP Theories*. Lecture Notes in Logic. Volume 44, Association for Symbolic Logic; Cambridge University Press, 2015.

www.ingramcontent.com/pod-product-compliance
Lightning Source LLC
Chambersburg PA
CBHW081128170426
43197CB00017B/2784